International Workshop on Physical and Chemical Processes in Atomic Systems 2015

Journal of Physics: Conference Series Volume 751

Moscow, Russia
15 – 17 November 2015

Editor:

Vladimir D. Borman

ISBN: 978-1-5108-3060-8
ISSN: 1742-6588

Printed from e-media with permission by:

Curran Associates, Inc.
57 Morehouse Lane
Red Hook, NY 12571

Some format issues inherent in the e-media version may also appear in this print version.

Printed by Curran Associates, Inc. (2016)

For permission requests, please contact the Institute of Physics at the address below.

Institute of Physics
Dirac House, Temple Back
Bristol BS1 6BE UK

Phone: 44 1 17 929 7481
Fax: 44 1 17 920 0979

techtracking@iop.org

Additional copies of this publication are available from:

Curran Associates, Inc.
57 Morehouse Lane
Red Hook, NY 12571 USA
Phone: 845-758-0400
Fax: 845-758-2633
Email: curran@proceedings.com
Web: www.proceedings.com

Table of contents

Volume 751

International Workshop on Physical and Chemical Processes in Atomic Systems
15–17 November 2015, Moscow, Russia

Accepted papers received: 2 September 2016
Published online: 28 September 2016

II. Extreme Gas Dynamics

III. Porous Medium, Surface And Colloids

This special volume contains the proceedings of "International Workshop on Physical and Chemical Processes in Atomic Systems". The conference was held in Moscow, Russian Federation in 15–17 November, 2015.

The organizers of the conference are

- National Research Nuclear University "MEPhI" (Moscow Engineering Physics Institute).
- SSC RF "Troitsk Institute of Innovative and Thermonuclear Research"

This newly organized international workshop consisted of oral invited lectures which contained detailed reviews on selected topics and numerous posters.

The wide range of fundamental and applied physical and chemical processes in various atomic systems has been considered during the meeting. The main topics of the conference were physics of separation processes, including separation of isotopes in rapidly rotating flows, selective transport phenomena in membranes, water treatment, physical phenomena connected with separation science such as physics of surface and colloids, extremal gas dynamics and porous medium.

As is known, the study of physical and chemical processes in atomic systems requires the use of diverse physical and chemical methods. So a fair amount of reports are focused on development of new instruments and techniques for the research of multiple phenomena in atomic systems.

Some reports presented in the proceedings have been considerably modified by their authors after discussion held during the workshop. Besides, it was decided to include in the proceedings a few additional posters corresponding to a scope of the meeting as a result of a round table final discussion.

The interest shown by Russian and foreign scholars either to the forum format or the subjects of reports, gives hope for a continuation of this meeting.

Special volume is supported by MEPhI Academic Excellence Project, contracts No 02.a03.21.0005, 27.08.2013.

Guest Editor
Professor Vladimir D. Borman
Head of the Department of Molecular Physics
National Research Nuclear University MEPhI, Russia

Journal of Physics: Conference Series **751** (2016) 011002 doi:10.1088/1742-6596/751/1/011002

Peer review statement

All papers published in this volume of *Journal of Physics: Conference Series* have been peer reviewed through processes administered by the proceedings Editors. Reviews were conducted by expert referees to the professional and scientific standards expected of a proceedings journal published by IOP Publishing.

Laser photochemical lead isotopes separation for harmless nuclear power engineering

P A Bokhan, N V Fateev, V A Kim and D E Zakrevsky

Rzhanov Institute of Semiconductor Physics of Siberian Branch, Russian Academy of Sciences, Pr.Lavrent'eva 13, Novosibirsk, 630090, Russia

E-mail: fateev@isp.nsc.ru

Abstract. The collisional quenching of the metastable $^3P_{1,2}$ and $^1D_2$ lead atoms is studied experimentally in the gas flow of the lead atoms, reagent-molecules and a carrier gas Ar. The experimental parameters were similar to the conditions that are required in the operation of the experimental setup for photochemical isotope separation. Excited atoms are generated under electron impact conditions created by a gas glow discharge through the mixture of gases and monitored photoelectrically by attenuation of atomic resonance radiation from hollow cathode ^{208}Pb lamp. The decay of the excited atoms has been studied in the presence various molecules and total cross section data are reported. The flow tube measurements has allowed to separate the physical and chemical quenching channels and measure the rates of the chemical reaction excited lead with N_2O, CH_2Cl_2, SF_6 and *CuBr* molecules. These results are discussed in the prospects of the obtaining isotopically modified lead as a promising coolant in the reactors on the fast-neutron.

1. Introduction

The great interest currently exists to isotopically modified lead as a promising coolant in fast-neutron nuclear power plants. It is assumed that the heavy metal coolant and the proton-to-neutron converter will be used (lead or lead–bismuth alloy). However, long-term irradiation of lead formed an extremely dangerous α- and γ- radionuclides, including ^{210}Po and ^{207}Bi. It was shown that a coolant enriched by isotope ^{208}Pb (up to the abundance of 99.0%) produces far less (by a factor of $10^3 \div 10^4$) polonium and bismuth radionuclides than natural lead [1].

The technologies to produce the superpure and isotopically modified lead are well developed. Unfortunately, the cost of lead isotopes depending on the production method is not suitable for modern power engineering. The necessity for isotopically modified lead can be thousands tons already within 10 years. Therefore, the development of alternative technologies leading to the production of a considerably (by orders) less cost product is actual. The laser isotope separation has the significant advantages [2, 3] compared with other known methods. They are as follows a high selectivity of an elementary separation act, the low energy expenditures, the possibility of the target isotope isolation, non-contact of the separation method. When two types of atoms of different isotopic composition (^{1}A and ^{2}A) have even one spectral line that does not overlap with others, it is possible to excite selectively with laser light an atom or molecule ("particle" ^{1}A) of a needed isotopic composition. The excitation of the ^{1}A particles changes their chemical and physical properties and, hence, may be used to separate substances by any method based on differences in the characteristics of excited and unexcited particles. Selective laser excitation of ^1A particles is followed by the chemical reaction of $^1A^*$ to give products designated ^1AR. The chemical reactivity of atoms or molecules is often increased by excitation. For laser isotope separation, long-living (metastable) states are preferred. Indeed, the chemical interaction probability for an excited atom should exceed the probability of radiation decay of the level, that is, the condition $k^*n > \tau_0^{-1}$ must be satisfied. Where k^* is the rate constant of the reaction, n is the concentration of reagent molecules and τ_0 is the radiation lifetime. Laser can be used to prepare reagents in particular quantum states. The excitation of a reagent can have several effects on a chemical reaction. Laser excitation of a reactant atom not only increases its internal energy, it also generally modifies its electronic state symmetry. It is well known that symmetry plays an important role in photon-induced but it can also play an important role in chemical reactivity. Electronic excitation invariably changes the shape and symmetry of the potential energy surface (PES) and it may induce a different reaction mechanism compared with that of the ground state. Possibility of the photochemical method of lead isotope extraction is based also on the peculiarities of a low electron level structure of lead. He has metastable states with the life time of $10^{-1} \div 10^{-2}$ s, which suits ideal for conducting the photochemical reactions. The second peculiarity is that the transitions inside the configuration are hit into radiation of the highly effective semiconductor and other lasers. It gives the perspective to simplify and to make cheaper the technology of lead isotope production.

Collisions of metastable lead atoms with various molecules investigated in this work. The use of the flow tube measurements [4-6] has allowed us to separate the physical and chemical quenching channels and measure the rates of these quenching processes. The scheme of these experiments is similar to the real system for laser isotope separation. A considerable amount of data on the quenching of he optically metastable states $Pb(6^1D_2)$, $Pb(6^1S_0)$, $Pb(6^3P_1)$ and $Pb(6^3P_2)$ has already been reported [7-10]. Total thermal deactivation rate constants, but not product channels, for all excited states have been measured with some molecules at room temperature. The main objective of this study is to justify a physical possibility of using the photochemical laser method for lead isotope separation. For this purpose some experimental work should be carried out to find the effective chemical reactions of the metastable *Pb* atoms with molecules.

2. Description of the method

The gas flow apparatus for the laser isotope separation is shown in Figure 1. Collisional reactions of the metastable lead atoms and reagent take place in the interaction chamber constructed of cylindrical tube with inner cross section area S and long *L* having two end windows.

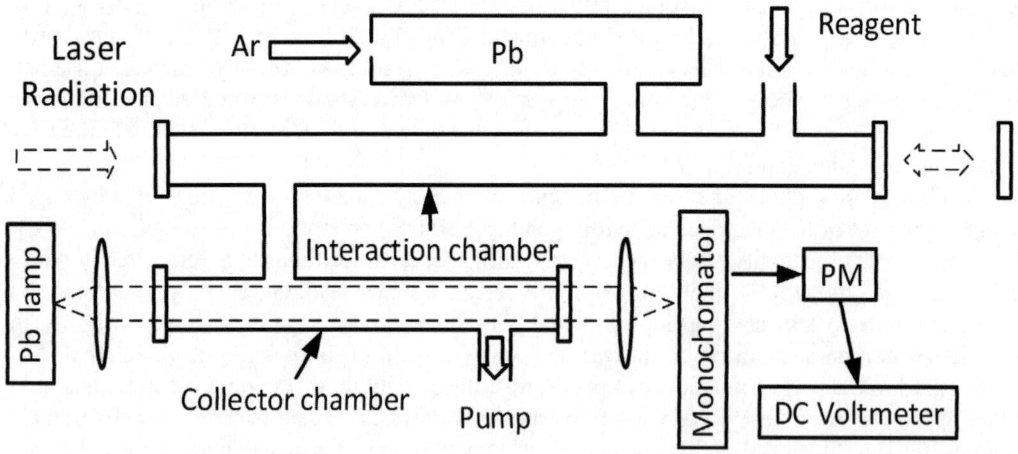

Figure 1. Experimental setup for photochemical isotope separation of lead atoms.

The section of the reactor on the reaction zone was heated by an external furnace to prevent the deposition of lead on tube walls. The lead atoms in a mixture with a carrier gas were introduced into the region of interaction from a heated reservoir. By varying the reservoir temperature, we could vary the concentration of lead atoms in the interaction zone. The carrier gas (Ar) is the major component in the flow tube and thus serves to define the physical properties of the gas flow. The reagent-molecules are mixed homogeneously with the carrier gas. The gas mixture was evacuated at the pumping rate G by a vacuum pump. The gas flow velocity is $v = G \cdot S^{-1}$. The propagation time of the lead atoms through the interaction area is $T = L \cdot v^{-1}$. To excite the lead atoms the monochromatic laser radiation propagates along the gas flow. Long-lived (metastable) states produced via the forbidden transitions. They can lose their excitation energy by collisions with molecules and other particles by the action many various physical and chemical processes in the interaction chamber. These processes we have combined only in three channels:

$$Pb^* \rightarrow Pb + h\nu \quad Pb^* + (wall) \rightarrow Pb \tag{1a}$$

$$Pb^* + XY \rightarrow Pb + (XY)^* \tag{1b}$$

$$Pb^* + XY \rightarrow PbX + Y \tag{1c}$$

Channel (1a) contains contributions to the lifetime τ_{0i} of metastable atoms in the absence of a reagent. Channel (1b) is the transfer of excitation energy to the internal and translational degrees of freedom of the colliding partner proceeding at a rate constant k_{ip} (physical quenching). Lead atoms in the ground state are formed by such processes. Channel (1c) is the collisions of the metastable atom Pb^* with XY molecules which leading to a chemical reaction. Lead atoms combine with molecule into stable chemical compounds with the value the rate constant k_{ic} (chemical quenching) and lifetime is more than T. The chemical compounds of *Pb* atoms with other particles have other chemical and physical properties. The stable reaction components are deposited on the walls of the interaction chamber or move along with the flow. In such a way, this process is characterized as depletion atoms in the reaction zone. From the interaction chamber the gas flow arrives in the collector chamber. By selecting the walls temperature can be done there the deposition of the reaction products. The time interaction of the *Pb* atoms with the laser radiation can be changed by varying the pumping rate. The flow tube technique is a good method to measure the kinetic parameters of the reactions. It allows determine the individual values for the rate constants of the physical and chemical quenching. The solution of the rate equation for the concentration of the lead atoms $N(T)$ in the interaction chamber is presented by the expression:

$$N(T) = N_0 \cdot e^{-T[k_0 n + \Sigma(W_i \theta_i)]}, \tag{2}$$

where T is the propagation time of the lead atoms through the interaction area, N_0 is the concentration of the lead atoms in the interaction chamber with $T = 0$, where the atoms and gas-reagent are mixed, n – concentration of reagent-molecule, k_0 - rate constant of chemical reaction between ground state lead atoms and reagent. The rate of atomic excitation W_i of the i metastable state per second under action laser radiation or other processes, the decay rate R_i of i metastable atoms in the interacting volume and the efficiency (probability) θ_i of the formation chemically stable product in process (1c) is given by the relations:

$$W_i = R_i \cdot \left(\frac{N_i(T)}{N(T)} \right), \; R_i = \left(\frac{1}{\tau_{0i}} + k_i \cdot n \right), \; \theta_i = \frac{k_{ic} \cdot n \cdot \tau_{0i}}{1 + (k_{ip} + k_{ic}) \cdot n \cdot \tau_{0i}} \tag{3}$$

where $N_i(T)$ is the stationary concentration of i metastable lead atoms in the interaction chamber, τ_{0i} is the lifetime of i metastable atoms in the absence of a reagent, $k_i = k_{ic} + k_{ip}$ is the total second-order rate constants for chemical reaction and physical quenching.

3. Selective excitation of the metastable lead atoms

The possible optical schemes and properties of a long lived level excitation are presented in Figure 2 and Table 1. The $6s^26p^2$ ground configuration of *Pb* consists of five levels: $^3P_{0,1,2}$, $^1D_2$ and $^1S_0$. Since electric-dipole (E_1) transitions between states of the same parity are forbidden, all levels of these configurations are metastable. In the second-order perturbation theory weak magnetic-dipole (M_1), electric-quadrupole (E_2), or mixed (M_1+E_2) type transitions between these levels are allowed. Lead occurs in nature in the form of four different stable isotopes of masses 204, 206, 207, and 208. For ^{207}Pb the nuclear spin quantum number is I=1/2, and only this isotope shows a hyperfine structure. The isotope shifts of multipole lines are relatively small in comparison with resonance on (λ = 283.3 nm) [11-13]. All known values are presented in Table 1. Figure 2 shows the possible excitation schemes of the metastable lead atoms by one and two step way. The rate atomic CW excitation of the atom W_i in the flow is determined by the equation:

$$W_i = \frac{\sigma \cdot P}{S_L \cdot h \cdot \nu_0}, \qquad (4)$$

where P is power, S_L is cross-sectional area and ν_0 is central frequency of radiation transition. The excitation cross sections for the metastable states are small and have values $\sigma \leq 10^{-17} \div 10^{-19}$ cm^2.

Table 1. Properties of optical transitions for selective excitation of metastable lead atoms.

Transition	λ, nm	$^a A_{ki}$, s^{-1}	$^b\sigma$, cm^2,	$^c\Delta\nu_\Delta$, MHz	$^d\Delta\nu_\iota$, MHz 208-207	$^d\Delta\nu_\iota$, MHz 208-206	$^d\Delta\nu_\iota$, MHz 208-204	$^d\Delta\nu_\iota$, MHz 207-208
$^3P_0 \to 7s^3P_1$	283,3	4,7×10^7	2,6×10^{-12}	1600	10277,9	2317,6	4292,5	2922,1
$^3P_0 \to {}^3P_1$	1279,3	7,9	4,1×10^{-17}	352	-1066,7	-220,4	-	2533,3
$^3P_0 \to {}^3P_2$	939,2	0,49	1,7×10^{-18}	478	-	-	-	-
$^3P_0 \to {}^1D_2$	446,2	0,17	7,3×10^{-20}	957	-	-	-	-
$^3P_2 \to {}^1S_0$	531,6	18,7	4,6×10^{-19}	846	-	-	-	-
$^3P_1 \to {}^1D_2$	733,4	14,5	8,1×10^{-18}	614	-1320	-184	334	2117
$^3P_2 \to {}^1D_2$	925,6	10,6	7,3×10^{-18}	480	-	-	-	-

[a] probabilities of spontaneous emission from level *k* to *i*.
[b] cross-section at center of the Doppler absorption contour.
[c] Doppler width of the absorption contour at 930°K.
[d] shift between two isotopes of *Pb*.

This circumstance necessitates the exciting at high laser power, the major part of which may be lost due to a relatively low absorption. A multipass system is used as a technical solution. Very attractive way of selective production the metastable excited state if the lead atoms were excited from the ground state $6p^2\ ^3P_0$ to the $6p7s\ ^3P_1^0$ resonance state under action monochromatic laser radiation at (λ= 283.3 nm) with lifetimes 5.8 ns. Then $6p7s\ ^3P_1^0$ resonance state decay into the ground and well into the metastable levels by the superluminescence. In this case the laser part of the experimental setup is more expensive than the other methods. For isotope separation by photochemical method the isotope atom concentration is restricted by the process of resonance energy transfer from an atom of one isotope to an atom of another one. This process obviously acts as the excitation selectivity deterioration. Excitation transfer cross-section in the approximation of the dipole-dipole interaction is in the inverse proportion to the life time of the upper level. The long-lived states have the small values of the dipole moment and a cross-section excitation $\sigma \leq 10^{-17}$ cm^{-2} (Table 1). The rate of energy transfer to another isotope ($\sigma \cdot v_a \cdot N$) must be less than (1/T). Here v_a is the arithmetic mean velocity of the lead atoms. This gives the upper limit of the atomic concentration ~ 10^{13} cm^{-3}. This value is by three orders of magnitude higher than the limiting atom concentration admitted in photoionization isotope separation method. The main peculiarity is that the transitions between metastable states are hit into radiation of the highly effective semiconductor and other lasers. It gives the perspective to simplify and to make cheaper the technology of lead isotope production. Laser isotope separation of lead atoms in the gas flow is effective even if the corresponding isotope structure is masked by the Doppler broadening. In this case, the selectivity of excitation increases at frequency detuning from the lines of all isotopes. This demonstrates Figure 3, which shows the total and Doppler absorption contour of each *Pb* isotope. The narrowband laser radiation can be used to provide velocity selective excitation of the inhomogeneously broadened absorption contours. For a mixture of isotopes, the absorption contour represents the sum of values with corresponding v_0 and weights proportional to the isotope percentages. If the laser frequency is tuned to the isotope absorption edge, the contribution from other isotopes becomes small. In this case, decreases sharply the total absorption coefficient, but the large interaction time *T* between the laser radiation and atoms provides 100% excitation efficiency for the desired isotope.

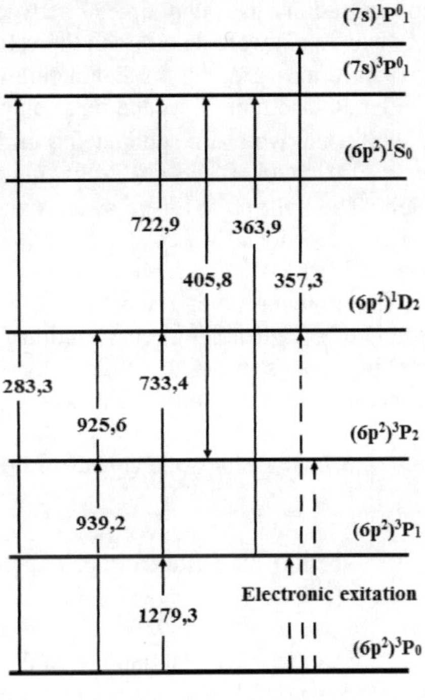

Figure 2. Energy levels of atomic lead revelant to this paper. All wavelength transition are given in nm.

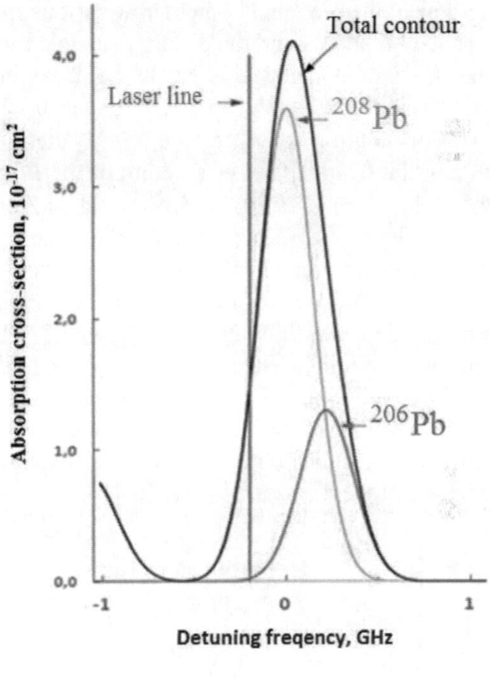

Figure 3. Lead atom absorption spectrum at the transition $6p^2\,{}^3P_0$-$6p^2\,{}^3P_1$ ($\lambda = 1279.3$ nm).

4. Experiment and results

The wall temperature of the interaction chamber was 930° K and argon concentration is $4{\cdot}10^{16}$ cm^{-3}. Gas mixture of the molecules, carrier gas and *Pb* atoms was evacuated at a rate $G = 0.6$ l·s^{-1} by a pump with velocity is v = 90 cm·s^{-1}. Lead atoms were excited under electron impact conditions created by DC or pulsed glow discharges through the mixture of gases. The length of the discharge gap is 10 cm. The reaction time T = 0.11 s. The concentration of the *Pb* atoms was determined by absorption measurements of the transmitted intensity of the resonant radiation from ^{208}Pb lamp through the interaction region obeyed the Beer's law. For the ground state measurements were used the $6p^2\,{}^3P_0 \rightarrow 7s\,{}^3P_1^0$ (λ = 283.3 nm) resonance transition. For measurements the populations of the excited states were used other spectral lines. The output radiation from ^{208}Pb lamp focused into the entrance slit of a monochromator. He was used as a band-pass filter and tuned to some emission lines of Pb. Measurements were carried out in two cases transverse direction to the gas flow. The first direction is located in the discharge region. The second direction is located beyond the region of the interaction of excited atoms and molecules. The light-induced fluorescence (LIF) method is used for the temporal behavior population of metastable lead atoms in the interaction chamber. Under action of resonance radiations the *Pb* atoms occur the optical transitions from the metastable states to the $7s\ ^3P_1^0$ level (Figure 1). The time dependent fluorescence $S_i(t)$ at 405.78 nm ($7s\ ^3P_1^0 \rightarrow 6p^2\ ^3P_2$) was recorded from illuminated area. The multiplier signal $S_i(t)$ is proportional intensity of this fluorescence and, accordingly, metastable state population determined by the equation:

$$S_i(t) = C \cdot e^{-(R_i t)}, \tag{5}$$

where C is a time independent constant. The study of the metastable state quenching in collisions with molecules were carried out using the afterglow pulsed discharge with a repetition rate of 100 Hz and current pulse with half-width of 4 μs. The output multiplier signal $S_i(t)$ processed by a box-car integrator and stored in a computer at various molecules concentrations and delay times t between the resonance radiations and the discharge pulse. From the decay curves (5) were determined the values of τ_{0i} and k_i. For $^1D_2$ metastable levels the decay curves of $S_i(t)$ were measured by the interaction of the *Pb* atoms with pulsed *Pb* vapor laser radiation (722.9 nm) with half-width of 10 ns and pulse energy of 10 μJ. For monitoring $6p^2\ ^3P_1$ and $6p^2\ ^3P_2$ metastable levels was used two sharp spontaneous emission lines emitted from the active medium of the *Pb* vapor laser at 363.96 and 405.78 nm, belonging to the $7sP_1^0 \rightarrow 6p^2\ ^3P_1$ and $7s\ ^3P_1^0 \rightarrow 6p^2\ ^3P_2$ transitions, respectively. The light pulses were synchronized in time with discharge pulses. Plots for the obtained quenching cross sections for $Pb(^3P_1, ^3P_2, ^1D_2)$ atoms versus the serial number of interacting molecules are shown in Figure 4. This figure does not reveal the correlated behavior of the quenching cross sections for different metastable. The data for 6^3P_2 state can be divided into three groups on the basis of the magnitude of the quenching cross section. Weak quenchers, including H_2O, NH_3, and C_2H_5OH all have quenching cross sections less than 1 Å^2. Strong quenchers, including $(C_2H_5)_2O$, CH_2Cl_2 and $CuBr$ have quenching cross sections in excess of 10 Å^2. Whereas the rest of the molecules have quenching cross sections values between 1 and 10 Å^2. This division might indicate in a general way that several different mechanisms must be considered for the quenching process.

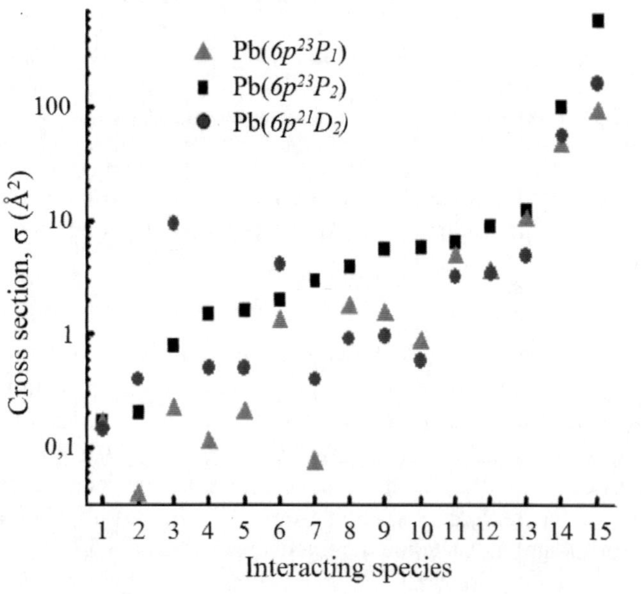

Figure 4. Quenching cross sections for excited Pb($^3P_1$, $^3P_2$, $^1D_2$) by various molecules collisions at 930 K. Serial number of interacting molecules:
1 - Water (H_2O).
2 - Ammonia (NH_3).
3 - Ethanol (C_2H_5OH).
4 - Propane (C_3H_8).
5 - Isobutanol ($C_4H_{10}O$).
6 - Benzene (C_6H_6).
7 - Carbon dioxide (CO_2).
8 - Methanol (CH_3OH).
9 - Ethyl bromide (C_2H_5Br).
10 - Nitrous oxide (N_2O).
11 - Acetone (CH_3COCH_3).
12 - Sulfur hexafluoride (SF_6).
13 - Diethyl ether ($(C_2H_5)_2O$).
14 - Methylene chloride (CH_2Cl_2).
15 - Copper bromide. (CuBr).

Chemical contributions to the total collisional quenching are investigated by the reduction of lead atom concentration due to chemical reaction for excited atoms with molecules. Relative atomic concentration measurements were carried in the second direction of the flow during the presence and absence of a glow discharge in the interacting region. In these measurements, the metastable lead atoms were excited in a CW glow discharge with current 50 mA. Simultaneously with photochemical reaction (1c) occur chemical reaction between of lead atoms in the ground state and added molecules with the rate constant k_0. For some molecules the reaction rates of these two processes can be compared with each other. Nearly all chemical reactions have the activation barrier. Table 2 lists the results of our measurements for the values which describe the depletion excited lead atoms in the interaction chamber by chemical reaction. Unexpectedly, but only for four of the fifteen investigated molecules the change in concentration of the *Pb* atoms was revealed.

Table 2. Experimental results for depletion of the $^3P_0$ and $^3P_2$ Lead atoms by molecules

Molecule	$k_0^{a,b}$, 10^{-13}	$\sum (W_i\theta_i), s^{-1}$	k_{ic}/k_{ip}	$k_{ic}^{a}, 10^{-12}$	Product	ΔH^0, *eV*
N_2O	0,01	14,3; $[2{\cdot}10^{-2}]^c$	0,28	0,93	*PbO*	-2,31
CH_2Cl_2	8,2	5,2; $[10^{-3}]^c$	0,043	24	*PbCl*	0,38
SF_6	0,0016	4,1; $[6{,}8{\cdot}10^{-2}]^c$	0,031	1,3	*PbF*	0,27
CuBr	4,1	3,3; $[4{,}5{\cdot}10^{-4}]^c$	0,025	70	*PbBr*	0,83

[a]Second-order rate constants, units $cm^3{\cdot}s^{-1}$.
[b]Reference 6.
[c]Pressure gas-reagent [Torr].

For eleven other molecules studied, the effect of depletion did not exceed measurement errors in spite of the high rate constants k_i for quenching. This means that the collisions of metastable lead atoms with these molecules caused effective quenching of electronic excitation energy without noticeable chemical reactions ($k_{ic} << k_{ip}$). The extent of chemical quenching is reflected in the depletion in the ground state lead atom concentration measured sufficiently long after collisions to allow for complete relaxation of all excited species. Our gas flow apparatus is allows studying chemical reactions with half-lives of $T \approx 0.11$ sec or longer. This time is longer than the time of the radial diffusion ($4{\cdot}10^{-2}$ s) for excited species on the walls of the interaction chamber where they effectively relax to the ground state before reaching the observation region. Very many lead atomic states are excited in the gas discharge, which participate in the collision process. To find the individual contribution of the each excited state, we consider the required conditions for efficient chemical reaction. According to equation (2) high efficiency depletion of the *i* lead atoms states is determined by the condition $(\theta_i{\cdot}W_i) \geq 1$. From the experimental data we can determine only total sum of values $(\theta_i{\cdot}W_i)$ for all possible states. The short-lived and high-lying states give a minor contribution to the depletion of lead atoms, because the condition $(k_i\ n\ \tau_{0i}) \leq 0.1$ is always satisfied for them. This condition requires another condition ($\theta_i << 1$), which follows from the expression (3). From the relation (2) follows, that under this condition the efficiency of the chemical reaction is low. The metastable *Pb* atom states produced in the discharge are mostly a combination of four $^3P_1$, $^3P_2$, $^1D_2$ and $^1S_0$ states. All considered chemical reactions are exothermic. It is believed that only the lowest metastable states are formed in significant amount, since all higher lying states will strongly radiate to these states. The minor contribution of the $^1D_2$ and $^1S_0$ states in a chemical reaction is related with exclusions by the Wigner rules [14]. Thus, only two metastable states $^3P_1$ and $^3P_2$ produced in the glow discharge can effectively react with molecules to yield stable products. They are effectively excited in a gas discharge (their stationary concentration amounts to 20% of the concentration of the ground

Journal of Physics: Conference Series **751** (2016) 012001 doi:10.1088/1742-6596/751/1/012001

state) and possess long lifetimes with respect to spontaneous decay. Between these two states should choose only ${}^3P_2$ state, for which the activation barrier is smaller and hence the chemical reaction is more efficient. Table 2 lists the values for rate constants for k_{ic} and k_{ip} at $W_i = 144$ s^{-1}, $\tau_{0i} = 10^{-3}$ s. More accurate data can give the experiments with selective excitation. The choice of molecules as reagent in the present experimental study was motivated by the low value of k_0 and by the ecological purity of those substances. These properties are necessary for laser isotope separation.

5. Discussion

The experimental results allowed us to forecast prospects of the photochemical lead atoms isotope separation. The appropriate power continuous wave laser radiation is $\lambda = 1.3$ μm, P = 10 W and a low line width of ≤ 10 MHz is use. The number of photons arriving per second is K $\approx 6\times10^{19}$ photon/s. This yields the rate of isotope separation H = 650 kg/year for the 100% quantum efficiency of the separation process (from one photon comes one atom of the desired isotope). In reality, this value is much less. It is defined the real physical and chemical processes that take place in the interaction chamber. We adopt the following parameters for the interaction chamber: the cross section area and length of the interaction chamber are $S = 10$ cm^2, $L = 10^2$ cm respectively, the drift velocity of the particles in the gas flow is v = 10^2 cm·s^{-1}. Then the interaction time of lead atoms with laser radiation is $T = 1$ s. The following concentrations values are acceptable in the gas flow: $N = 10^{13}$ cm^{-3}, $n = 10^{13} \div 10^{15}$ cm^{-3}, 5×10^{16} cm^{-3} for the lead atoms, reagent-molecules and argon atoms respectively. The number of atoms per second in the gas flow moving through the cross-section S is equal to $F = N{\cdot}v{\cdot}S = 10^{16}$ atoms·s^{-1}. At 100% excitation of lead atoms we obtain the rate of isotope separation H = 0.1 kg·year^{-1}. Let us consider excitation of lead atoms at the optical transition $6p^{2\,3}P_0 \rightarrow 6p^{2\,3}P_1$ ($\lambda = 1.279$ μm). From relation (4) we find the rate of atomic excitation under action resonance laser radiation $W_i = 2.5{\cdot}10^2$ s^{-1} at the center of the absorption line with $\sigma = 4.1\times10^{-17}$cm^2 (Table 1). This implies that during the time of the movement $T = 1$ s of the atoms into the interaction region of the laser excitation probability is $W_i{\cdot}T = 2.5{\cdot}10^2 >> 1$. That is, the atom can be excited multiple times in the interaction chamber. Even for the detuning laser frequency at the edge of the absorption line (Figure 3), the probabilities of atomic excitation remain a large enough. The quantity θ_i in the relation (2) is defined as the ratio of the decay rate of chemical quenching of metastable atoms to the overall quenching decay rate. This quantity can assume values in the range $0 \leq \theta_i \leq 1$. The results of the Table 2 show that the most effective chemical reaction is realized with N_2O molecule, for which $\theta_i \approx 0.1$. At 100% efficiency of the laser excitation for the metastable lead atoms in the flow and the same efficiency of the chemical reaction can be realized the production rate of the necessary isotopes. The obtained results allow one to define the range of the possible chemical reagents for laser isotope separation of lead by photochemical method. The $6p^2\ {}^3P_2$ lead state can be excited isotopically selectively in $6p^2\ {}^3P_0 \rightarrow 6p^2\ {}^3P_2$ scheme by radiations from semiconductor laser, operating at 939 nm. Gas flow conditions allow multiple radiation interactions with atoms to be performed with high efficiency in spite of small cross-section transition and low laser power. The obtained results allow one to define the range of the possible chemical reagents for laser isotope separation of lead by photochemical method.

References

[1] Khorasanov G L, Korobeynikov V V, Ivanov A P and Blokhin A I 2009 *Nucl. Eng.Des.* **239** 1703-7

[2] Letokhov V S 1977 *Annu. Rev. Phys. Chem.* **28** 133-59

[3] Bokhan P A, Buchanov V V, Fateev N V, Kalugin M M, Kazaryan M A, Prokhorov A M and Zakrevskii D E 2006 *Laser isotope separation in atomic vapor* (Wiley-VCH Verlag GmbH: Weinheim) p 185

[4] Howard C J 1979 *J. Phys. Chem.* **83** 3

[5] Bokhan P A, Fateev N V, Kim V A and Zakrevskii D E 2010 *Chem. Phys.* **372** 6

[6] Bokhan P A, Zakrevskii D E, Kim V A, Fateev N V and Yudin N A 2007 *Rus. J. Phys. Chem. B* **1** 521-7

[7] Husain D and Littler J G F 1972 *J. Chem. Soc.-Faraday Trans. 2* **68** 2110-6

Journal of Physics: Conference Series **751** (2016) 012001 doi:10.1088/1742-6596/751/1/012001

[8] Husain D and Littler J G F 1972 *Chem. Phys. Lett.* **16** 145
[9] Ewing J J, Trainor D W and Yatsiv S. *J. Chem. Phys.* **61** 4433
[10] Husain D and Littler J G F 1974 *Int. J. Chem. Kinet.* **6** 61
[11] Wasowicz T J and Kwela J 2008 *Phys. Scr.* **77** 025301
[12] Wasowicz T J, Werbowy S, Kwela J and Drozdowski R. 2010 *J. Opt. Soc. Am. B* **27** 2628
[13] Reeves J M and Forston E N 1991 *Phys. Rev. A* **44** R1439
[14] Shuler K E 1953 *J. Chem. Phys.* **21** 624

Regenerated Uranium Separation in Matched Abundance Ratio Cascade with Additional Product Flow

E V Maslyukov and V A Palkin

Ural Federal University named after the first President of Russia B.N. Yeltsin, Institute of Physics and Technology, 19 Mira Street, Yekaterinburg, Russian Federation

E-mail: eugene_v_m@mail.ru

Abstract. This paper addresses known methods used to purify regenerated uranium in single and double cascades. A new method for separating regenerated uranium has been developed that enables a significant reduction of the concentration of $^{232, 234}$U in the additional product flow. Matched abundance ratio cascades (M*-cascades) with different key components and additional product flow are used in the new method. Main product flow of the M*-cascade contains low enriched regenerated uranium. It can be used for reactor fuel production. Purified product can be enriched in the ordinary cascade in compliance with the requirements of ASTM C 996–10 set for isotopes $^{232, 234}$U in low enriched commercial uranium, which is usually produced from the natural one. Computer experiment based on the new method has been performed. The experiment shows that the best cascade with the maximum flow of the enriched purified product is M*-cascade with key components ^{232,236}U.

1. Introduction

Regenerated uranium is produced from irradiated nuclear reactors fuel. It plays an important role in the nuclear fuel cycle. It contains isotope ^{235}U, which has a concentration higher than that in the natural uranium. However, there are significant amounts of impurity isotopes $^{232, 234, 236}$U in the regenerated uranium. It is impossible to separate them chemically. The impurity isotopes cause the main difficulties for using regenerated uranium in the production of reactor fuel [1].

The most significant problem is the presence of the isotopes ^{232}U and ^{234}U because of their high radioactivity. The isotope ^{236}U captures neutrons in a nuclear reactor. Its presence leads to an increase of the concentration of the fissile isotope ^{235}U in the fuel. This problem affects the economic indicators of a nuclear reactor. To minimize radiation hazards and improve fuel quality it is necessary to reduce the concentration of $^{232, 234, 236}$U. For this purpose cascade separation technique of the uranium hexafluoride (UF_6) and dilution operations can be used.

There are different ways of regenerated uranium enrichment with the reduction of $^{232, 234, 236}$U concentration in a separate cascade. One of these is its enrichment to a concentration of the ^{235}U up to 10–90% in the ordinary cascade and subsequent dilution down to 2–7% [2]. Despite the good quality of the product obtained, this method has significant drawbacks: ^{235}U isotope with a concentration over 5% is formed as an intermediate product of the purification process. Moreover, this method involves using natural uranium.

Also, there are methods of regenerated uranium purification in double cascades [3–5]. The first ordinary cascade enriches the uranium hexafluoride up to 20% and above for ^{235}U. The product flow of this cascade is fed the second cascade, where $^{232,\,234}$U are removed from the waste flow. The purified wasted uranium hexafluoride of the second cascade is diluted to obtain the desired concentration of ^{235}U. When the additional carrier gas [5] is used in the second cascade feed flow the content of ^{232}U is reduced down to 3×10^{-8} – 4×10^{-9} % without using special diluents. The only disadvantage of the double cascades [3–5] is obtaining a highly enriched uranium with ^{235}U concentration more than 20% at some stages of separation.

Another method uses additional feed flow with natural uranium in the cascade [6]. When using this method, the regenerated uranium is enriched to a concentration less than 5% ^{235}U while the content of $^{232,\,234,\,236}$U is decreased. The consumption of natural uranium can be efficiently reduced by utilizing one more feed flow with depleted waste uranium [7]. However, this leads to an increase in the expenses for separation work. Another method involves using a cascade with an additional product flow and two feed flows with regenerated and with natural uranium [8]. In this case, additional expenses for separation work are practically nonexistent. However, with a significant reduction in the concentration of $^{232,\,234,\,236}$U in the intermediate product flow, their content increases in the final commercial product flow of the low enriched uranium.

The alternative is a double cascade, where ^{235}U concentration is limited and the pattern is changed so as to remove $^{232,\,234}$U in the first ordinary cascade. Thus, the feeding point is significantly shifted toward the product flow. Its purified depleted stream is fed to the second ordinary cascade, where the desired commercial product of low enriched uranium is generated in the product flow [9]. Studies have shown that reducing the concentration of ^{232}U to $(3\text{-}4)\times10^{-9}$ % in the final product is possible with ^{235}U enrichment to 5 % in the cascades [10]. However, this method features high expenses for separation work and does not provide for significant removal of $^{234,\,236}$U isotopes. More efficient removal of ^{234}U is possible using the method described in [11].

This paper deals with the purification of the regenerated uranium in the additional product flow of the matched abundance ratio cascade (M*-cascade) [12] and subsequent enrichment of the purified product in the ordinary cascade. Calculations are based on the methodology [13] for the M*-cascades for various key components.

2. Calculations

The scheme consisting of two cascades is shown in Figure 1.

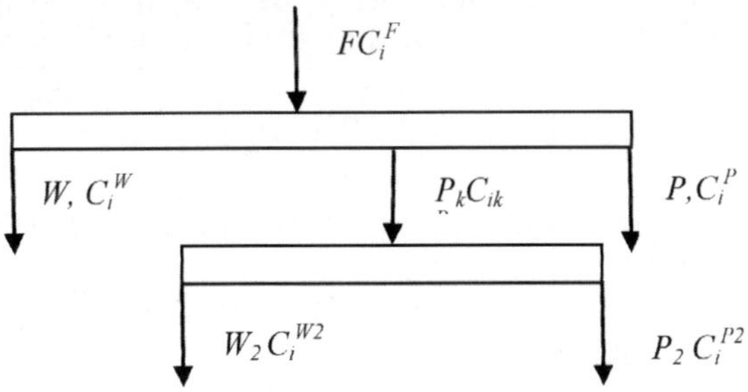

Figure 1. Double cascade scheme

Let F, W, and P denote feed, waste, and primary product flows of the first cascade, respectively; let P_k denote additional product flow, and C_i^F, C_i^W, C_i^P, C_{ik}^P denote corresponding concentrations (mass fractions) of the i-th component. Additional flow P_k is taken from the waste flow of stage k of the first cascade. W_2, P_2 are waste and product flows of the second cascade; and C_i^{W2}, C_i^{P2} are the corresponding concentrations of the i-th component.

Matched abundance ratio cascades with one additional product flow were calculated for various key components. Feed flow F was 10 g/s and consisted of a mixture of isotopes: ^{232}U – 1.5×10^{-7} %, ^{234}U – 1.6×10^{-2} %, ^{235}U – 0.85 %, ^{236}U – 0.35 %.

The selected concentrations of feed correspond to the characteristic isotopic composition of regenerated uranium [14]. The cascades consisted of 51 stages with the feed in the stage number 45. This configuration has been chosen based on the given separation factors and the need to achieve a concentration of the ^{235}U not exceeding 5% in the primary product flow and 0.04 – 0.3% in the waste. Under these conditions, the primary product flow meets the requirements of ASTM C 996–10 for low enriched regenerated uranium and can be used to produce the corresponding reactor fuel.

The number of the additional product flow stage k varied from 2 to 45, its flow quantity P_k ranged from 1 g/s to 5 g/s in steps of 0.1 g/s. The separation factor per unit mass difference was assumed to be 1.091.

For each calculation, the additional product flow of the M*-cascade fed the ordinary cascade. The parameters of the latter were assessed by analytical formulas [15]. The concentration of the ^{235}U at this cascade was set equal to 0.3% for the waste and 4.4% for the product. The concentration of the ^{235}U in the product flow of the ordinary cascade was taken on the basis of its required content of 3.2 – 3.6% in the fuel without regard for the parasitic neutron capture by ^{236}U isotope. The compensation factor in the product was 0.2 – 0.3. The content of the impurity isotopes in the product was limited to 1×10^{-8} % for the ^{232}U and 4.8×10^{-2} % for the ^{234}U. These concentrations of impurity isotopes meet the ASTM C 996-10 specification for low enriched uranium with a concentration of commercial ^{235}U equal to 4.4%.

3. Results

It was found that the concentration of the components $^{232\text{-}236}U$ decreases with the increase of the additional product flow (Table 1).

Table 1. Concentration of components in function of additional product flow for stage No. 14 (k = 14) of M*-cascade with key components $^{232,\,236}U$.

P_k(g/s)	C_{ik}^P(%)			
	^{232}U	^{234}U	^{235}U	^{236}U
1.7	6.45×10^{-10}	7.03×10^{-3}	0.8498	0.4057
2.0	6.32×10^{-10}	6.70×10^{-3}	0.8193	0.3978
2.5	6.12×10^{-10}	6.21×10^{-3}	0.7731	0.3853
3.0	5.94×10^{-10}	5.79×10^{-3}	0.7318	0.3736
3.5	5.76×10^{-10}	5.42×10^{-3}	0.6946	0.3626
4.0	5.60×10^{-10}	5.09×10^{-3}	0.6610	0.3521
4.4	5.47×10^{-10}	4.86×10^{-3}	0.6364	0.3442

As the additional product stage number increases, the $^{232\text{-}236}U$ concentrations increase too (Table 2). It follows from these that an optimal stage number exists for each selected value of the additional product flow P_k in the matched abundance ratio cascade. The product flow of the ordinary cascade $P2$ will be maximal for this stage number taking into account the limits assumed.

Table 2. Concentration of components in function of additional product flow stage number of M*-cascade with key components $^{232,236}U$ when P_k=2.7 g/s.

k	C_{ik}^P(%)			
	^{232}U	^{234}U	^{235}U	^{236}U
2	5.58×10^{-11}	2.66×10^{-3}	0.469	0.287
5	1.07×10^{-10}	3.54×10^{-3}	0.568	0.324
10	2.87×10^{-10}	4.96×10^{-3}	0.691	0.364
14	6.05×10^{-10}	6.03×10^{-3}	0.756	0.381

It should be noted that when the restrictions on the concentration of $^{232,234}U$ are met the ^{236}U content in the product flow of the ordinary cascade increases as P_k grows (Figure 2).

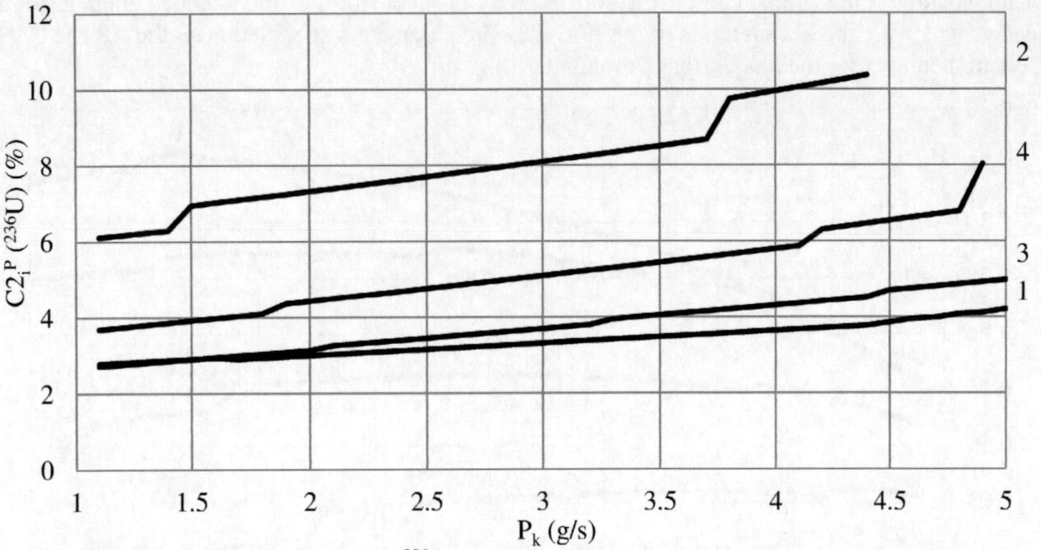

Figure 2. Concentration of the ^{236}U in product flow of ordinary cascade in function of additional product flow of M*-cascade: 1 – key components $^{232,236}U$; 2 – key components $^{232,238}U$; 3 – key components $^{234,235}U$; 4 – key components $^{235,236}U$.

Thus, it is possible to reduce the product flow of the ordinary cascade to decrease the ^{236}U concentration. For example, a decrease of the flow by 4 % from the optimal value for the matched abundance ratio cascade with key components $^{232,238}U$ leads to a decrease of the ^{236}U concentration by more than 16 % (Table 3).

Table 3. Variation of ^{236}U concentration in product flow of ordinary cascade

K	P_k(g/s)	C_{ik}^P(%)				$P2$ (g/s)	$C2_i^P$(%)		
		^{232}U	^{234}U	^{235}U	^{236}U		^{232}U	^{234}U	^{236}U
30	1.8	2.26×10^{-9}	4.40×10^{-3}	0.654	0.390	0.156	9.93×10^{-9}	3.76×10^{-2}	4.10
29	2.7	1.74×10^{-9}	3.50×10^{-3}	0.523	0.340	0.162	8.98×10^{-9}	3.96×10^{-2}	5.10

When the P_k flow increases the feed flow of the ordinary cascade also increases. At the same time, the ^{235}U concentration in the feed flow decreases. These two factors have opposite effects on generation of the product flow $P2$ at the given concentrations for the product and waste of the ordinary cascade. The first of them results in the increase of $P2$ and the second one leads to its decrease. As a result, the maximum $P2$ is observed at certain optimum value of P_k (Figure 3).

Table 4 presents examples of additional product flows of the M*-cascades that correspond to the maximal values of $P2$ with the restrictions set for $^{232,234}U$ met.

Table 4. Parameters of additional product flows of M*-cascades and ordinary cascades with maximal product flows $P2$.

MARC	k	P_k (g/s)	ΣL (g/s)	C_{ik}^P (%)				C_i^{P235} (%)	$P2$ (g/s)	$C2_i^P$ (%)		
				^{232}U	^{234}U	^{235}U	^{236}U			^{232}U	^{234}U	^{232}U
$^{232,236}U$	14	4.4	2357	5.5×10^{-10}	4.9×10^{-3}	0.64	0.344	4.95	0.36	2.5×10^{-9}	4.3×10^{-2}	3.78
$^{232,238}U$	29	3.0	2821	1.7×10^{-9}	3.5×10^{-3}	0.52	0.339	3.94	0.16	9.0×10^{-9}	4.0×10^{-2}	5.10
$^{234,235}U$	26	3.2	2576	2.0×10^{-9}	5.2×10^{-3}	0.66	0.362	4.50	0.28	8.8×10^{-9}	4.4×10^{-2}	3.80
$^{235,236}U$	32	1.4	3094	1.8×10^{-9}	2.9×10^{-3}	0.49	0.368	3.41	0.06	9.9×10^{-9}	3.5×10^{-2}	6.29

These examples show that the best matched abundance ratio cascade is a cascade with key components $^{232,236}U$. Moreover, the product flow of the ordinary cascade contains the minimal quantity of the impurity component ^{236}U. The total flow ΣL in this M*-cascade is significantly lower as compared to other cases.

Table 5 shows the isotopic compositions of primary product flows of the matched abundance ratio cascades, which are the best in terms of the flow *P*2. The parameters specified meet the ASTM C 996–10 requirements set for the low enriched uranium.

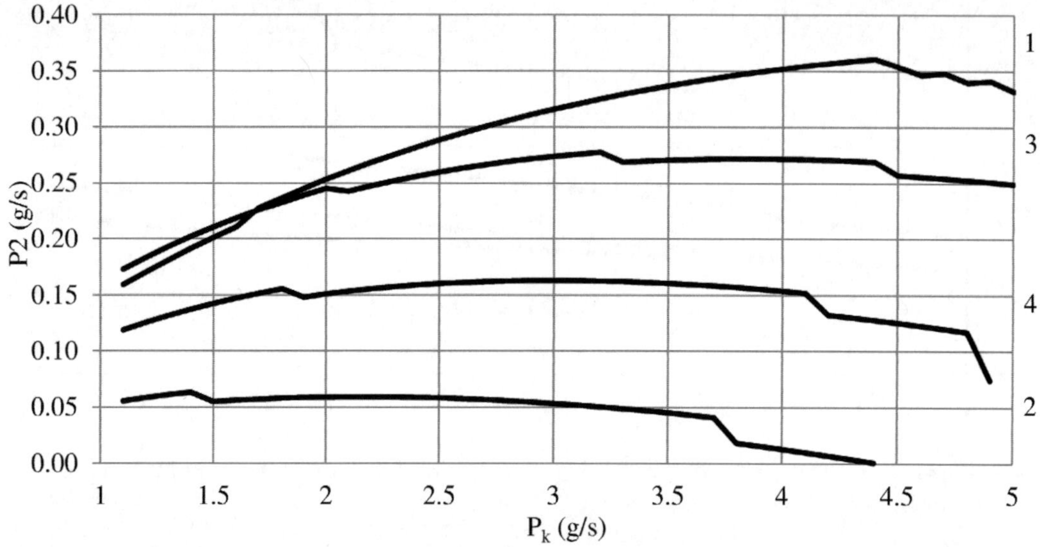

Figure 3. Product flow of ordinary cascade in function of additional product flow of M*-cascade: 1 – key components $^{232,236}U$; 2 – key components $^{232,238}U$; 3 – key components $^{234,235}U$; 4 – key components $^{235,236}U$.

Table 5. Parameters of the primary product flows of M*-cascades with maximal product flows *P*2.

MARC	C_i^P(%) ^{232}U	^{234}U	^{235}U	^{236}U	*P*(g/s)	*P*2 (g/s)
$^{232,236}U$	1.65×10^{-6}	0.91	0.36	1.128	1.65×10^{-6}	0.146
$^{232,238}U$	9.04×10^{-7}	1.65	0.16	1.068	9.04×10^{-7}	0.090
$^{234,235}U$	1.22×10^{-6}	1.23	0.28	1.109	1.22×10^{-6}	0.115
$^{235,236}U$	6.73×10^{-7}	2.22	0.06	1.034	6.73×10^{-7}	0.070

4. Conclusion

The method developed allows purifying the regenerated uranium to remove the impurity isotopes $^{232,234}U$ from the additional product flow of the matched abundance ratio cascade with a concentration of the ^{235}U in the product flow up to 5 %. The purified product can be then enriched in another cascade while meeting the requirements of ASTM C 996–10 set for these isotopes in the low enriched commercial uranium. As for achieving the maximum flow of the enriched and purified uranium, the best cascade is the M*-cascade with key components $^{232,236}U$.

References

[1] Nikipelov B V and Nikipelov V B 2002 *At. Energy Bull.* **9** 34

[2] Vlasov A A *et al* 2004 *Russian Pat. 2236053* Method of the isotopic restoration of regenerated uranium

[3] Vlasov A A *et al* 2004 *Russian Pat. 2242812* Method of the isotopic restoration of regenerated uranium

[4] Vodolazskih V V *et al* 2006 *Russian Pat. 2282904* Method of the isotopic restoration of regenerated uranium

[5] Prusakov V N, Sazykin A A, Sosnin L Yu, Utrobin D V and Chel'tsov A N 2008

J. Atomic Energy **105** 194
[6] Sulaberidze G A et al 2006 *Theor. Found. of Chem. Eng.* **40** 5
[7] Smirnov A Yu and Sulaberidze G A 2014 *J. Atomic Energy* **117** 44
[8] Palkin V A 2010 *Perspektivnye materialy* **8** 11
[9] Sulaberidze G A, Borisevich V D and Xie Quanxin 2004 *Proc. of IX Russian sci. conf. "Fiziko-himicheskie processy pri selekcii atomov i molekul" (Zvenigorod) p 78*
[10] Palkin V A 2013 *J. Atomic Energy* **115** 32
[11] Palkin V A 2015 *J. Atomic Energy* **118** 130
[12] A De La Garza, G A Garrett and J E Murphy 1961 *Cascades J. Chem. Eng. Sci.* **15** 188
[13] Palkin V A and Maslyukov E V 2012 *J. Atomic Energy* **112** 389
[14] Zhurin V A *at al* 2010 *Russian Pat. 2399971* Method of the isotopic restoration of regenerated uranium
[15] Palkin V A and Maslyukov E V 2012 *Proc. of 12th Int. Workshop on Separation Phenomena in Liquids and Gases p 21*

Journal of Physics: Conference Series **751** (2016) 012003 doi:10.1088/1742-6596/751/1/012003

Influence of parameters of a Q-cascade with a local flow "extension" on mass-transfer of components along the cascade length

A Yu Smirnov, G A Sulaberidze

National Research Nuclear University MEPhI (Moscow Engineering Physics Institute), Kashirskoe highway 31, Moscow, 115409, Russia

E-mail: AYSmirnov@mephi.ru

Abstract. The presented results complement the theory of separation cascades having a local increase of working substance flow at its internal stages (cascades with the flow "extension") used to obtain isotopic products with relatively high concentrations of intermediate components. Based on the Q-cascade model theoretical dependences of the maximum possible concentration of the target component in the end product flow of the cascade, the maximum achievable concentration of the target component inside the cascade and the value of the cascade total relative flow versus the length of the cascade section between the end and intermediate product flows - $\varepsilon_0 S_E$ were obtained. Comparison of these dependences allows to approximately determine the optimal value $\varepsilon_0 S_E$ for the given value of the target component concentration in one of the cascade product flows.

1. Introduction

In the separation of multicomponent isotope mixtures enrichment of intermediate isotopes causes particular difficulties [1]. This is due to fact that in the case of multi-isotope mixtures separation, the intermediate components are produced at both ends of a cascade (conventionally designated as "heavy" and "light" ones) together with the utmost components, the components with the smallest and largest mass numbers. As a result, increase in the concentrations of the intermediate components in the light and heavy fractions of a three-flow (one ingoing and two outgoing flows) cascade is limited by the maximum achievable values [2].

There exist a number of specific methods for enrichment of intermediate components of multi-isotope mixtures designed to achieve maximum concentrations in the outgoing flows of a three-flow cascade. These methods are based on enrichment of a mixture in a target isotope by a series of sequential operations in the same cascade or by use of the so-called dual cascade, where the product flow from the first cascade serves as a feed flow of the second one [3]. Alternatively, one can use the multi-flow cascades [4-7].

In the papers [6–7] it was demonstrated the fundamental possibility of obtaining relatively high concentrations of intermediate components in the additional product flow within the cascade having an "extension" of a working substance flow at its internal stage. The basic principle of such a cascade work can be explained by the following statement: changing a gradient of a working substance flow at one of the internal stages of a cascade leads to change of the concentration gradients of all mixture components over cascade stages. It was shown that increasing the concentration of the intermediate target component can be achieved at the internal cascade stages by selection of a certain gradient flow function. Besides, an additional product flow at one of these cascade stages enables to achieve a higher target component concentration as compared to the maximum achievable one [5-7].

However, the results of [5-7] do not answer the question of the maximum possible effect of increasing the concentration of the intermediate component in the cascade with the "extension" as well as about the influence of the parameters of " extension " (the "flow jump", etc.) on the mass transfer characteristics of the desired intermediate component in such a cascade.

The aim of this work is to study the mass transfer regularities of the intermediate target component in the cascade with the flow "extension" depending on the length of the section with the flow "extension". As the object of the study the Q-cascade [8] with the flow "extension" is considered. Analytical dependences to assess the maximum possible concentration of the target component in the end product flow of the cascade with flow "extension" and its internal stages are obtained.

2. Mathematical model

The detailed mathematical model of the *Q*-cascade with the flow "extension" is presented in [6-7]. In this paper its brief description is presented.

Consider the cascade, external parameters which are power flow F with concentrations of the components C_i^F , and three exit streams: an end product flow P with concentrations of the components C_i^P, an additional product flow E withdrawn from an intermediate stage of the cascade with concentrations C_i^E and a waste flow W with concentrations C_i^W (Figure 1.). The number of stages in the enriching and depleted sections of the cascade is denoted, respectively, S_P and the S_W, the number of stages between the two product flows is denoted S_E. One supposes that the flow of the working substance $L(s)$ is continuous over the length of the cascade, with the exception of the points in which the additional product is withdrawn and the cascade feed flow is injected. Also, suppose that that the flow of the working substance vanishes at the cascade ends [6, 8]. In the point where the additional product is withdrawn the flow of the working substance increases abruptly in the K-times.

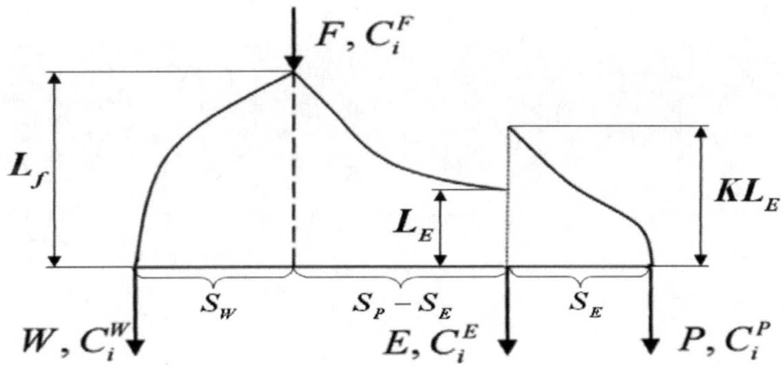

Figure 1. Schematic drawing of a cascade with extension of a flow profile.

Expressions for the calculation of external parameters and total flow of this cascade in the steady-state operation are as follows [6-7]:
for the enriching section:

$$\frac{P}{F}=\sum_{j=1}^{m}\left\{\frac{C_j^F\cdot\dfrac{e^{Q_j^{\mathrm{I}}S_W}-1}{Q_j^{\mathrm{I}}}}{\left[\dfrac{1-e^{-Q_j^{\mathrm{II}}S_E}}{K\cdot Q_j^{\mathrm{II}}\cdot e^{Q_j^{\mathrm{I}}(S_P-S_E)}}+\dfrac{1-e^{-Q_j^{\mathrm{I}}(S_P-S_E)}}{Q_j^{\mathrm{I}}}+\dfrac{e^{Q_j^{\mathrm{I}}S_W}-1}{Q_j^{\mathrm{I}}}\right]+\dfrac{E}{P}\dfrac{C_j^E}{C_j^P}\cdot\left[\dfrac{1-e^{-Q_j^{\mathrm{I}}(S_P-S_E)}}{Q_j^{\mathrm{I}}}+\dfrac{e^{Q_j^{\mathrm{I}}S_W}-1}{Q_j^{\mathrm{I}}}\right]}\right\},\quad(1)$$

$$C_j^P=\frac{F}{P}\cdot\frac{C_j^F\cdot\dfrac{e^{Q_j^{\mathrm{I}}S_W}-1}{Q_j^{\mathrm{I}}}}{\left[\dfrac{1-e^{-Q_j^{\mathrm{II}}S_E}}{K\cdot Q_j^{\mathrm{II}}\cdot e^{Q_j^{\mathrm{I}}(S_P-S_E)}}+\dfrac{1-e^{-Q_j^{\mathrm{I}}(S_P-S_E)}}{Q_j^{\mathrm{I}}}+\dfrac{e^{Q_j^{\mathrm{I}}S_W}-1}{Q_j^{\mathrm{I}}}\right]+\dfrac{E}{P}\dfrac{C_j^E}{C_j^P}\cdot\left[\dfrac{1-e^{-Q_j^{\mathrm{I}}(S_P-S_E)}}{Q_j^{\mathrm{I}}}+\dfrac{e^{Q_j^{\mathrm{I}}S_W}-1}{Q_j^{\mathrm{I}}}\right]}\quad(2)$$

for the depleted section:

$$\frac{W}{F}=\sum_{j=1}^{m}\left\{\frac{C_j^F\cdot\left\{\left[\dfrac{1-e^{-Q_j^{\mathrm{II}}S_E}}{K\cdot Q_j^{\mathrm{II}}\cdot e^{Q_j^{\mathrm{I}}(S_P-S_E)}}+\dfrac{1-e^{-Q_j^{\mathrm{I}}(S_P-S_E)}}{Q_j^{\mathrm{I}}}\right]+\dfrac{E}{P}\dfrac{C_j^E}{C_j^P}\cdot\dfrac{1-e^{-Q_j^{\mathrm{I}}(S_P-S_E)}}{Q_j^{\mathrm{I}}}\right\}}{\left[\dfrac{1-e^{-Q_j^{\mathrm{II}}S_E}}{K\cdot Q_j^{\mathrm{II}}\cdot e^{Q_j^{\mathrm{I}}(S_P-S_E)}}+\dfrac{1-e^{-Q_j^{\mathrm{I}}(S_P-S_E)}}{Q_j^{\mathrm{I}}}+\dfrac{e^{Q_j^{\mathrm{I}}S_W}-1}{Q_j^{\mathrm{I}}}\right]+\dfrac{E}{P}\dfrac{C_j^E}{C_j^P}\cdot\left[\dfrac{1-e^{-Q_j^{\mathrm{I}}(S_P-S_E)}}{Q_j^{\mathrm{I}}}+\dfrac{e^{Q_j^{\mathrm{I}}S_W}-1}{Q_j^{\mathrm{I}}}\right]}\right\},\quad(3)$$

$$C_j^W=\frac{F}{W}\cdot\frac{C_j^F\cdot\left\{\left[\dfrac{1-e^{-Q_j^{\mathrm{II}}S_E}}{K\cdot Q_j^{\mathrm{II}}\cdot e^{Q_j^{\mathrm{I}}(S_P-S_E)}}+\dfrac{1-e^{-Q_j^{\mathrm{I}}(S_P-S_E)}}{Q_j^{\mathrm{I}}}\right]+\dfrac{E}{P}\dfrac{C_j^E}{C_j^P}\cdot\dfrac{1-e^{-Q_j^{\mathrm{I}}(S_P-S_E)}}{Q_j^{\mathrm{I}}}\right\}}{\left[\dfrac{1-e^{-Q_j^{\mathrm{II}}S_E}}{K\cdot Q_j^{\mathrm{II}}\cdot e^{Q_j^{\mathrm{I}}(S_P-S_E)}}+\dfrac{1-e^{-Q_j^{\mathrm{I}}(S_P-S_E)}}{Q_j^{\mathrm{I}}}+\dfrac{e^{Q_j^{\mathrm{I}}S_W}-1}{Q_j^{\mathrm{I}}}\right]+\dfrac{E}{P}\dfrac{C_j^E}{C_j^P}\cdot\left[\dfrac{1-e^{-Q_j^{\mathrm{I}}(S_P-S_E)}}{Q_j^{\mathrm{I}}}+\dfrac{e^{Q_j^{\mathrm{I}}S_W}-1}{Q_j^{\mathrm{I}}}\right]}.\quad(4)$$

Also, the following relations are taken place [6-7]:

$$C_j^E=C_j^P\cdot\frac{1-e^{-Q_j^{\mathrm{II}}S_E}}{Q_j^{\mathrm{II}}}\Bigg/\sum_{i=1}^{m}\left(C_i^P\cdot\frac{1-e^{-Q_i^{\mathrm{II}}S_E}}{Q_i^{\mathrm{II}}}\right),\quad(5)$$

$$\frac{C_j^E}{C_j^P}=\frac{C_n^E}{C_n^P}\cdot\frac{Q_n^{\mathrm{II}}}{Q_i^{\mathrm{II}}}\cdot\frac{1-e^{-Q_i^{\mathrm{II}}S_E}}{1-e^{-Q_n^{\mathrm{II}}S_E}}. \tag{6}$$

The total relative flow of the cascade can be calculated as follows:

$$\begin{aligned}\frac{\sum L}{2P}=\sum_{j=1}^{m}\Bigg\{\Bigg[&\frac{W}{P}C_j^W\left(e^{Q_i^I S_W}-1-Q_i^I S_W\right)+C_j^P\left(e^{-Q_i^I(S_P-S_E)}-1+Q_i^I(S_P-S_E)\right)+\\&+\frac{E}{P}C_j^E\left(e^{-Q_i^I(S_P-S_E)}-1+Q_i^I(S_P-S_E)\right)\Bigg]\Big/\left(Q_i^I\right)^2+\\C_j^P&\left[\frac{e^{-Q_i^{II}S_E}-1+Q_i^{II}S_E}{\left(Q_i^{II}\right)^2}+\frac{\left(e^{-Q_i^{II}S_E}-1\right)\left(1-e^{Q_i^I(S_P-S_E)}\right)+Q_i^{II}S_E}{KQ_i^IQ_i^{II}e^{Q_i^I(S_P-S_E)}}\right]\Bigg\}.\end{aligned} \tag{7}$$

where $Q_i^{I,II}$ – are constants which role will be considered below.

From the theory of the Q-cascade it is followed that the values Q_i connected with each other by conditions [8]:

$$Q_i-Q_k=\varepsilon_{ik}. \tag{8}$$

For the molecular-kinetic separation methods [8]:

$$Q_i=\varepsilon_0(M-M_i), \tag{9}$$

where M_i – the mass number of the i-th component, ε_0 – enrichment coefficient per unit mass difference, M – the parameter which choice in the range $M_1\le M\le M_m$ allows to calculate all values Q_i and, consequently, the flow distribution $L(s)$ [8].

Parameter M plays an important role: setting value M determines the flow distribution $L(s)$ and the ratio of the product flow to the waste one. All this causes that components with mass numbers $M_i<M$ enriched with the lightest component to "light" end of the cascade, and components with $M_i>M$ - enriched with the heaviest component to the "heavy" end of the cascade. Within the Q-cascade model by varying of the M one can achieve accumulation of the intermediate component in the internal stages of the cascade. For example, by setting M, so that in one section - $M_n<M$ (n-target component number), and in the other - $M_n>M$. Under this condition the flow distribution $L(s)$ is such that the n-th component will begin to enrich in opposite directions in different sections of the cascade. In this case, the n-th component will be "locked" on intermediate stages of the cascade. As a result, the concentration of n-th component in the maximum will increase, in comparison with the Q-cascade without extension. Thus, suppose that in the sections between the waste flow blade and the additional product flow the parameter M is set to $M^{\mathrm{I}}>M_n$ with its corresponding constant Q_i^I ($i=\overline{1,m}$), and in the rest section M takes the value M^{II} with corresponding constants Q_i^{II} ($i=\overline{1,m}$).

3. Results and discussion

Consider the effect of some parameters of the Q-cascade with the "extension' on the mass transfer of components along its length. Let us assess the impact of the value of S_E (the length of the section located between the end and additional product flows) to mass transfer of the target intermediate component along the length of the cascade. Detailed analysis of these relationships allows to define the maximum possible value of the target component concentrations in the end product flow and the maximum concentration of a given component within the cascade.

As follows from the analysis of relations connecting the external and internal parameters of the cascade, for the study the influence of the S_E on mass transfer of the target component, the most interesting are two cases:

Case 1: $\varepsilon_0 S_P \gg 1$, $\varepsilon_0 S_P \gg \varepsilon_0 S_E$;

Case 2: $\varepsilon_0 S_P \gg 1$, $\varepsilon_0 S_P \sim \varepsilon_0 S_E$.

The first case corresponds to the "long" cascade in which the length of the end section is small compared to the entire length of the enriching section (see Fig. 1). This cascade corresponds to small influence of "extension" on the component mass transfer along its length. The second case corresponds to a cascade in which a significant part of the enriching section coincides with the section with "extension". Note also that in both cases the condition $\varepsilon_0 S_W \gg 1$ is taken. For small values $\varepsilon_0 S_W$ and $\varepsilon_0 S_P$ (in the case of "short" cascades) the concentration distribution of the intermediate component may not maximum at internal cascade stages which makes such cascades inappropriate to obtain relatively high concentrations of intermediate components [2, 7].

After limiting transitions in (2) it is possible to obtain formulas for the maximum achievable concentration of the target component with n number at the end product flow of the cascade for the cases 1 and 2, respectively:

$$\left(C_n^P\right)_{\max} = \frac{C_n^F \Big/ \left(1+\dfrac{E}{P}\dfrac{C_i^E}{C_i^P}\right)}{\sum\limits_{j=1}^{n} C_j^F \Big/ \left(1+\dfrac{E}{P}\dfrac{C_j^E}{C_j^P}\right)}, \tag{10}$$

$$\left(C_n^P\right)_{\max} = \frac{C_n^F \Big/ \left(\dfrac{Q_n^I}{KQ_n^{II}}\dfrac{1-e^{-Q_n^{II}S_E}}{e^{Q_n^I S_W}}+1+\dfrac{E}{P}\dfrac{C_n^E}{C_n^P}\right)}{\sum\limits_{j=1}^{n-1} C_j^F \Big/ \left(1+\dfrac{E}{P}\dfrac{C_j^E}{C_j^P}\right)+C_n^F \Big/ \left(\dfrac{Q_n^I}{KQ_n^{II}}\dfrac{1-e^{-Q_n^{II}S_E}}{e^{Q_n^I S_W}}+1+\dfrac{E}{P}\dfrac{C_n^E}{C_n^P}\right)}. \tag{11}$$

From these ratios it is evident that their use for assessment is difficult due to the fact that unknown values are E/P, C_j^E, C_j^P explicitly included in them. However, relations (10), (11) can be used for practical assessments with condition $E/P \to 0$, provided that corresponds to an absence of the additional product flow or case of its negligible smallness in relation to other external cascade flows. Using the condition $E/P \to 0$ the above relations (10) and (11) can be rewritten as:

$$\left(C_n^P\right)_{\max} = \frac{C_n^F}{\sum\limits_{j=1}^{n} C_j^F}, \tag{12}$$

$$\left(C_n^P\right)_{\max} = \frac{C_n^F}{\sum\limits_{j=1}^{n-1} C_j^F + C_n^F \Big/ \left(\dfrac{Q_n^I}{KQ_n^{II}}\dfrac{1-e^{-Q_n^{II}S_E}}{e^{Q_n^I S_W}-1}+1\right)}. \tag{13}$$

It can be seen that the expression (12) coincides with the corresponding expression obtained for monotonic profile cascades [2]. Thus, in case of smallness of the section with "extension" with respect to the entire length the cascade mass transfer regularities in cascade with the flow "extension" will be

Journal of Physics: Conference Series **751** (2016) 012003 doi:10.1088/1742-6596/751/1/012003

similar to those for cascade of monotonic flow profile. With extending of the section with "extension" concentration distribution of the target component will be more different from the corresponding distribution in the cascade of monotonic flow profile that will cause changes in the maximum achievable concentration of the target component in the end product flow of the cascade.

Fig. 2 presents us a graph showing the analytical dependence of the maximum concentration of the isotope ^{183}W on the length of the cascade at the section with the “extension” - $\varepsilon_0 S_E$. The following parameters were given: $\varepsilon_0 S_W = 5{,}94$, *K*=1,5, $M^I = 183{,}5$, $M^{II} = 182{,}5$. From the analysis of the character of the curve on Fig. 2 it follows that with increasing of the length of the section with “extension” the target component concentration in the end product decreases. Obviously, this is due to the fact that the target component begins to actively accumulate inside the cascade.

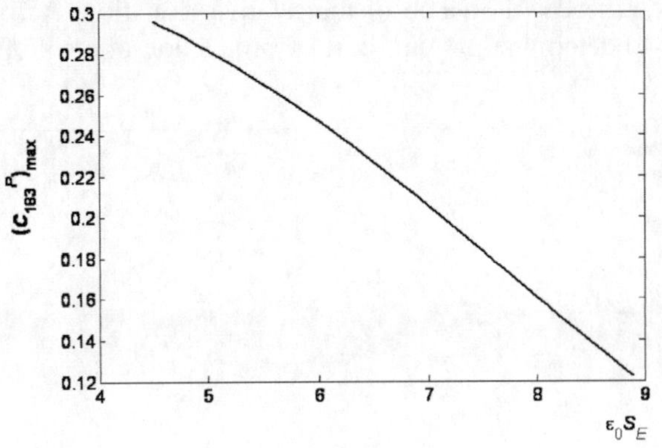

Figure 2. Analytical dependence of the limiting concentration of ^{183}W in the end product flow versus length of the utmost section of the cascade with “extension” $\varepsilon_0 S_E$.

The conclusions based on analysis of the dependence shown in Fig. 2, fully confirmed by Fig. 3, which shows the concentration dependence of ^{183}W in the end product flow of the cascade with “extension” derived from direct calculation of the cascade with different $\varepsilon_0 S_E$ which were varied from $\varepsilon_0 S_P/2$ to $\varepsilon_0 S_P$. In the calculations, the following values were given: $\varepsilon_0 S_W = 5{,}94$, $\varepsilon_0 S_P = 9$, *K*=1,5, $M^I = 183{,}5$, $M^{II} = 182{,}5$, $E/P = 0$.

From a comparison of Fig. 2 and 3 it is evident that both dependences consistent qualitatively and quantitatively. This fact indicates the correctness of the (12) formula.

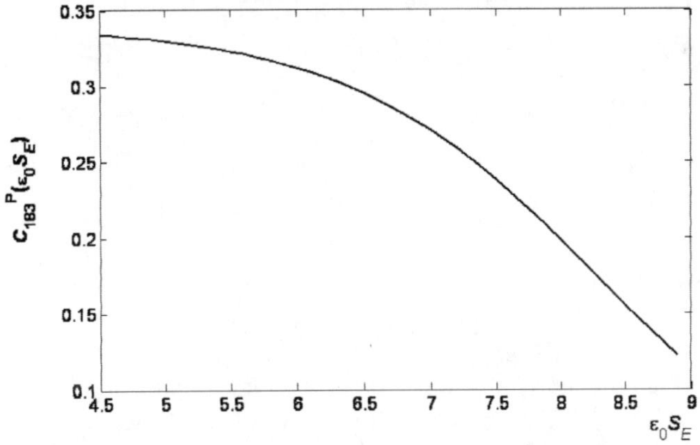

Figure 3. Dependence of the concentration of ^{183}W in the end product flow versus length of the utmost section of the cascade with “extension” $\varepsilon_0 S_E$.

On Fig. 4, 5 are shown dependences of the maximum concentration of the target component (target component concentration at the point of its maximum inside the cascade) and relative total flow of the cascade versus the length of end section $\varepsilon_0 S_E$. It can be seen that both of the curves has maximum values. However, it should be noted that the maxima of both dependencies are achieved at different values of $\varepsilon_0 S_E$. Comparison of the curves in Fig. 3-5 allows to determine approximately the optimum value $\varepsilon_0 S_E$ for the given (or approximately defined) values of the target component concentration in one (or both) of the cascade product flows. Found in this way the value $\varepsilon_0 S_E$ can then be used as an initial approximation of this parameter in the strict solution of the problem of optimization of Q-cascade with “extension”. Furthermore, approximately the results for a given value $\varepsilon_0 S_E$ of the optimal parameters of the cascade (the value of the relative total flow etc.) can be used during the estimations, for example, to determine the unit cost of production of stable isotopes in cascade with “extension”.

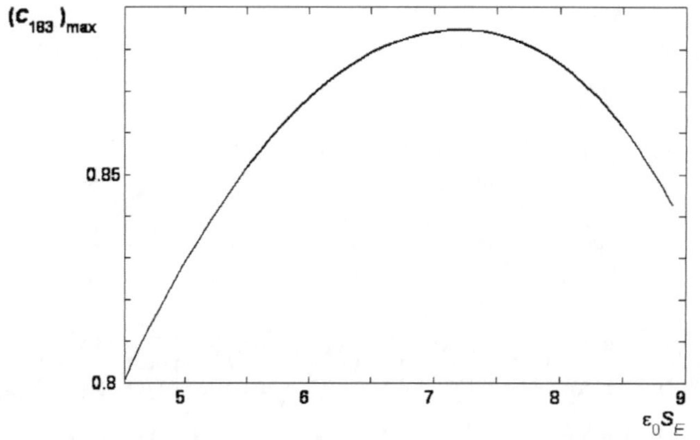

Figure 4. Dependence of the maximum concentration of ^{183}W inside the cascade versus length of the utmost section of the cascade with “extension” $\varepsilon_0 S_E$.

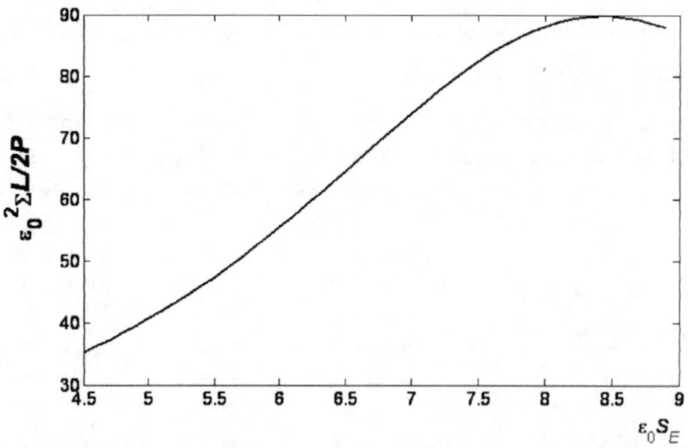

Figure 5. Dependence of the cascade relative total versus length of the utmost section of the cascade with “extension” $\varepsilon_0 S_E$.

It is quite obvious that the mass transfer regularities described above in the absence of the additional product flow will not differ from the relevant laws in cascades with the “extension” in case if the additional product will be quite small $E/P \ll 1$. Therefore, obtained relations can be used to find the "good" initial approximations in solving problems of calculation and optimization of the parameters of cascades with "extension" with the additional product flow.

Conclusion

The presented results complement the theory of separation cascades having a local increase of working substance flow at its internal stages (cascades with the flow "extension") used to obtain isotopic products with relatively high concentrations of intermediate components. Based on the Q-cascade model theoretical dependences of the maximum possible concentration of the target component in the end product flow of the cascade, the maximum achievable concentration of the target component inside the cascade and the value of the cascade total relative flow versus the length of the cascade section between the end and intermediate product flows - $\varepsilon_0 S_E$ were obtained. Comparison of these dependences allows to approximately determine the optimal value $\varepsilon_0 S_E$ for the given value of the target component concentration in one of the cascade product flows.

Acknowledgement

This research is carried out in the framework of the joint Sino-Russian project "Separation of multicomponent isotope mixtures by the method of centrifugation to obtain new functional materials for nuclear energy, fundamental physics and medicine" supported by the Russian Fund for Basic Research (RFBR) under the contract no. 16-58-53058 GFEN_a and the National Natural Science Foundation of China (NSFC) under the contract no. 1151101155 (Sulaberidze G.A.), and the grant No.14.Y30.16.6284-MK of the President of the Russian Federation to support young scientists (Smirnov A.Yu.).

References

[1] Sulaberidze G A, Borisevich V D 2001 *Sep. Sci. Technol.* **36** 1769–817
[2] Minenko V P 1972 *At. Energy* **33** 814–6
[3] Vetsko V M, Laguntsov N I, Levin E V et al 1987 *At. Energy* **63** 692-7
[4] Sulaberidze G A, Borisevich V D, Xie Q X 2006 *Theor. Found. Chem. Eng.* **40** 5-14
[5] Smirnov A Yu, Sulaberidze G A, Xie Q, Borisevich V D 2015 *Chem. Eng. Res. Des.* **95** 47
[6] Smirnov A Yu, Sulaberidze G A, Borisevich V D 2010 *Theor. Found. Chem. Eng.* **44** 888-96
[7] Smirnov A Yu, Sulaberidze G A 2013 *Theor. Found. Chem. Eng.* **47** 375-80
[8] Zeng S, Sulaberidze G A, Jiang D, Borisevich V D, Smirnov A Yu 2012 *Sep. Sci. Technol.* **47** 1591–5

Enhancing the performance of Q-cascade for separating intermediate components

S Zeng[1], D J Jiang[1], Y N Zhang[1], V D Borisevich[2], G A Sulaberidze[2], A Yu Smirnov[2]

[1] Department of Engineering Physics, Tsinghua University, Beijing 100084, China
[2] National Research Nuclear University MEPhI (Moscow Engineering Physics Institute), Kashirskoe highway 31, Moscow, 115409, Russia

E-mail: zengs@tsinghua.edu.cn

Abstract. It has been shown recently that Q-cascade with an expansion of the entering flow at the intermediate withdrawal point is able to obtain a concentration of an intermediate component far exceeding the concentration limit available from an end withdrawal. To enhance the applicability of this approach, it is necessary to reduce the relative total flow while maintaining the concentration of the intermediate target component unchanged. Optimization is carried out by using the technique of cascade segmentation, and using the mass numbers of the virtual components in the segments and the lengths of the segments as decision variables. The results demonstrate that the relative total flow is considerably reduced through optimization.

1. Introduction

Because the concentration is limited of an intermediate component available from an end withdrawal of a cascade, the separation has to be performed successively two or more times, with a withdrawal of the previous separation as the feed of the next to obtain its concentration to the required level. The whole separation process may involve one or more cascades, possibly, of different configurations. This definitely increases the complexity of separation, and consequently increases the cost. Therefore, it is of great interest to explore new separation cascades or new ways of separation to enrich more efficiently an intermediate component to the required concentration.

The model cascade with continuous profile, Q-cascade, has been recently investigated quite extensively. It has also been applied to obtaining an intermediate component by employing a technique which may be referred to as profile broadening [1][2]. With this technique the concentration of an intermediate component can be enriched to any level. However, as we known, for a separation task, there are a number of cascades that are able to satisfy the requirement. Taking into account practical applications, of all cascades in fulfilling the same separation task, a cascade that has a smaller relative total flow is considered to have a better separation performance and is more competitive. Here we present a way of enhancing the performance of a Q-cascade.

2. Q-cascade with profile broadening

Figure 1 illustrates a Q-cascade profile with profile broadening in the enriching section. The enriching section is split into two segments, whose lengths are $S_{\mathrm{P}}^{(1)}$ and $S_{\mathrm{P}}^{(2)}$, respectively. The location of a stage in a segment is indicated by l, and counted in the way as shown in the figure. The entering flow profiles of the two segments are $L_{\mathrm{P}}^{(1)}$ and $L_{\mathrm{P}}^{(2)}$, respectively. For the stripping section, the length is $S_{\mathrm{W}}^{(1)}$, and the profile is $L_{\mathrm{W}}^{(1)}$. F is the feed of the cascade, and P and W are, conventionally, referred to as the product and the waste. The intermediate withdrawal W_{m} takes place at the border of the first and the second segments, where is the location of the profile broadening.

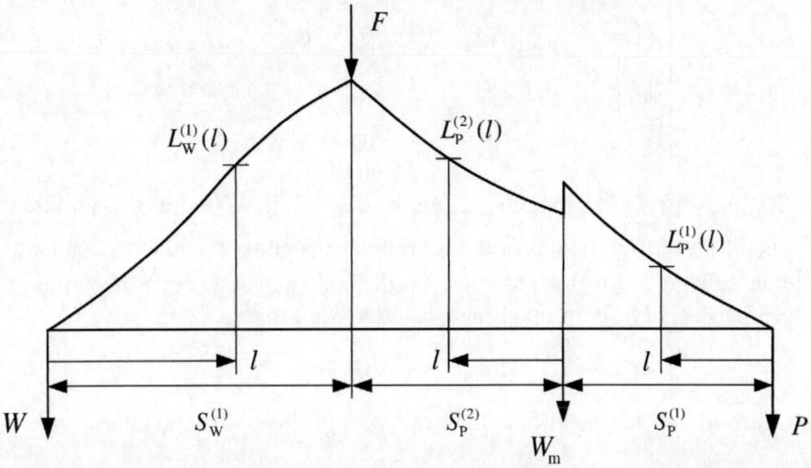

Figure 1. Illustration of the profile of a Q-cascade with flow broadening at the intermediate withdrawal.

According to the Q-cascade theory [3], for the stripping section and the enriching section, we have:

$$C_{i,\mathrm{W}}^{(n)}(l)L_{\mathrm{W}}^{(n)}(l)=2WC_i^{\mathrm{W}}f_{i,\mathrm{W}}^{(n)}(l) \tag{1}$$

$$C_{i,\mathrm{P}}^{(n)}(l)L_{\mathrm{P}}^{(n)}(l)=2PC_i^{\mathrm{P}}f_{i,\mathrm{P}}^{(n)}(l)+\begin{cases}0 & (n<m)\\ 2W_{\mathrm{m}}C_i^{\mathrm{m}}f_{i,\mathrm{P}}^{n,m}(l) & (n\ge m)\end{cases} \tag{2}$$

Here, $C_{i,\mathrm{W}}^{(n)}(l)$ $(n=1,2,\ldots,N_{\mathrm{W}})$ is the concentration of the i-th component at stage l of the n-th segment in the stripping section, and similarly, $C_{i,\mathrm{P}}^{(n)}(l)$ $(n=1,2,\ldots,N_{\mathrm{P}})$ the concentration of the i-th component at stage l of the n-th segment in the enriching section. The component index for the isotope mixture $i=1,2,\ldots,N_{\mathrm{C}}$, with N_{C} the number of components in the multi-component isotope mixture to be separated. N_{W} and N_{P} are the numbers of segments in the stripping and enriching sections, respectively. The intermediate withdrawal takes place at the beginning of the m-th segment. C_i^{W}, C_i^{P}, and C_i^{m} are respectively the concentrations of the i-th component in the waste, product and intermediate withdrawals. Obviously, for the cascade shown in Figure 1, $N_{\mathrm{W}}=1$, $N_{\mathrm{P}}=2$, and $m=2$. In equations (1) and (2),

$$f_{i,\mathrm{W}}^{(n)}(l)=\begin{cases}(\mathrm{e}^{Q_{i,\mathrm{W}}^{(n)}l}-1)(Q_{i,\mathrm{W}}^{(n)})^{-1}+\mathrm{e}^{Q_{i,\mathrm{W}}^{(n)}l}f_{i,\mathrm{W}}^{(n-1)}(S_{\mathrm{W}}^{(n-1)}) & (n\ge 1)\\ 0 & (n=0)\end{cases} \tag{3}$$

$$f_{i,\mathrm{P}}^{(n)}(l)=\begin{cases}(1-\mathrm{e}^{-Q_{i,\mathrm{P}}^{(n)}l})+\mathrm{e}^{-Q_{i,\mathrm{P}}^{(n)}l}f_{i,\mathrm{P}}^{(n-1)}(S_{\mathrm{P}}^{(n-1)}) & (n\neq m)\\ (1-\mathrm{e}^{-Q_{i,\mathrm{P}}^{(n)}l})(Q_{i,\mathrm{P}}^{(n)})^{-1}+\mathrm{e}^{-Q_{i,\mathrm{P}}^{(n)}l}f_{i,\mathrm{P}}^{(n-1)}(S_{\mathrm{P}}^{(n-1)})J_{\mathrm{L}} & (n=m)\\ (1-\mathrm{e}^{-Q_{i,\mathrm{P}}^{(n)}l})(Q_{i,\mathrm{P}}^{(n)})^{-1} & (n=1)\end{cases} \tag{4}$$

$$f_{i,\mathrm{P}}^{n,m}(l)=\begin{cases}(1-\mathrm{e}^{-Q_{i,\mathrm{P}}^{(n)}l})(Q_{i,\mathrm{P}}^{(n)})^{-1}+\mathrm{e}^{-Q_{i,\mathrm{P}}^{(n)}l}f_{i,\mathrm{P}}^{n-1,m}(S_{\mathrm{P}}^{(n-1)}) & (n>m)\\ (1-\mathrm{e}^{-Q_{i,\mathrm{P}}^{(n)}l})(Q_{i,\mathrm{P}}^{(n)})^{-1} & (n=m)\\ 0 & (n<m)\end{cases} \tag{5}$$

Note that $n=1,2,\ldots,N_{\mathrm{W}}$ for the stripping section, and $n=1,2,\ldots,N_{\mathrm{P}}$ for the enriching section. In equation (4), J_{L} is the parameter, here referred to as the broadening parameter, that takes into account the flow profile broadening. At the common border of two segments, if without flow profile broadening, the profile should be continuous across the border, that is,

$$L_{\mathrm{P}}^{(n-1)}(S_{\mathrm{P}}^{(n-1)})=L_{\mathrm{P}}^{(n)}(0) \tag{6}$$

If a broadening is carried out, for instance at the beginning of the b-th segment,

$$L_{\mathrm{P}}^{(b-1)}(S_{\mathrm{P}}^{(b-1)})=J_{\mathrm{L}}L_{\mathrm{P}}^{(b)}(0) \tag{7}$$

We let $b=m$, which means that the location where there is the intermediate withdrawal is the location where the broadening happens. It is clear that the meaning of the broadening parameter is actually the ratio of the flows at the border of two connecting segments. The quantities $Q_{i,\mathrm{W}}^{(n)}$ and $Q_{i,\mathrm{P}}^{(n)}$ are the Q-parameters of the n-th segment in the stripping and enriching sections, respectively, and defined as

$$Q_{i,\mathrm{W}}^{(n)}=\varepsilon_0(M_{\mathrm{W}}^{*(n)}-M_i),\quad Q_{i,\mathrm{P}}^{(n)}=\varepsilon_0(M_{\mathrm{P}}^{*(n)}-M_i) \tag{8}$$

where $M_{\mathrm{W}}^{*(n)}$ and $M_{\mathrm{P}}^{*(n)}$ are the mass numbers of the virtual component in the corresponding segments, M_i is the mass number of the i-th component, satisfying $M_i<M_j$ if $i<j$, and ε_0 is the enrichment separation factor for unit mass difference. When $n=m$, we have

$$C_i^{\mathrm{m}}=C_i^{\mathrm{P}}f_{i,\mathrm{P}}^{(m-1)}(S_{\mathrm{P}}^{(m-1)})\left[\sum_{j=1}^{N_{\mathrm{C}}}C_j^{\mathrm{P}}f_{j,\mathrm{P}}^{(m-1)}(S_{\mathrm{P}}^{(m-1)})\right]^{-1} \tag{9}$$

The relative total flow of the cascade is defined as:

$$G_{\mathrm{RT}}=\frac{\varepsilon_0^2}{2W_{\mathrm{m}}}\left(\sum_{n=1}^{N_{\mathrm{W}}}\int_0^{S_{\mathrm{W}}^{(n)}}L_{\mathrm{W}}^{(n)}(l)\mathrm{d}l+\sum_{n=1}^{N_{\mathrm{P}}}\int_0^{S_{\mathrm{P}}^{(n)}}L_{\mathrm{P}}^{(n)}(l)\mathrm{d}l\right) \tag{10}$$

which is the cascade total flow G divided by $2W_{\mathrm{m}}/\varepsilon_0^2$, and stands for the total flow needed to obtain a unit quantity of product with the required concentration. In multi-component isotope separation, so far we have no unanimously accepted definition of value function and separative power, as in the case of a binary separation, to evaluate the performance of a separation facility. Therefore, it is reasonable to think that, to fulfill a separation task, the cascade with a minimum relative total flow has the best performance, because a less relative total flow means that a smaller number of separation units will be used. Note that in [2], the relative total flow is defined as the cascade total flow divided by $2P/\varepsilon_0^2$. We think the definition in equation (10) is more appropriate to reflect the economic consideration, as the actual product, which is the intermediate withdrawal W_{m} other than P, is directly taken into account.

3. Enhancing cascade performance

In [2], separation is carried out to separate the 4-th component of the Krypton isotopes. Table 1 gives the components of Krypton, and their mass numbers and natural concentrations.

Table 1. Krypton's components, their mass numbers and natural concentrations.

i	1	2	3	4	5	6
M_i	78	80	82	83	84	86
C_i^{F}	0.0035	0.0227	0.1156	0.1152	0.5690	0.1740

Here, we take the last separation task in [2] as the example here to demonstrate how to enhance the cascade performance. In this separation task, $C_k^{\mathrm{m}} = 0.72$ $(k = 4)$, $\varepsilon_0 S_{\mathrm{W}} = 6$, $J_{\mathrm{L}} = 1.4^{-1}$, $W_{\mathrm{m}} / P = 0.1$ $(W_{\mathrm{m}} / F \approx 2.285 \times 10^{-2})$, and the mass numbers of the virtual component $M_{\mathrm{W}}^{*(1)} = M_{\mathrm{P}}^{*(2)} = 83.5$, $M_{\mathrm{P}}^{*(1)} = 82.5$. To fulfill this task, $G_{\mathrm{RT}} = 548.72$ in [2]. However, this is not the best cascade for this separation task. Without stating this clearly, people might mistake this cascade as the cascade of choice and compare it with other cascades. Then the conclusion would be misleading. In order to compete with other ways of separation, it is necessary to make G_{RT} minimum. For this purpose, we set up the following separation task:

$$\begin{aligned} &\text{given:}\, W_{\mathrm{m}} / F,\, J_{\mathrm{L}} \\ &\min G_{\mathrm{RT}}(M_{\mathrm{W}}^{*(1)}, M_{\mathrm{P}}^{*(1)}, M_{\mathrm{P}}^{*(2)}, \varepsilon_0 S_{\mathrm{W}}^{(1)}, \varepsilon_0 S_{\mathrm{P}}^{(1)}, \varepsilon_0 S_{\mathrm{P}}^{(2)}) \\ &\text{s.t. } C_k^{\mathrm{m}} = 0.72 \end{aligned} \tag{11}$$

4. Results and discussions

Four separation tasks are considered, and the results are presented in Table 2.

Table 2. Results for 4 separation tasks.

Task	1	2	3	4
W_{m} / F	0.02285	0.02285	0.15	0.15
J_{L}	1.4^{-1}	1.4^{-1}	1.4^{-1}	1
$M_{\mathrm{W}}^{*(1)}$	83.5	85.0000	83.4996	83.4929
$M_{\mathrm{P}}^{*(2)}$	83.5	83.7387	83.5428	83.5621
$M_{\mathrm{P}}^{*(1)}$	82.5	82.5240	82.5360	82.5554
$\varepsilon_0 S_{\mathrm{W}}^{(1)}$	6	.202688	5.70738	5.71418
$\varepsilon_0 S_{\mathrm{P}}^{(2)}$	13.4050	8.38396	6.90026	6.92507
$\varepsilon_0 S_{\mathrm{P}}^{(1)}$	2.35987	9.10625	12.0108	12.3014
G_{RT}	548.770	86.5874	82.0462	83.3437

Task 1 is the task mentioned above and computed in [2]. To allow readers to make later comparisons, in the table six digits are given for the values obtained from computation; the rest values are the specified ones. The cascade flow profiles as well as the concentration distributions of the target component C_k, $C_k^{(n)}(l)$, are plotted in Figure 2 and Figure 3, respectively, where s is defined as:

$$s = \begin{cases} l & (0 \le l \le S_{\mathrm{W}}^{(1)}, \text{in the stripping section}) \\ S_{\mathrm{W}}^{(1)} + S_{\mathrm{P}}^{(2)} - l & (0 \le l \le S_{\mathrm{P}}^{(2)}, \text{in segment 2 of the enriching section}) \\ S_{\mathrm{W}}^{(1)} + S_{\mathrm{P}}^{(2)} + S_{\mathrm{P}}^{(1)} - l & (0 \le l \le S_{\mathrm{P}}^{(1)}, \text{in segment 1 of the enriching section}) \end{cases} \quad (12)$$

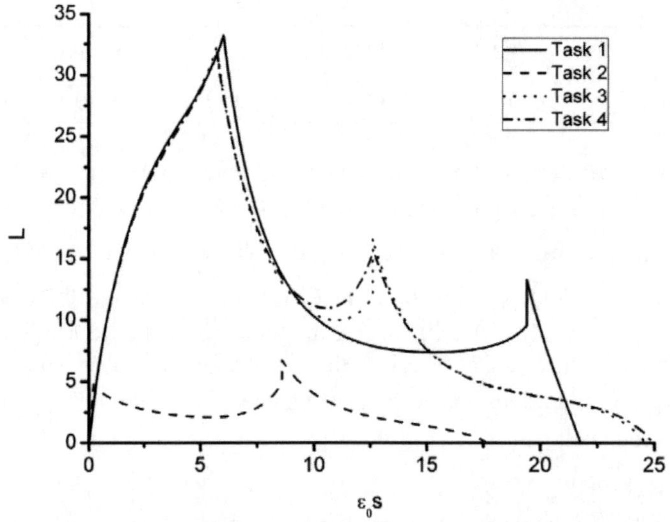

Figure 2. Flow profile distributions of the 4 tasks along cascades.

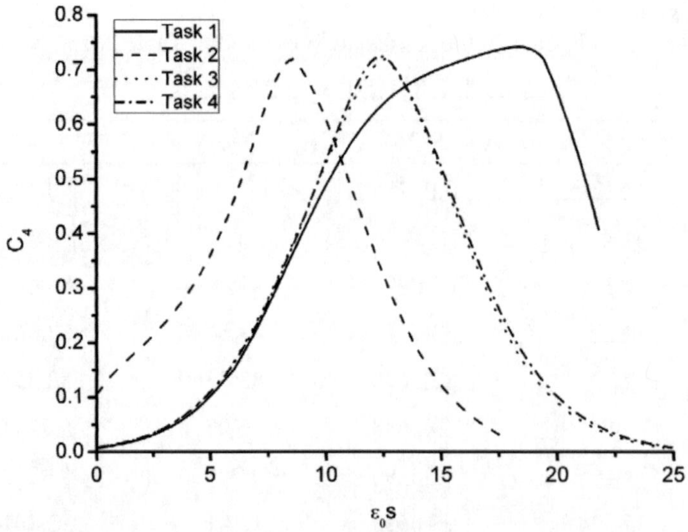

Figure 3. Concentration distributions of C_k along cascades.

The results show that taking some measures to enhance the separation performance is absolutely necessary. One can see that by employing minimization of the relative total flow, there is a reduction of at least 6 times in the relative total flow, which is very significant. In other words, to obtain the same quantity of product, the cascade can be about 6 times smaller in scale, or to use the same scale of cascade, about 6 times more product can be produced.

G_{RT} of Task 2 is larger than that of Task 3, which is surprising, because W_{m}/F of Task 2 is much smaller than that of Task 3. We see that $M_{\mathrm{W}}^{*(1)}$ has a value of 85, which is the upper bound set up for the minimization. So it is understandable that the result of Task 2 has not yet reached the real minimum. However, the larger the upper bound for $M_{\mathrm{W}}^{*(1)}$ is, the smaller $\varepsilon_0 S_{\mathrm{W}}^{(1)}$, and the smaller G_{RT}; the minimum is always obtained when $M_{\mathrm{W}}^{*(1)}$ takes its upper bound. The tendency seems to be that when $M_{\mathrm{W}}^{*(1)}$ tends to infinity, the real minimum is obtained. In this case, $\varepsilon_0 S_{\mathrm{W}}^{(1)}$ becomes zero, that is, the cascade has no stripping section. That a cascade without a stripping section can fulfill the task suggests that such a cascade has some surplus separation capability for the specified separation task, which is one of the reasons why a larger value of W_{m}/F is taken in Task 3.

It is interesting to note that the shapes of the concentration distributions in Tasks 2-4 are alike. Also note that for the 3 tasks the locations of the intermediate withdrawal are very close to or coincident with the peak locations of the concentration, which indicates that the separation capability of the cascades are made full use, whereas for Task 1 the location of the intermediate withdrawal is somewhat far away from the peak location of the concentration, which suggests that the cascade is not configured to its best separation state and some separation capability is wasted.

In Task 4, the only difference from Task 3 is $J_{\mathrm{L}} = 1$, that is, there is no cascade flow profile broadening specified. Observing Figure 2, we see clearly that the broadening is automatically incorporated into the flow profile. However, with J_{L} specified, G_{RT} is reduced slightly, which indicates that J_{L} can also be used as a parameter to further enhance the cascade performance.

5. Conclusions

The performance of Q-cascades in separating an intermediate component is enhanced significantly through optimization of the relative total flow. The optimization is minimizing the relative total flow by using the mass numbers of the virtual component in and the lengths of the cascade segments as the decision variables.

In comparison with a cascade in literature for separating the ^{83}Kr isotope, it is achieved that a cascade with a relative total flow of being approximately six times less is able to produce the same amount of the target component of the required concentration, or a cascade with about the same amount of relative total flow to produce approximately six times more product.

The flow profile broadening is automatically taken into account in the optimization. However, the benefit of the broadening can be further promoted to reduce the relative total flow by using the broadening parameter.

Acknowledgment

This research is supported by National Natural Science Foundation of China (Grant No.11575097), National Key Basic Research and Development Program of China (Grant No.2014CB744100), and the Russian Fund for Basic Research (Grant No.16-58-53058 GFEN_a).

References

[1] Smirnov A Yu and Sulaberidze G A 2013 *Theor. Found. Chem. Eng.* **47** 75–380.
[2] Smirnov A Yu, Sulaberidze G A and Borisevich V D 2010 *Theor. Found. Chem. Eng.* **44** 888.
[3] Kolokol'tsov N A, Minenko V P, Nikolaev B I et al 1970 *At. Energy* **29** 1193–8.
[4] Zhang Y N, Zeng S, Jiang D J, Borisevich V D and Sulaberidze G A 2015 *Chem. Eng. Res. Des.* **100** 509–17.

Evaluating the effectiveness of dilution of the recovered uranium with depleted uranium and low-enriched uranium to obtain fuel for VVER reactors

A Yu Smirnov[1], G A Sulaberidze[1], A A Dudnikov[2] and V A Nevinitsa[2]

[1] National Research Nuclear University MEPhI (Moscow Engineering Physics Institute), Kashirskoe highway 31, Moscow, 115409, Russia
[2] National Research Center "Kurchatov Institute", Moscow, Russia

E-mail: AYSmirnov@mephi.ru

Abstract. The possibility of the recovered uranium enrichment in a cascade of gas centrifuges with three feed flows (depleted uranium, low-enriched uranium, recovered uranium) with simultaneous dilution of U-232,234,236 isotopes was shown. A series of numerical experiments were performed for different content of U-235 in low-enriched uranium. It has been demonstrated that the selected combination of diluents can simultaneously reduce the cost of separative work and the consumption of natural uranium, not only with respect to the previously used multi-flow cascade schemes, but also in comparison to the standard cascade for uranium enrichment.

1. Introduction

According to experts, despite the projected development and commissioning of fast reactors, almost inevitable imbalance between the estimated reserves of natural uranium and its consumption to fuel the current amount of reactors on thermal neutrons in the foreseeable future [1]. This demonstrates the necessity of using fissile materials extracted from spent nuclear fuel for supply such reactors. One of the possible materials is recovered uranium, the use of which will reduce the requirements for natural uranium and significantly reduce the amount of radioactive wastes, as compared with an open fuel cycle [2].

The recovered uranium can be used as raw material for production of low-enriched uranium in the cascades of gas centrifuges [2-7]. However, it is difficult to enrich such material due to the presence in its composition of isotopes ^{232}U and ^{236}U [2]. To the content of these isotopes in the low-enriched uranium (LEU) impose strict requirements related to the need to perform radiation safety conditions in the manufacture of fuel elements and save the neutron-physical characteristics of the nuclear fuel.

At the present, there is a number of ways to enrich recovered uranium in gas centrifuge cascades. Many of them are based on a simple cascade which has one feed flow, one product flow (commercial product) and one waste flow (depleted uranium). Such cascade may be used, for [8]:

- enrichment of pre-diluted with natural uranium recovered uranium;
- production of enriched recovered uranium followed by dilution with natural uranium or another diluent;

- to enrich natural uranium to the content of ^{235}U slightly above the desired value for the fuel and consequent mixing of the product with the recovered uranium.

The main disadvantage of using such methods of recovered uranium enrichment are losing of the separation work due to the mixing of uranium with different contents of ^{235}U isotope.

Besides the mentioned above methods there other ways to enrich recovered uranium which are based on the use multi-flow separation cascades. Such cascades may have, for example, two feed flows (recovered uranium and natural uranium) [4] or an additional product flow [5]. The main advantage of using these cascades in comparison with three-flow cascades is minimization of separation work losses. However, a significant reduction of $^{232,234,236}U$ concentrations in low-enriched uranium is achieved, primarily due to dilution of recovered uranium by natural one. This fact leads to small savings of natural uranium (no more than 15%) [6].

More complex modification of multi-flow cascades to enrich the recovered uranium is a cascade with three feed flows, the first one is recovered uranium, the second one is depleted uranium, as the third feed it is possible to use either natural uranium or a small amount of low-enriched uranium [7, 9]. In the case of use of natural uranium and depleted uranium as diluents cascade with three feed flows provides a distinct advantage (over 50%) in the consumption of natural uranium, compared with the standard cascade for natural uranium enrichment. However, in this case, the loss of separation work can also be ~ 50% [7]. In case of using as the second diluent low-enriched uranium with ^{235}U content from 1.0 to 2.0% (in weight fractions) instead of natural uranium, it is possible to obtain the desired product with an economy both in separation work and consumption of natural uranium [9]. However, the question whether the use of such cascades effective or not in terms of production remains open.

This work is a continuation of the study [9]. The main purpose of the work is to evaluate the efficiency of the cascade with three feed flows to dilute the recovered uranium with low-enriched uranium and depleted uranium, as compared to the simple cascade for enrichment of natural uranium.

2. Mathematical model

As the object of study Matched Abundance Ratio Cascade (MARC) is considered. MARC is a model cascade in which relative concentrations of the selected pair of components are constant along the cascade length. This model is widely used for the analysis of the physical laws of mass transfer in cascades for separation of multicomponent isotope mixtures [10-12]. The schematic drawing of the cascade under consideration is presented on Fig. 1. The cascade has three feed flows (F_1, F_2, F_3) with concentrations $C_i^{F_1}, C_i^{F_2}, C_i^{F_3}$, one product flow ($P$) and one waste flow ($W$) with concentrations C_i^P, C_i^W, where i - the number of shared component mixture; m - total number of components in the mixture. Hereinafter component concentration expressed in weight fractions. The F_1 is the flow of the waste uranium, F_2 is the flow of the recycled uranium and F_3 is the flow of the LEU-diluent. Cascade consists of N stages connected symmetrically-countercurrent manner and numbered from the waste end of the cascade to the product one [11]. Flows F_1, F_2, F_3 entered stages f_1, f_2 and f_3, respectively. The mathematical model of such a cascade is described in [7].

The calculations were performed for the problem statement as follows: 1) given: concentrations C_n^P, C_n^W (n - number of the target component - ^{235}U), relations F_2/F_3, F_1/F_3; 2) cascade parameters to vary: f_1, N; 3) to be determined: C_i^P, C_i^W $(i \neq n), W/F_3, P/F_3$ and all internal parameters of the cascade. Such a problem could be solved by means of numerical method used in [7].

The stated problem is interesting from a practical point of view, because in this case is uniquely defined quality of the product that allows to compare the effectiveness of different ways of production of equivalent isotopic materials.

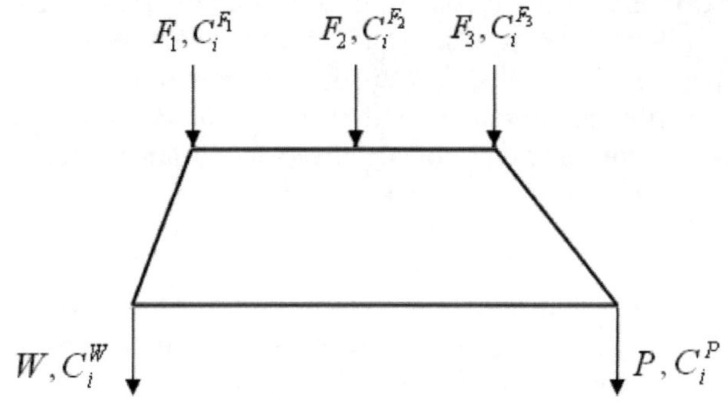

Figure 1. The schematic drawing of the cascade with three feed flows to enrich recycled uranium

3. Results and discussion

Let us consider the enrichment of the recovered uranium with the following components concentrations: ^{233}U – 2.98·10^{-7}%, ^{233}U – 5.81·10^{-7}%, ^{234}U – 1.91·10^{-2}%, ^{235}U – 0.901%, ^{236}U – 0.573%. This uranium mixture corresponds to spent fuel of Russian reactor VVER-1000 [9].
The MARC with no-mixing in the relative concentrations of the isotopes ^{235}U and ^{238}U is considered. The following cascade parameters were given: concentration of ^{235}U in the product was 4% (excluding additional enrichment in order to compensate the effect of ^{236}U presence), concentration of ^{235}U in the waste flow was 0.1%, the reactivity compensation coefficient was 0.29. The concentration of ^{232}U in the product was limited by value 2·10^{-7}% [2]. As diluents were considered the depleted uranium containing 0.25% ^{235}U and LEU with ^{235}U concentration varied from 1.0% to 2.0%. For each case the ratio of the natural uranium to the depleted one was varied. The results of calculation of cascade parameters were compared with the analogue values of the three-flow MARC to enrich natural uranium to the same concentrations of ^{235}U in the product and waste flows.
To compare the effectiveness of compared cascades as a criterion selected ratio of values S which were calculated for the simple cascade and cascade with three feed flows by means of the following formula:

$$S = A_C \cdot Z + A_{F_1} \cdot F_1 + A_{F_2} \cdot F_2 + A_{F_0} \cdot F_0, \qquad (1)$$

where Z - total number of separating elements (gas centrifuges) needed to enrich uranium at given concentrations of ^{235}U in outgoing flows, for the cascade with three feed flows this value also takes into account the number of separating elements needed for production of LEU-diluent; A_C - coefficient that takes into account the costs associated with the work of separating elements; F_0 - natural uranium consumption for commercial LEU production; A_{F1}, A_{F2}, A_{F0} –costs of the depleted, recovered and natural uranium, respectively.

This value largely determines the unit cost of production of the isotope product [13]. The first term in (1) is proportional to the cost of separation work, and the second one takes into account material cost to obtain the product. The calculations assumed that the average cost of natural uranium of about nine times more than the average cost of depleted uranium [14], and three times less than the cost of the uranium hexafluoride obtained from reprocessed uranium [15]. In the case of calculating the value of S for the simple cascade, the value of F_1 and F_2 were set to zero.

Fig. 2 shows the ratio of S parameter of the cascade with three feed flows and the simple cascade for cases with concentrations of ^{235}U in the LEU-diluent 1.1%, 1.5% and 2.0%. As can be seen from the curves on Fig. 2 unit cost of production of the commercial LEU decrease with increasing of concentration of ^{235}U in the LEU-diluent. Only at low concentration of ^{235}U in LEU-diluent unit costs will be higher than for the simple cascade. The greatest gain is obtained when the diluent has the highest concentration of ^{235}U.

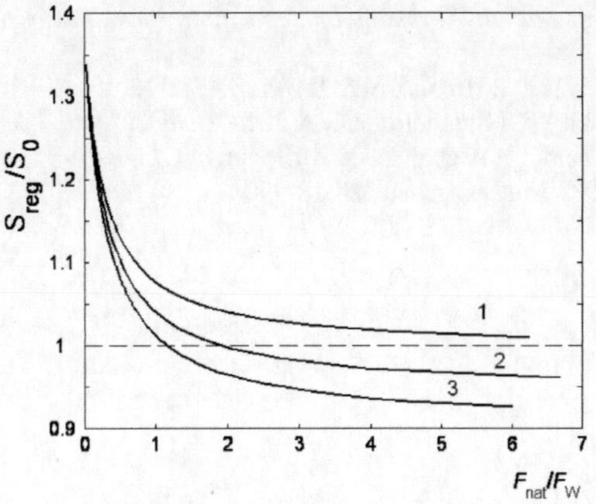

Figure 2. The dependence of the ratio for the *S* parameter of the cascade with three feed flows (*S*reg) and the simple cascade (S_0) versus the ratio between depleted uranium and LEU-diluent at various concentrations of ^{235}U in it (curves 1,2 and 3 correspond to concentrations of ^{235}U 1,1, 1 5 and 2.0% in LEU-diluent)

At the end, let us note that the use of cascade with three feed flows may allow to reduce the amount of storage of depleted uranium and spent fuel. For example, according to the data for 2009 in the USA alone amounts of storage of depleted uranium make up 708189 tons, and according to the work [1] to store the present amount the world's spent fuel it is necessary to have from 3 to 5 storages commensurate with the Yucca Mountain (USA). Obviously, when calculating the overall effect of the involvement of depleted and recovered uranium in the fuel cycle, one must consider reduced storage costs of these materials.

4. Conclusion

The scheme of the five-flow cascade (three feed flows, product and waste flows) of gas centrifuges to enrich recovered uranium with simultaneous dilution of its by low-enriched uranium and the depleted uranium is proposed. Based on the results of computational experiments the benefits of this cascade in comparison to the simple cascade in the value of the unit cost of production of the commercial LEU were shown.

Acknoledgement

This research is supported by the grant No.14.Y30.16.6284-MK of the President of the Russian Federation for young scientists.

References

[1] Andrianova E A, Davidenko V D, Tsibul'skii V F 2015 *At. Energy* **118** 301–6
[2] Smirnov A Yu, Sulaberidze G A, Alekseev P N, Dudnikov A A, Nevinitsa V A, Proselkov V N and Chibinyaev 2012 *Phys. At. Nuclei* **75** 1616-25
[3] Palkin V A 2013 *At. Energy* **115** 32–7
[4] Sulaberidze G A, Borisevich V D, Xie Q X 2006 *Theor. Found. Chem. Eng.* **40** 5-14
[5] Palkin V.A. 2010 *Prosp. Mater.* **8** 11–4
[6] Smirnov A Yu et al 2012 *Nucl. Phys. Eng.* **3** 396–403
[7] Smirnov A Yu, Sulaberidze G A 2014 *At. Energy* **117** 44–51
[8] Sulaberidze G A, Borisevich V D, Xie Q X 2004 *In: Proc. of IX All-Russia (International) Scientific Conference "Physical and Chemical Processes on Selection of Atoms and Molecules"* (Russia, October 4-8, 2004) p 70
[9] Smirnov A Yu, Sulaberidze G A 2015 *Phys. Proc.* **72** 132-5
[10] De la Garza A, Garrett G A and Murphy J E 1961 *Chem. Eng. Sci.* **15** 188–209

Journal of Physics: Conference Series **751** (2016) 012005 doi:10.1088/1742-6596/751/1/012005

[11] von Halle E. 2003 *In: Proc. of 8th Workshop on Separation Phenomena in Liquids and Gases,* (Oak Ridge, TN, USA)
[12] Song T, Shi Z, Borisevich V D 2010 *Sep. Sci. Technol.* **45** 2113–8
[13] Smirnov A Yu, Borisevich V D, Sulaberidze G A 2012 *Theor. Found. Chem. Eng.* **46** 373-8
[14] Varela C, Akapo D, Schneider E A et al. 2009 *In: Proc. of GLOBAL-2009* 9463
[15] Butler G G, Wilcox P 1988 *Nucl. Engineer* **28** 186-90

Peculiarities of the transient processes in cascades for separation of isotope mixtures with various numbers of components

A Yu Smirnov[1], A K Bonarev[1], G A Sulaberidze[1], V D Borisevich[1], S Zeng[2], D Jiang[2] and Y N Zhang[2]

[1] National Research Nuclear University MEPhI (Moscow Engineering Physics Institute), Kashirskoe highway 31, Moscow, 115409, Russia
[2] Tsinghua University, Engineering Physics Department, Beijing, P.R.China

E-mail: AYSmirnov@mephi.ru

Abstract. The peculiarities of the transient processes in separation of isotope mixtures with various numbers of components are considered. A model *Q*-cascade without its depletion part is taken as an object of the research. It is shown that before entering the stationary mode of operation in the cascade the concentrations of the intermediate multicomponent mixture components in the product flow demonstrate the nonmonotonic behavior when they pass through the maximum and the minimum before the subsequent achievement of their steady-state values. It is revealed during calculations that the "abnormal" behavior of the end concentrations is primarily caused by the initial composition of a separated mixture.

1. Introduction

In recent decades, there has been substantial growth in demand for the isotopically modified products. This is the result of intensive development of science, technology, and medicine which requires the use of high purity isotopes of various chemical elements. The most methods of isotope separation are characterized by the use of multistage separation installations (cascades), consisting of series-connected separation stages, which contain parallel connected separating elements (for example, gas centrifuges).

For a theoretical description of isotope separation processes in cascades it was developed various models of molecular-selective mass transfer in multistage installations as well as related methods to calculate their parameters [1-9]. An important role in theory of cascades for isotope separate plays the so-called model cascades [1, 2, 4, 8-12] that allow to simplify the analysis of the mass transfer laws in a cascade and corresponding computational procedures for calculation and optimization of its parameters. At the same the results obtained correspond to the separation process in the cascades used in practice. One of the widely used for this purpose model cascade is a *Q*-cascade [4, 11-18].

The duration of transient processes largely determines the efficiency of the cascade work, is one of its important characteristics. The main way to solve the problem of describing the non-stationary mass transfer in cascades for the separation of multicomponent mixtures is the numerical calculation of parameters for the symmetrically countercurrent cascade [19-23]. In the recent paper [24] Recently the mathematical model of a Q-cascade without its depleted part has been offer to study the physical laws of the transient processes in cascades for separation of multicomponent mixtures.

For the first time it was obtained the solution allowed to calculate analytically the values of the partial flows and concentrations for all components of a separated mixture at any time at any stage of the cascade. It was found that before entering a steady state operation mode of the cascade, the concentrations of some components with intermediate mass numbers in the product flow of the cascade demonstrate nonmonotonic behavior, passing through the maximum and minimum values before reaching their steady-state values. It was found in the conducted computational experiments that such the "abnormal" behavior is determined by the initial composition of a separated mixture, i.e. by the relations between the concentrations of its various components.

The present work is an extension of the research which results are published in [24]. It is devoted to an in-depth analysis of the physical reasons of the anomalous behavior of intermediate component concentrations when a cascade entering its stationary mode.

2. Mathematical model

As the object to study we use a Q-cascade without a depletion part which mathematical model is described in detail in [24]. We consider that this Q-cascade is connected to the infinite volume reservoir (V) filled with a mixture of the component concentrations C_i^F where i is a number of an arbitrary component of a mixture consisting of the m components (see figure 1). The product flow P is withdrawn from the opposite end of the cascade, where the light components of the mixture are enriched (so-called the "light" end of a cascade). We believe that at the initial time moment the cascade is also completely filled the homogeneous mixture whose composition corresponds to that in the reservoir. The length of the cascade (the number of stages in it) is denoted as S_P. The stage number is counted from the stage where the product flow P is withdrawn to the stage where a feed flow enters the cascade.

Figure 1. Schematic drawing of a Q-cascade with the infinite reservoir at its heavy end.

The partial flow of the ith component in the non-stationary regime of the Q-cascade under consideration is presented as follows [24]:

$$G_i(y_i,\tau_i) = 2P \cdot C_i^F \frac{1-\exp\left(y_i - y_i^P\right)}{1-\exp\left(-y_i^P\right)} \sum_{j=1}^{m} \frac{C_j^P}{Q_j}\left[1-\exp\left(-y_i^P\right)\right] + \sum_{l=1}^{\infty} A_{l,i} \sin\left(\frac{\pi l y_i}{y_i^P}\right) \exp\left[\frac{1}{2} y_i - \left(\frac{1}{4} + \left(\frac{\pi l}{y_i^P}\right)^2\right)\tau_i\right] \quad (1)$$

where $y_i = Q_i s$, $\tau_i = \frac{2Q_i^2}{\omega} t$ are the new variables; $y_i^P = Q_i S_P$; ω is the residence time of a separated mixture at a cascade stage which value is considered for all cascade stages; Q_i are the constants connected by the condition $Q_i - Q_j = \varepsilon_{ij}$, which for the molecular kinetic separation methods are transformed to the following relationship $Q_i = \varepsilon_0 (M - M_i)$ [4, 12]; M_i is a mass number of the

component with the i number; M is a parameter which allows to calculate the Q_i constants for all components of the separated mixture; ε_{ij} are the relative enrichment factors for the pair of components with the i and j numbers; ε_0 is the enrichment factor for the unit mass difference.
The $A_{l,i}$ are expressed as follows

$$A_{l,i} = \frac{4P \cdot C_i^F}{S_P}\sum_{j=1}^{m}\frac{C_j^P}{Q_j}\left\{\left[1-\frac{1-\exp(-Q_jS_P)}{1-\exp(-Q_iS_P)}\right]\frac{S_P/\pi l\left[1-\exp(-1/2Q_iS_P)\cos\pi l\right]}{1+(Q_iS_P/2\pi l)^2}-\right.$$
$$-\exp(-Q_jS_P)\frac{S_P/\pi l\left[1-\exp((Q_j-1/2Q_i)S_P)\cos\pi l\right]}{1+((Q_j-1/2Q_i)S_P/\pi l)^2}+ \tag{2}$$
$$\left.+\exp(-Q_iS_P)\frac{1-\exp(-Q_jS_P)}{1-\exp(-Q_iS_P)}\frac{S_P/\pi l\left[1-\exp(1/2Q_iS_P)\cos\pi l\right]}{1+(Q_iS_P/2\pi l)^2}\right\}$$

The values of the component concentrations at the product end of the cascade in the stationary mode of its exploitation can be calculated by the following formula:

$$C_i^P = \frac{Q_iC_i^F}{1-\exp(-y_i^P)} \Bigg/ \sum_{j=1}^{m}\frac{Q_jC_j^F}{1-\exp(-y_j^P)} \tag{3}$$

Thus, after specifying the number of stages in the cascade S_P, the initial composition of a separated mixture as well as the values of the parameters M by the formulas (1) - (3) one can carry out the analytical calculation of the non-stationary concentration distributions over cascade length at various moments of time.

It should be emphasized that despite of the fact that the mathematical model of a Q-cascade is developed for the case of the so-called "low enrichment", the results obtained using this model, valid also in the case of arbitrary enrichments at cascade stages [12, 24].

3. Results and discussion

The model of non-stationary mass transfer in a Q-cascade briefly described above was used to study the basic laws of transient processes in multistage installations for separation of multicomponent isotope mixtures. The calculations were performed for isotopic mixtures with the various numbers of components (from 3 to 5). The isotopic compositions of examined mixtures are presented in tables 1-3. The details of the algorithms to calculate the parameters of the transient processes in a Q-cascade may be found in [24].

In the analysis of the results of our computational experiments we based on the assumption that the process of reaching the stationary mode of operation of a cascade can be conditionally divided into 2 parts [25, 26]. At first one the change in concentration of the mixture components occurs independently. At the second part there is the mutual influence of the components on each other, associated with the process of their accumulation on the cascade stages. If one would like to consider the specifics of the initial part of the process in detail, it can be assumed that at this stage the dynamics of concentration change is determined by the below system of partial differential equations, which describes the unsteady separation process in the case of the "weak" enrichment [25-26]:

$$\frac{\partial C_i}{\partial \tau} = \frac{\partial^2 C_i}{\partial y^2} - \frac{\partial C_i}{\partial y}\sum_{j=1}^{m}(M_j - M_i)\cdot C_j - C_i\sum_{j=1}^{m}(M_j - M_i)\frac{\partial C_j}{\partial y} - \frac{2T}{\varepsilon_0 \cdot L}\frac{\partial C_i}{\partial y} \tag{4}$$

$$y = \varepsilon_0 s,\quad \tau = \frac{\varepsilon_0^2 t}{\omega},\quad \omega = \frac{H(s)}{L(s)} = const\,.$$

where $H(s), T(s), L(s)$ are the stage holdup (the amount of a working substance that is held by a separation stage of a cascade), the mass transfer and the flow of a working substance at the entrance of the cascade stage with the s number, respectively.
The nonlinear term in equation (4) can be replaced by the following linear approximation

$$C_i \sum_{j=1}^{m} (M_j - M_i)\cdot C_j \approx b_i \cdot C_i \tag{5}$$

where $b_i = \overline{M} - M_i$; $\overline{M} = \sum_{j=1}^{m} M_j C_j^F$ is the average mass number of the initial mixture of isotopes.

Three multi-isotope mixtures were investigated in the computing experiments (see tables 1-3). The first one was the 5-component natural mixture of the tungsten isotopes. The intermediate component with the mass number 184 was chosen as a target component; the separation factor for the unit mass difference was equal to q_0=1.16 [9]; the parameter M was equal to be 184.5, which corresponds to the preferential enrichment of the key isotope in the flow of the light fractions. In addition, two model 3 and 4-component mixtures which compositions were chosen so that the content of the intermediate components was comparable with that for the utmost components were investigated. This suggests that the impact of these components on the mass transfer laws in the transient regimes is comparable to the effect of utmost isotope masses.

The mass numbers and concentrations of isotopes, as well as the number of the target component for the model mixtures were chosen so that the b_i parameter for the key isotope was close to zero. As is shown in [25, 26], it is crucial for the emergence of the "abnormal" behavior of a component. The M parameter for each of the model mixtures corresponded to enrichment the targeted "intermediate" isotopes at the light end of the cascade. The separation cascade parameters for each case under consideration are listed in table 4.

The dependences of the component concentrations for the intermediate masses in the considered multi-component mixtures on the relative time of the cascade work are shown in figures 2-6. As the relative time we understand the ratio of the current time of the cascade work (t) to the residence time of a separated mixture at a cascade stage (ω=const). The quantity t/ω was varied in the limits from 0 to a value which is corresponded to the steady state mode of a cascade exploitation.
The analysis of the computational experiment results allowed confirming the following fact. The concentrations of some intermediate components in the transient mode in the product flow from the cascade successively pass through the maximum and minimum values that resemble a kind of a swing and then reaching their steady-state values. The results are qualitatively consistent with that of [25, 26] in which the problem of the non-stationary mass transfer in cascades for the separation of multicomponent isotope mixtures is solved numerically.
As it follows from the analysis of the initial compositions of the separated mixtures considered in this research and the features of the curves in figures 2-6, the "abnormal" change of concentration takes place for those components of the mixture, in which the absolute value of the b_n coefficient is the smallest one or close to zero.

From the physical point of view, such a non-monotonic behavior of the component concentrations in the separated mixture can be explained by the proximity of an average mass number and the mass of the component. It means that this component has almost no the strongly marked direction of enrichment in a cascade. In other words, the concentration gradient of this isotope is the alternating one within a cascade, including the area next to its product end. This fact is illustrated by the curves in figure 7, which demonstrate the dependence of the b_n parameter for the ^{183}W isotope. It can be seen that the magnitude of b_n in the transient process takes both positive and negative values. When

comparing the curves in figures 5 and 7 one can notice that the area of the "abnormal" behavior corresponds to that, in which the value of b_n changes its sign. In addition, for the components $^{182,184}W$ the b_n parameter had the permanent signs throughout all transition process. Note that such kind of instability in the enrichment of the intermediate components can be observed in inside cascade either in the transient [21] or in the steady regime of a cascade work [27].
Calculations performed for the 3 and 4-component mixtures have led to the similar results. At the initial stage of the transient process there were observed the maxima and minima ("a swing") for the end concentrations of the intermediate components, for which the value of the b_i parameter was close to zero (see figures 2 and 3).

Thus, the "abnormal" change of the concentrations of the isotope with the intermediate masses in the product flow from a cascade is mostly due to the initial composition of the separated mixture.

Table 1. Composition of the three-component model mixture

Number of component i	Mass number of isotope	c_{iF},%
1	100	20.0
2	101	40.0
3	102	40.0

Table 2. Composition of the four-component model mixture

Number of component i	Mass number of isotope	c_{iF},%
1	298	10.0
2	299	30.0
3	300	30.0
4	301	30.0

Table 3. Composition of the natural mixture of tungsten isotopes

Number of component i	Mass number of isotope	c_{iF},%
1	180	0.12
2	182	26.50
3	183	14.31
4	184	30.64
5	186	28.43

Journal of Physics: Conference Series **751** (2016) 012006 doi:10.1088/1742-6596/751/1/012006

Table 4. Parameters of a *Q*-cascades for separation of multicomponent mixture

	Model 3-component mixture	Model 4-component mixture	Natural abundance of the 5-component mixture of the tungsten isotopes
Cascade length, S_p	18.5	18.5	52.5
Parameter M	101.5	299.5	184.5
Separation factor, q_0	2.00	2.00	1.16
Number of a target component, n	2	2	4

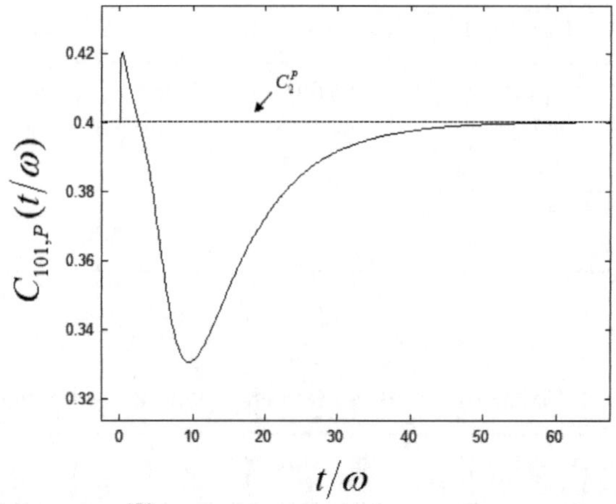

Figure 2. Variation of the ^{101}X component concentration of the model 3-component mixture in the product flow of the cascade on the relative time of its work.

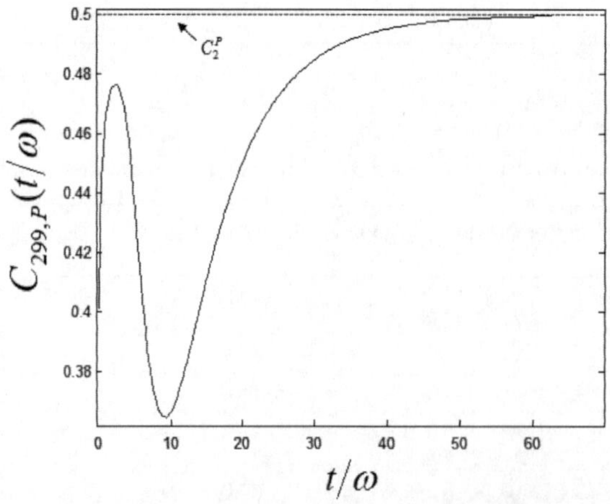

Figure 3. Variation of the ^{299}X component concentration of the model 4-component mixture in the product flow of the cascade on the relative time of its work.

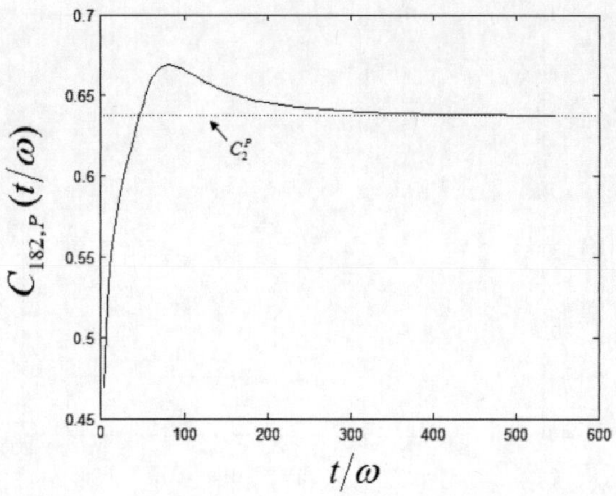

Figure 4. Variation of the ^{182}W component concentration of the tungsten isotopes of a natural abundance in the product flow of the cascade on the relative time of its work.

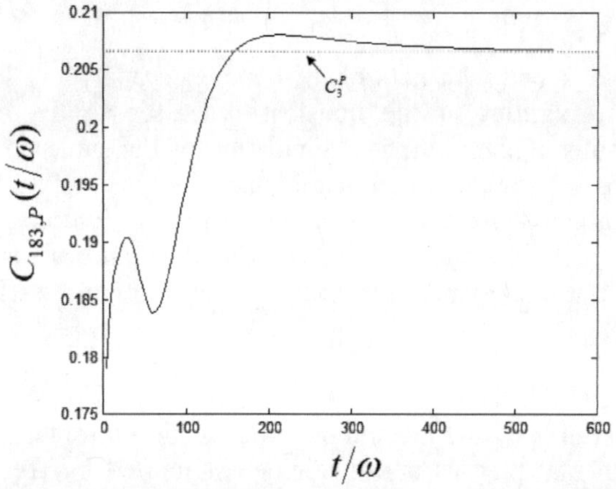

Figure 5. Variation of the ^{183}W component concentration of the tungsten isotopes of a natural abundance in the product flow of the cascade on the relative time of its work.

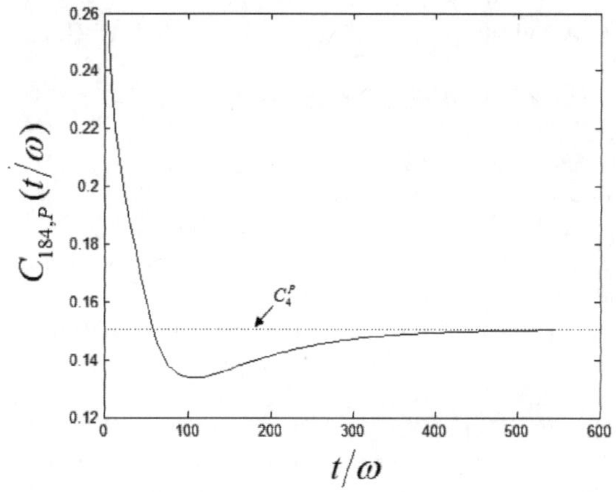

Figure 6. Variation of the ^{184}W component concentration of the tungsten isotopes of a natural abundance in the product flow of the cascade on the relative time of its work.

Journal of Physics: Conference Series **751** (2016) 012006 doi:10.1088/1742-6596/751/1/012006

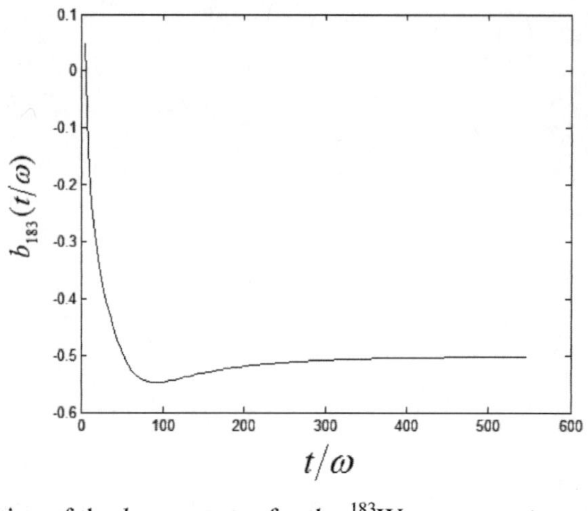

Figure 7. Variation of the b_i parameter for the ^{183}W component concentration of the tungsten isotopes of a natural abundance on the relative time of the cascade work.

4. Conclusion

The results of calculations obtained with help of the developed analytical solution demonstrated the qualitative agreement in regularities of the transient processes in the model cascades and the previously accumulated results of the numerical simulation of the same processes in the symmetric countercurrent cascades for separation of multicomponent isotope mixtures. In particular, it was detected the "abnormal" variation the end concentrations for the intermediate component (passing through the maximum and minimum values before entering the stationary one), which is due both to the initial composition of the separated mixtures, and the parameters of the separation cascade.

Acknowledgement

This research is carried out in the framework of the joint Sino-Russian project "Separation of multicomponent isotope mixtures by the method of centrifugation to obtain new functional materials for nuclear energy, fundamental physics and medicine" supported by the Russian Fund for Basic Research (RFBR) under the contract no. 16-58-53058 GFEN_a and the National Natural Science Foundation of China (NSFC) under the contract no. 1151101155.

References

[1] De la Garza A, Garrett G A and Murphy J E 1961 *Chem. Eng. Sci.* **15** 188–209
[2] De la Garza A 1963 *Chem. Eng. Sci.* **18** 73–82
[3] Levin S 1963 *J. Chem. Phys. Phys.-Chem. Biol.* **60** 218-23
[4] Kolokoltsov N A, Minenko V P, Nikolaev B I, et al 1970 *At. Energy* **29** 1193-8
[5] Kolokoltsov N A and Sulaberidze G A 1971 *At. Energy* **31** 954-7
[6] Kolokoltsov N A, Laguntsov N I and Sulaberidze G A 1973 *At. Energy* **35** 749-50
[7] Yamamoto I and Kanagawa A J 1978 *J. Nucl. Sci. Technol.* **15** 28–32
[8] Sazykin A A 2000 *ISOTOPES. Properties. Production. Application (the book),* ed V Yu Baranov (Moscow) pp 72-108
[9] Halle E 1987 *Proc. 1st Workshop on Separation Phenomena in Liquids and Gases* pp 325-55
[10] Borisevich V, Sulaberidze G and Zeng S 2011 *Chem. Eng. Sci.* **66** 393–6
[11] Zeng S, Jiang D J, Borisevich V D and Sulaberidze G A 2011 *Chem. Eng. Sci.* **66** 2997–3002
[12] Zeng S, Sulaberidze G A, Jiang D et al 2012 *Sep. Sci. Technol.* **47** 1591–5
[13] Tolstoy V V 2010 *Proc. of 11th Workshop on Separation Phenomena in Liquids and Gases* 114–21
[14] Zeng S, Smirnov A Yu, Jiang D, et al 2013 *Sep. Sci. Technol.* **48** 15–21
[15] Smirnov A Yu and Sulaberidze G A 2013 *Theor. Found. Chem. Eng.* **47** 375-80
[16] Raichura R C, Al-Janabi M A M and Langbein G M 1991 *Ann. Nucl. Energy* **18** 327-56
[17] Frolov V V, Chuzhinov V A and Sulaberidze G A 1986 *At. Energy* **61** 457-9

Journal of Physics: Conference Series **751** (2016) 012006 doi:10.1088/1742-6596/751/1/012006

[18] Muller G 1965 *Kernenergie* **8** 226-36
[19] Agostini J-P 1994 *Report CEA-R-5666*
[20] Higashi K 1975 *J. Nucl. Sci. and Techn.* **12** 243-9
[21] Xie Q, Wang L and Lei Z 2010 *Prospective Materials* 25-28
[22] Vetsko V M, Devdariani O A, Levin E V, et al 1982 *Isotopenpraxis* **18** 288-93
[23] Laguntsov N I, Levin E V, Nikolaev B I, et al 1987 *At. Energy* **62** 452–8
[24] Smirnov A Yu, Sulaberidze G A, Borisevich V D, Zeng S and Jiang D 2015 *Chem. Eng. Sci.* **127** 418-24
[25] Vetsko V M, Devdariani O A, Levin E V, Sulaberidze G A and Partsahashvili G L 1982 *Isotopenpraxis* **18** 288-93
[26] Laguntsov N I, Levin E V, Nikolaev B I and Sulaberidze G A 1987 *At. Energy* **62** 452–8
[27] Smirnov A Yu and Sulaberidze G A 2014 *Theor. Found. Chem. Eng.* **48** 629-36

Determination of the optimal mesh parameters for Iguassu centrifuge flow and separation calculations

S M Romanihin and I V Tronin

National Research Nuclear University MEPhI (Moscow Engineering Physics Institute), Kashirskoe highway 31, Moscow, 115409, Russia

E-mail: IVTronin@mephi.ru

Abstract. We present the method and the results of the determination for optimal computational mesh parameters for axisymmetric modeling of flow and separation in the Iguasu gas centrifuge. The aim of this work was to determine the mesh parameters which provide relatively low computational cost whithout loss of accuracy. We use direct search optimization algorithm to calculate optimal mesh parameters. Obtained parameters were tested by the calculation of the optimal working regime of the Iguasu GC. Separative power calculated using the optimal mesh parameters differs less than 0.5% from the result obtained on the detailed mesh. Presented method can be used to determine optimal mesh parameters of the Iguasu GC with different rotor speeds.

1. Introduction

Currently, the most effective method for the separation of a large number of isotope mixtures is the gas centrifuge (hereafter GC) [1]. Experimental investigation of flow and diffusion in a rotor of a gas centrifuge is a serious and unsolved problems due to the disturbances which are brought into a rotating flow by the measuring equipment. This makes the theoretical studies of the flow in the gas centrifuge an important tool for finding ways to improve the efficiency of centrifugal separation.

At present, the most effective approach to study high-speed centrifuges is a numerical solution of the system of differential equations describing the flow of gaseous mixture inside the rotor of GC. The accuracy of the results obtained by these calculations depends on the quality of the mesh [3]. To obtain reliable numerical solutions it is necessary to use a computational mesh with small spatial steps due to the complicated flow inside GC with large gradients of hydrodynamic parameters. This mesh discretization leads to compaction of the cells and increase the number of elements of the computational mesh. In this case significantly increases the computational cost and the duration of calculation of the given problem.

Studying the flow inside the rotor of a GC requires to carry out a large number of calculations. In particular, the determination of the most effective operating regime of a certain GC, one need to carry out up to several hundreds of simulations. Such calculations significantly increase time and computational cost of such investigations.

Thus, obtaining an optimal mesh to achieve high accuracy with an acceptable temporal and computational costs is an important issue. The aim of this study is to search for such a computational mesh that can provide reliable numerical solution that does not require much time and computational costs.

2. Model and mesh generation

We use a standard model of a Iguasu GC [7], which has a rotor diameter d=12 cm and the rotor length L=1 m. The modeling was performed in a rotating frame system in a two-dimensional axisymmetric approximation. The computational domain contains working camera and waste chambers located between two coaxial cylinders. The outer cylinder corresponds to the wall of the rotor. The inner cylinder corresponds to the boundary, where the hydrodynamic approximation is not valid [8]. Product chamber is not included in the computational domain. Its influence is simulated by imposing pressure on the wall of the rotor near the lower baffle of the working chamber. The effect of the waste scoop is modeled as a sources of mass, momentum and energy, distributed in small toroidal region [9]. The speed of rotation of the rotor in the range from 600m/s to 900m/s.

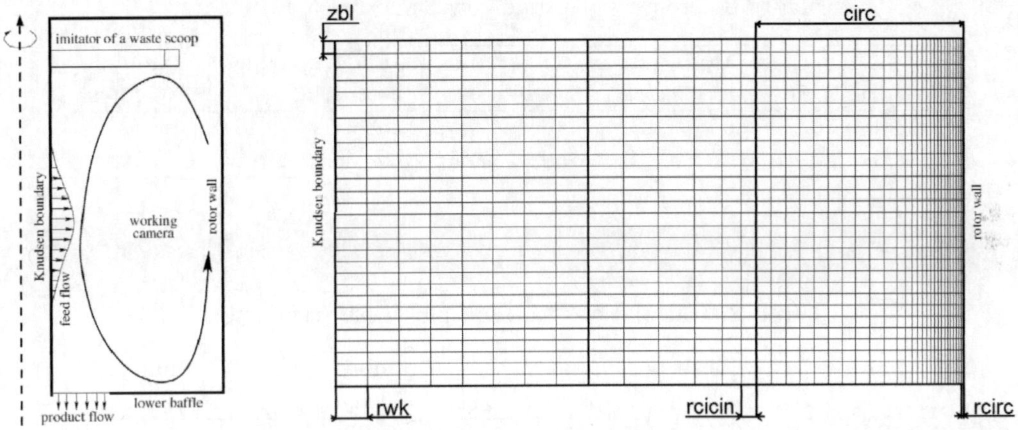

Figure 1. Scheme of the computational domain and part of mesh near lower baffle.

We use a mesh with a concentration of elements near the wall of the rotor to obtain a reliable solution in the boundary layer. Simultaneously, to ensure correct solutions and proper resolution of the gradients of the variables the refinement of the mesh is smooth, so that the sizes of neighboring cells differ not more than twice. This approach provides a reduction of errors of the interpolation scheme in case of the large gradients of density and pressure. In our problem, mesh is characterized by five parameters.

3. Optimization objective function

To find the optimal mesh parameters it is nessesary to determine maximum of the objective function that reflects precision and computational cost of numerical solution. Since the main parameter characterizing the work of GC is separative power, we use the deviation of the separation power from the value obtained using the most detailed mesh as a criterion of accuracy. Computational cost is characterized by number of cells in the computational mesh. This value directly affects the duration of the calculation and the required computing resources. These considerations were drawn up the objective function of our problem:

$$f(x_i) = \frac{N_{eta} - N(x_i)}{N_{eta}} \cdot \frac{\delta U_{eta} - \delta U_{res}(x_i)}{\delta U_{eta}}, \tag{1}$$

where x_i- parameters describing the mesh, $N(x_i)$- the number of elements and $\delta U_{res}(x_i)$- the deviation from the reference values of the separative power, δU_{eta}- the reference value of the separative power, N_{eta}- the number of elements in the mesh for reference solution.

To determine the maximum of the objective function (1) we use BOBYQA [4] direct search method, which requires only calculations of the objective function at certain points [5]. This algorithm is included in a library of nonlinear optimization algorithms, NLopt [6], freely distributed in the Internet.

4. Results and discussion

To obtain the optimal mesh parameters we needed to minimize the number of elements and deviation from the reference values of the separation powers. The reference separative power value was obtained using the most detailed mesh, which apparently provides good accuracy. Reference separative power, mesh parameters and computational costs are shown in tab. 1.

Table 1. Reference separative power and mesh parameters.

	Mesh parameters					Number	Time, s	Number
v, m/s	*circ*	*rwk*	*rcirc*	*rcircin*	*zbl*	of cells		of cores
600	0.1	0.05	0.01	0.04	0.5	1268000	4545.0	72
900	0.1	0.05	0.01	0.02	0.5	625738	2622.7	72

Table 2. Calculations with optimal mesh parameters.

v, m/s	Mesh parameters					Number	Time, s	Number
	circ	rwk	rcirc	rcircin	zbl	of cells		of cores
600	0.1	0.7	0.01	0.06	6.0	56342	455.1	4
900	0.1	0.2	0.01	0.03	6.0	89080	459.8	9

Using the reference values shown in tab. 1 and the objective function (1) we determine the optimal mesh parameters. Results are shown in tab. 2. It is clear that the number of cells in the computational mesh as well as the number of cores and calculation time using the optimal mesh significantly decrease.

In order to test the accuracy of the calculations using the obtained mesh parameters, we perform the following procedure. On the first step, we determine the optimal separative power of the Iguasu GC using the mesh generated with the optimal parameters on each step of the optimization procedure. The results of the first step are shown in tab. 3. On the second step we calculate the separative power of Iguasu GC in the optimal regime using the most detailed mesh. The results of the second step are shown in tab. 4.

Table 3. The results of Iguasu GC optimization using the optimal mesh parameters.

v, m/s	Operating parameters of GC Iguasu				δU, c.u.
	P, mmHg	F, mg/s	δT, K	W_{torm}, W	
600	154.0	66.9	16.5	9.4	40.4
900	709.9	146.8	18.5	25.0	117.4

It is clear that the results on the most detailed mesh differ less than 0.5% from the results obtained on the optimal mesh. Thus, obtained optimal mesh parameters can be used in precise

Table 4. Verification of the optimum on the most detailed mesh.

v, m/s	Number of elements	Calculation time, s	Operating parameters of GC Iguasu				δU, c.u.
			P, mm.Hg	F, mg/s	δT, K	W_{torm}, W	
600	1268000	6480.6	154.0	66.9	16.5	9.4	40.4
900	625738	2622.7	709.9	146.8	18.5	25.0	116.9

calculations of the Iguasu GC separative power. Obtained parameters allows to reduce the computational costs of the separative power calculations and optimization procedure of the Iguasu GC.

5. Conclusions

The results of the optimal mesh parameters for Iguassu GC flow and separation calculations are presented. Optimal parameters were determined for two rotor speeds – 600 m/s and 900 m/s. Calculations using the optimal mesh require significantly less computational costs compared to the calculations on the detailed mesh without loss of accuracy. Obtained mesh parameters can be further used to study hydrodynamic flow inside the rotor to determine the most effective working regime of the Iguasu GC.

Acknowledgements

The work has been performed under the support of the Ministry of education and science of Russia, grant no. 3.726.2014/K. The computer simulations have been partially performed on the computer cluster of National research nuclear university(MEPhI), Basov farm. This work was also performed within the framework of the Center "Physics of nonequilibrium atomic systems and composites" supported by MEPhI Academic Excellence Project (contracts No. 02.a03.21.0005, 27.08.2013).

References

[1] Borisevich V D et al 2011 *Fizicheskie osnovy razdeleniya isotopov v gazovoi centrifuge (Physical basics of the isotope separation in gas centrifuge)* (Izdatel'skii dom MEI, Moscow) (in Russian)
[2] Bogovalov S V et al 2013 *Computers and Fluids* **86** 177
[3] Godunov S K, Zabrodin A V and Ivanov M Ya 1976 *Nauka* (in Russian)
[4] Powell M J D 2009 *Technical report NA2009/06* (Department of Applied Mathematics and Theoretical Physics, Cambridge, England)
[5] Powell M J D 1998 *Acta Numerica* **7** 287
[6] Johnson S G, The NLopt nonlinear-optimization package URL: http://ab-initio.mit.edu/nlopt.
[7] 1996 *Proc. of the Fifth workshop on Separation Phenomena in Liquids and Gases*, Sep. 22-26, 1996. ed. C Schwab, N A S Rodrigues and H G Wood
[8] Von Halle E, Wood H G and Lowry R A 1983 *Nucl. Technol.* **62** 325
[9] Bogovalov S V et al 2015 *Conference of Physics of Non-equilibrium Atomic Systems* **72** 283

Dependence of optimal separative power of the "high-speed" Iguasu centrifuge on pressure of working gas

S V Bogovalov, V D Borman, V D Borisevich, O V Davidenko, I V Tronin and V N Tronin

National Research Nuclear University MEPhI (Moscow Engineering Physics Institute), Kashirskoe highway 31, Moscow, 115409, Russia

E-mail: `IVTronin@mephi.ru`

Abstract. The results of optimization calculations of the separative power of the "high-speed" Iguasu gas centrifuge are presented. Iguasu gas centrifuge has the rotational speed of 1000 m/s, the rotor length of 1 m. The dependence of the optimal separative power on the pressure of the working gas on the rotor wall was obtained using the numerical simulations. It is shown, that maximum of the optimal separative power corresponds to the pressure of 1100 mmHg. Maximum value of separative power is 31.9 SWU.

1. Introduction

Isotope separation by gas centrifuges (hereafter GC) receieving increasing attention during the last years both from theoretical and practical points of view. One of the main problems arising in the design of the gas centrifuges is the problem of increasing the efficiency of GC, i.e. increasing the separative power δU of single GC. One way is to increase the velocity of the rotor of GC [1, 2]. However, the characteristic thickness of the separative layer of gas inside the rotor GC reduces as rotor velocity increases. It can lead to a decrease of the separative power. In addition, the question arises: what are flow parameters inside GC rotor must be to achieve optimal separative power. The answers of these questions require long-term and expensive experimental studies.

In recent years the numerical methods of the GC separative power caclulation based on numerical simulations are becoming an increasingly popular [3, 4, 5]. These methods allows one to calculate the separative power using the numerical solution of the hydrodynamics equations and convective diffusion inside the rotor of GC. This approach allows to significantly reduce the cost of research and to answer to two basic questions: what is the maximum separative power of GC with the specified velocity of rotor and the rotor length, and what are flow parameters must be to achive this separative power.

In this paper we present the results of numerical calculation of flow and separation in "high-speed" Iguasu GC (rotational velocity of 1000 m/s). The main attention is focused on the calculation of the optimal separative power and its dependence on pressure of working gas.

2. Model and method

The modelling of the flow and separation in the Iguasu GC is performed in axisymmetric approximation in rotating coordinate system for Iguasu GC (diameter of rotor of GC $d = 0.12$ cm, linear velocity of rotation $v = 1000$ m/s and length of rotor of $L = 1$ m). The computational domain represents the cylindrical sector with a span angle 1°. Influence of a waste scoop was modelled by the sources of mass, momentum and energy, which are uniformly distributed on small toroidal region [6]. The mass sink value was fixed according to the feed calue and feed cut $\theta = 0.5$. The product chamber of the centrifuge was absent in the computational domain. Influence of the product chamber was modelled by pressure imposed to the rotor wall near the lower baffle. The mass flux at the outlet has been specified by the conservation of mass.

A linear temperature profile has been specified at the rotor wall. The minimal temperature T_0 has been specified above the temperature of sublimation of the working gas. At every step of the optimization procedure of the GC the temperature of the gas has been calculated at the pressure exceeding on 10% the current pressure at the wall. This temperature has been taken as T_0. The sublimation of the gas is avoided due to this procedure.

Internal boundary of the computational domain corresponds to the Knudsen zone were the gas pressure is of the order of 1 Pa. The feed flux has been specified at the surface of this boundary.

Optimization procedure was carried out by the direct search method without use of information about derivatives of separative capacity by the optimization parameters. The algorithm BOBYQA [7] was used in operation. This algorithm is included in a library of nonlinear optimization algorithms, NLopt [8]. In the begining of optimization procedure we fix the length of the rotor, its radius and rotational speed, and the pressure on a rotor wall. Thus, the optimization procedure was performed in three-dimensional space of the following list of working parameters: feed flux F, temperature drop ΔT and power of the braking of the gas by the scoop W. Calculations were performed for different values of the pressure p in the range of 15 $mmHg$ to 1500 $mmHg$ and the dependence of the optimal separative power on the pressure was obtained.

3. Results and discussion

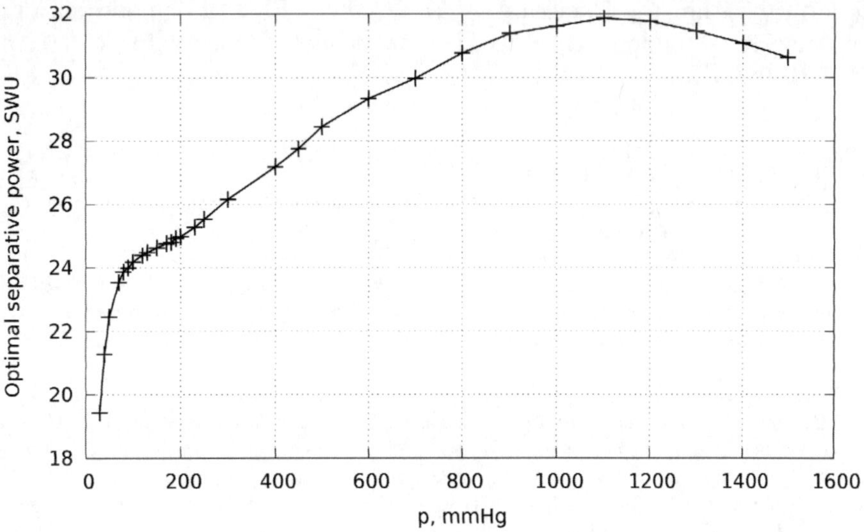

Figure 1. Dependence of the optimal separative power on the pressure at the wall of rotor.

The dependence of the optimal separative power on pressure at the wall of rotor is shown on fig. 1. Optimal separative power reaches its maximum value $\delta U_{max} = 31.9\ SWU$ at a pressure of $p = 1100\ mmHg$. It is necessary to emphasize that the obtained value of the optimal separative power is much higher than the value predicted by the empirical formula from the work [4]:

$$\delta U_{empirical} = 12L \cdot \left(\frac{v}{700\ m/s}\right)^2 \left(\frac{d}{12\ cm}\right) = 24.4\ SWU. \tag{1}$$

This distinction is connected with the fact that dependence of optimal separative power on the speed of rotor does not satisfy the quadratic law. As shown in [9], optimal separative power of the Iguasu GC grows with the speed of a rotor as $\sim v^{2.6}$. At the same time, from a figure 1 it is clear that optimal separative power has a small "plateau" in the pressure range from 100 $mmHg$ to 200 $mmHg$. In this pressure interval separative power δU changes in the range of 4%. Separative power value in this eange corresponds to the formula (1). Thus, if we introduce the top limit for the pressure, optimal separative power value would satisfy the law $\sim v^2$. It confirms the conclusions of the work [9].

4. Conclusions

The results of the optimization studies of the "high-speed" Iguasu GC with a rotor length $L = 1$ m and rotational velocity $v = 1000$ m/s are presended. The dependence of optimal separative power on a pressure of the working gas is calculated. Is is shown, that optimal separative power has a maximum $\delta U_{max} = 31.9\ SWU$ at the pressure of $p = 1100$ mmHg. It corresponds to dependence of the optimal separative power on velocity of the rotor $\sim v^{2.6}$. The dependence of the optimal separative power has a "plateau" the range of pressures from 100 $mmHg$ to 200 $mmHg$. Optimal separative power on this "plateau" changes slightly around the value 24.4 SWU, which corresponds to the dependence $\sim v^2$.

5. Acknowledgements

The work has been performed under the support of the Ministry of education and science of Russia, grant no. 3.726.2014/K. The computer simulations have been partially performed on the computer cluster of National research nuclear university(MEPhI), Basov farm. This work was also performed within the framework of the Center "Physics of nonequilibrium atomic systems and composites" supported by MEPhI Academic Excellence Project (contracts No. 02.a03.21.0005, 27.08.2013).

References

[1] Kemp R S 2009 *Sci. Glob. Security* **17** 119

[2] Cohen K 1951 *The theory of isotope separation as applied to the large-scale production of* U^{235} ed. G M Murphy, (McGraw-Hill: New York)

[3] Doneddu F, Roblin P and Wood H G 2000 *Sep. Sci. Technol.* **35** 1207

[4] Senchenkov A P, Senchenkov S A and Borisevich V D 2005 *Gas centrifuges.* In: ISOTOPES. Properties. Production. Application. ed. Baranov V Yu, (Fizmatlit: Moscow) p. 168 (in Russian).

[5] 1996 *Proc. of the Fifth workshop on Separation Phenomena in Liquids and Gases*, Sep. 22-26, 1996. ed. C Schwab, N A S Rodrigues and H G Wood

[6] Matsuda T, Sakurai T and Takeda H 1975 *Journal of Fluid Mechanics* **69** 197

[7] Powell M J D, The BOBYQA algorithm for bound constrained optimization without derivatives, Department of Applied Mathematics and Theoretical Physics, Cambridge England, technical report NA2009/06 (2009).

[8] Johnson S G, The NLopt nonlinear-optimization package URL: http://ab-initio.mit.edu/nlopt.

[9] Bogovalov S V, Borman V D, Borisevich V D, Tronin I V and Tronin V N *International Journal Numerical methods for Heat & Fluid flow*, in press

Two optimal working regimes of the "long" Iguasu gas centrifuge

V D Borman, S V Bogovalov, V D Borisevich, I V Tronin and V N Tronin

National Research Nuclear University MEPhI (Moscow Engineering Physics Institute), Kashirskoe highway 31, Moscow, 115409, Russia

E-mail: svbogovalov@mephi.ru

Abstract. We argue on the basis of the results of optimization calculations that the dependence of the optimal separative power of the Iguasu gas centrifuge with 2 m rotor has two local maxima,corresponding pressures of $p_{max1} = 35\ mmHg$ and $p_{max2} = 350\ mmHg$. The optimal separative power values in these maxima differ by the value of 0.6%. Low pressure maximum is caused by the thermal drive, whereas high pressure maximum is caused by both thermal and mechanical drives. High pressure maximum is located on wide "plateau" from $p_1 = 200\ mmHg$ to $p_2 = 500\ mmHg$, where the optimal separative power changes in the range of 0.7%. In this way, Iguasu gas centrifuge has two optimal working regimes with different sets of working parameters and close slightly different values of the separative power. Calculations show that high pressure regime is less sensitive to the parameters change than low pressure one.

1. Introduction

Nowadays, gas centrifuges (hereafter GC) are the main tool for the industrial production of enriched uranium for nuclear power stations. One of the most important problems in the GC design is the determination of the optimal working regime of the GC, i.e. determination of maximal separative power and flow parameters required to achieve this separative power. The existing estimates of the optimized separative power are based on the empirical relationships. Although these estimates can give correct predictions of the optimal separative power, they cannot predict flow parameters of the optimal working regime. Theoretical models give incorrect predictions of the optimal separative power as well as optimal flow parameters [2]. At present time, determination of the optimal working regime and the optimal separative power can be performed using the computer simulations.

In this work we present the results of the optimization calculations of the "long" Iguasu GC, i.e. Iguasu GC with the length of the rotor $L = 2\ m$ focusing on the dependence of the optimal separative power on the pressure of the working gas at the wall of rotor.

2. Model and method

The modelling of the flow and separation in the Iguasu GC is performed in the rotating frame system. The diameter of the rotor of the GC is $d = 0.12\ cm$, linear velocity of rotation $v = 600\ m/s$, the length of the rotor is $L = 2\ m$. The modelling is performed in axisymmetric approximation. The affect of the waste scoop has been modelled as a source of mass, momentum and energy distributed over a small toroidal region. The computational domain consists of the working chamber and waste chamber located between two coaxial cylinders. The outer cylinder corresponds to the rotor wall. The inner cylinder is the artificial wall introduced into the model to avoid the simulation in the rarefied gas were the hydrodynamical approximation is not valid. Then product chamber was not included into the computational domain. Its influence was modelled by the pressure imposed at the rotor wall near the low baffle of the working chamber. The mass flux at the outlet has been specified by the conservation of mass.

A linear temperature profile has been specified at the rotor wall. The minimal temperature T_0 has been specified above the temperature of sublimation of the working gas. At every step of the optimization procedure of the GC the temperature of the gas has been calculated at the pressure exceeding on 10% the current pressure at the wall. This temperature has been taken as T_0. The sublimation of the gas is avoided due to this procedure.

Internal boundary of the computational domain corresponds to the Knudsen zone were the gas pressure is of the order of 1 Pa. The feed flux has been specified at the surface of this boundary.

To determine the maximal separative power and optimal flow parameters we use BOBYQA [4] direct search method, which requires only calculations of the objective function at certain points [5]. This algorithm is included in a library of nonlinear optimization algorithms, NLopt [6], freely distributed in the Internet. In the beggining of each optimixation calculation we fix the pressure at the wall of rotor p. Optimization provides us the values of feed flux F, temperature drop along the wall of the rotor ΔT and power of the braking of the gas by the scoop W at which the optimized value of the separative power δU_{opt} of the GC is achieved. The calculations were performed for the pressure from $p_{min} = 15\ mmHg$ to $p_{max} = 900\ mmHg$.

All calculations of the gas flow, diffusion of the mixture and optimization were performed at fixed feed cut $\theta = 0.5$, rotor diameter $d = 0.12\ cm$ and rotor length $L = 2\ m$.

3. Results and discussion

Fig. 1 shows dependence of optimized separative power δU_{opt} on the pressure at the wall of rotor p. This dependence has tow maxima. The first one cooresponds to the pressure $p_{max1} = 35\ mmHg$ and the second one to the pressure $p_{max2} = 350\ mmHg$. It means that "long" Iguasu GC has tow optimal working regimes. Optimized separative power values in these regimes differ by 0.6%.

Table 1. Optimal working parameters of the "long" Iguasu GC.

p, mmHg	F, mg/s	δT, K	W, W	δU_{opt}
35	33.4	58.7	0.5	16.64
350	178.6	20.8	21.0	16.75

Calculated optimal working regimes differ in their optimal parameters. Sets of the optimal parameters are shown in tab. 1. It is clear that all the parameters are significantly different. Another important conclusion is that flow circulation in the first optimal regime is determined by the thermal drive. Indeed, the power W is relatively low, whereas temperature drop ΔT is relatively high. Additional calculations shows that all 95% of the separative power at the first regime is caused by the thermal drive and only 5% by the mechanical drive.

Second regime is characterized by higher feed flux, lower temperature drop and higher breaking power. In this regime thermal and mechanical drives give 50% of the optimal separative power.

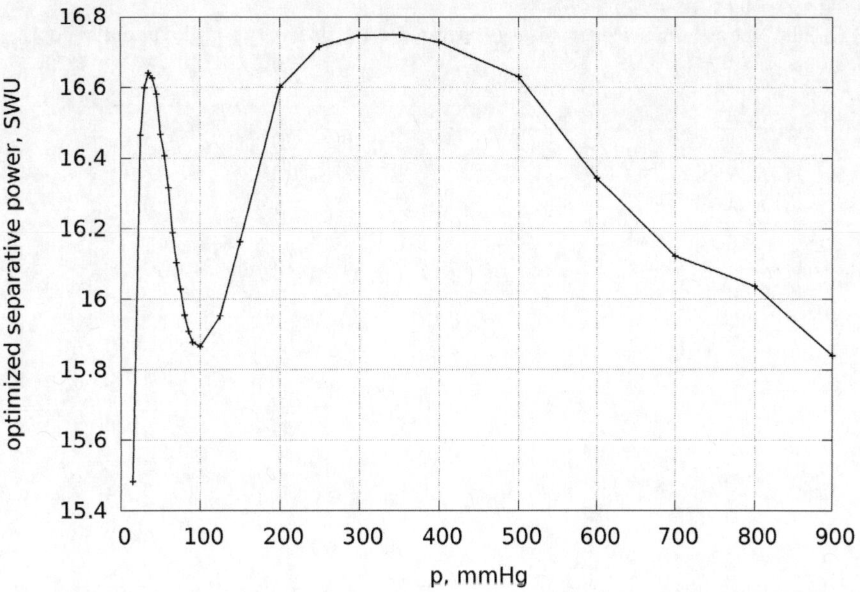

Figure 1. Dependence of the optimized separative power on the pressure at the wall of rotor.

It is worth to mention that the second regime is less sensitive to the working parameters changing comparing to the first one. A small change in pressure causes a small change in the separative power. Furthermore, there is a "plateau" in the dependence of the optimized separative power on a pressure starting from $p_1 = 200\ mmHg$ and to $p_2 = 500\ mmHg$. Optimized separative power on this "plateau" changes in the range of 0.7%.

4. Conclusions

The results of the optimization studies of the "long" Iguasu GC are presented. It is shown that Iguasu GC with 2 m rotor has two optimal working regimes with different flow parameters. The first one (low pressure regime) is reached at low pressure $p_{max1} = 35\ mmHg$ and determined mostly by the thermal drive. The second one (high pressure regime) is reached at $p_{max2} = 350\ mmHg$, less sensitive to the change of the parameters comparing to the first one and determined by both mechanical and thermal drives.

Acknowledgements

The work has been performed under the support of the Ministry of education and science of Russia, grant no. 3.726.2014/K. The computer simulations have been partially performed on the computer cluster of National research nuclear university(MEPhI), Basov farm. This work was also performed within the framework of the Center "Physics of nonequilibrium atomic systems and composites" supported by MEPhI Academic Excellence Project (contracts No. 02.a03.21.0005, 27.08.2013).

References

[1] Kai T 1977 *J Nucl. Sci. Technol* **14** 267
[2] Doneddu F, Roblin P and Wood H G 2000 *Sep. Sci. Technol.* **35** 1207
[3] Matsuda T, Sakurai T and Takeda H 1975 *J. Fluid Mech.* **69** 197
[4] Powell M J D 2009 *Technical report NA2009/06* (Department of Applied Mathematics and Theoretical Physics, Cambridge, England)

[5] Powell M J D 1998 *Acta Numerica* **7** 287
[6] Johnson S G, The NLopt nonlinear-optimization package URL: http://ab-initio.mit.edu/nlopt.

Engineering model of the electric drives of separation device for simulation of automatic control systems of reactive power compensation by means of serially connected capacitors

V M Juromskiy

National Research Nuclear University MEPhI (Moscow Engineering Physics Institute), Kashirskoe highway 31, Moscow, 115409, Russia

E-mail: juromskiy@mail.ru

Abstract. It is developed a mathematical model for an electric drive of high-speed separation device in terms of the modeling dynamic systems Simulink, MATLAB. The model is focused on the study of the automatic control systems of the power factor (Cosφ) of an actuator by compensating the reactive component of the total power by switching a capacitor bank in series with the actuator. The model is based on the methodology of the structural modeling of dynamic processes.

1. Introduction

As is known the energy efficiency of the multi-motor synchronous hysteretic electric drive of separation device is defined as the ratio of the active power *P* (which performs the mechanical work) to the total (apparent) power $S = \sqrt{P^2 + Q^2}$ that leads to the relation as follows Cos$\varphi = P/S$. Here $P = U\,I\,\mathrm{Cos}\varphi$, φ is the phase shift between the voltage *V* and the current *I* in an AC circuit, and $Q = U\,I$ Sin φ is the reactive power creates the electromagnetic field in a windings of an electric motor. The reactive power does not perform the useful work and additionally loads the power circuit.

The more power is, the less Cosφ takes place, and the more energy is lost in distribution systems, generators and other equipment of the power system. Also is well-known that the widely used in industry Cosφ is equal to 0.9 – 0.95.

Compensation of the reactive power in an electric drive of the separation device, which today are considered as the most advanced technology leads to a real possibility to reduce the cost of electricity supplying by nuclear power plants [1].

2. Statement of the problem

The synchronous hysteresis electric drives (HED) of high speed separation devices connected in parallel is powered by a static frequency converter (SFC), forming a grid "HED-SFC" with the non-standard frequency power and the low enough Cosφ (power factor, PF) = 0.2 – 0.3 [2].

The increase in a PF is possible as by means of the traditional inclusion of capacitors for compensation of the reactive power of HED connected in parallel [2] or in series with the HED [3] (figure 1a). The sequential capacitive compensation of the resonant frequency *R-L-C* circuit is not

Journal of Physics: Conference Series **751** (2016) 012010 doi:10.1088/1742-6596/751/1/012010

dependent on the active resistance *R* and significantly reduced the voltage *U* at the output of the frequency converter (FC) ensuring the nominal value of the voltage U_s for the separation device.
The initial task for the manual or automatic control of PF in the power system as a function of setting up the serial power loop *L*-*R*-*C* for a frequency close to the resonance f_0 by means of the capacitance change is stabilization of the nominal voltage supply of U_s of the actuator by the change of the output signal U_s of a FC.
Only in this way it is possible to identify the resonant (static) characteristics of the dependence of the voltage on the capacity *U*(*C*) in the HED-SFC grid in the $\{U; C\}$ coordinate space.

3. The construction of the model

To make simulation the energy system separation device- frequency converter (GC-FC) as the control object the structure presented in figure 1a is transformed into a dynamic system shown in figure 1*b* acceptable for presentation in the terms of the automatic control theory and the Simulink Toolbox of MATLAB [4] as follows:

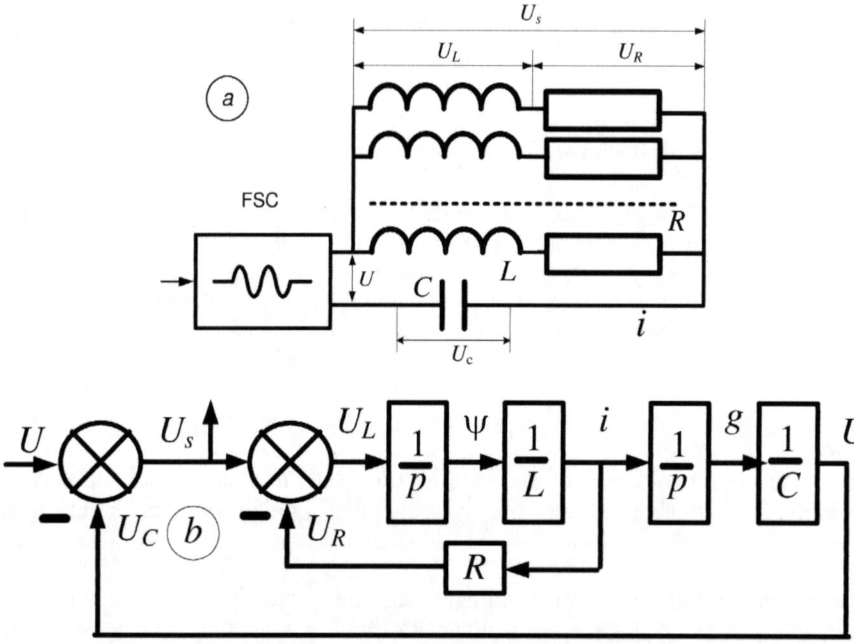

Figure 1. Compensation of the reactive component of the total power of HED in a case of the series configuration of capacitors *C*: *L*-*R* are the inductance and Ohmic resistance of the parallel-connected HED; 1*a is* the wiring diagram; 1*b* is the structural dynamic model; 1/*R* are the blocks of integration, the *p* is the Laplace operator.

In the structure of figure 1*a* the capacitor *C* has a charge calculated as $q = \int idt$, where *i* is the current in the circuit and the voltage on the capacitor is equal to $U_c = q/C$. The supply voltage U_s defined as $U_s = U - U_c$ is the sum of the voltages where U_L is related to the inductor *L* and U_R corresponds to the resistor *R*. If $U_R = iR$ is subtracted from U_c (the right adder in figure 1b), the voltage on inductance *L* is remains. The latter determines the flux linkage $\psi = \int_0^t U_L(t)dt$ which in turn generates the current $i(t) = \psi(t)\frac{1}{L}$ or in another form of writing $U_L = L\frac{di}{dt}$) [4]. The transfer function model for the channel input of the voltage U at the output SFC, the output voltage U_s is $W(p)=\frac{U_s(p)}{U(p)}=\frac{R(pL/C+1)pC}{T^2p^2+2\xi Tp+1}$ where $T=\sqrt{LC}$ is the time constant, $\xi = RC/2L$ is the rate of oscillation, p is the Laplace operator.

4. Model system in the terms of the Simulink Toolbox of MATLAB

Organization of the automatic control processes using the alternative current and their modeling in terms of the Simulink, MATLAB Toolbox requires the introduction of the high-frequency variables U and U_s (see figure 1a) which are proportional to their amplitude values, for example by means of the full wave rectification with following filtration (see blocks, "Abs", “Transfer Fsn” [5]). Control of exposure and changes of parameter are executed by the multiplying devices (see block «Product») [5]), one input of which receives the value of a parameter, and other deals with control action.
Then the engineering model of the electric drive of GCs for the purposes to control the power factor has the form shown in figure 2, where to be more definite the following parameters are specified: the supply frequency $f_0 = 1800\ Hz$ ($\omega_0 = 11280\ s^{-1}$), $U_s = 380\ V$, and the aggregate of GCs connected in parallel has the resulting parameters: $L = 50\ \mu H$, $R = 0.15\ \Omega$ and $C_0 = 157\ \mu F$ corresponding to the resonance *R-L-C* circuit. Under these conditions $\mathrm{Cos}\varphi \sim 0.25$ and the quality factor $Q \sim 3.8$.

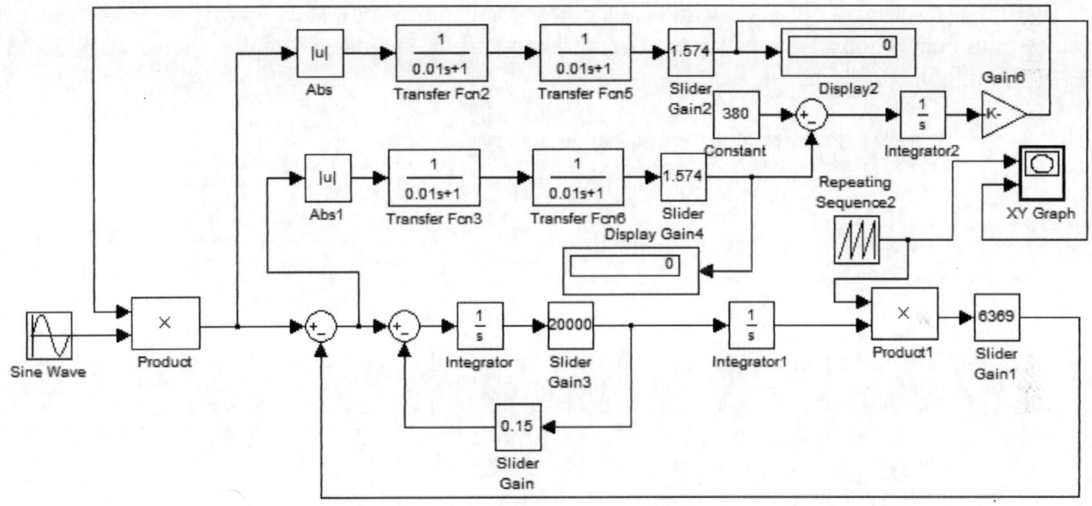

Figure 2. The engineering model of HED for the sequential reactive power compensation in conjunction with the system of automatic stabilization the U_s voltage for supplying HED in the terms of the Simulink Toolbox of MATLAB. Here “Slider Gain3” corresponds to $1/L$, “Slider Gain1” matches $1/C$, and “Slider Gain” meets R.

The structure of figure 1*a* is represented by the lower part of figure 2. It includes the blocks "Slider Gain". The feature of the "Slider Gain” block is its ability to visualize the adjustable parameters of the *L-R-C* circuit and to make the “online” changes them in the process of the computing code execution by computer simulation [5].

The automatic stabilization of the *Us* voltage (the upper part of figure 2) is performed by the control circuit consisting of special means to measure the high-frequency variables U, U_s ((detecting takes place in the block "Abs1"; filtration is realized in the block "Tranfer Fcn3"; 6" and the block "Slider Gain4" is intends to adjust an exact match of the unbent values of $|U_s|$, $|U|$ to their amplitude values);as well as the block for comparison “Constant” (U_{spec} = 380 V) and $|U_s(t)|$, the block of integration “Integrator2” with the transfer coefficient k = 0.02 (“Gain6”) and the model of the manged SFC on the basis of the multiplication unit (“Product”). The control signal is given to one of the inputs of the "Product" block and on the other is set the harmonic signal ("Sine Wave") with a frequency of ω_0 and amplitude, which value is wittingly greater than U_s/Q. The time-to-steady-state operation in the system of the U_s voltage stabilization in this case under consideration is equal to $t \sim 0.3$ s.

The static characteristics of the power resonant circuit when changing the *L-R* parameters of are determined at slow (relatively to the performance of the stabilization system of the U_s) change the parameter $1/C$ in the range (0.8 – 1.2) $/C_0$, where C_0 = 157 μF is the "resonant" value of the capacitance *C* of the capacitor bank at normal condition of the GC separation cascade. The capacitance variation (0.8 – 1.2) $/C_0$ corresponds to the change of *C* within the range (196 – 131) μF. The change in the magnitude of $1/C$ is carried out as multiplication of the $1/C$ value on the variable transmission coefficient of the managed block "Product". is The signal from the "Integrator1"goes to one input of the "Product" block, and the other receive a periodic linearly varying signal from the "Repeting Seguence2" block with the settings of the output changes from "0.8 to 1.2" for the period of time $T = 5$ s. The display changes to $|U|(1/C)$ is executed by the "XY Graf" block.

The external time of calculation of the structure presented in figure 3 using the PC is equal to a few minutes In the process of simulation by a computer it is possible every 5 "machine" seconds make drastically changes of the *R-L* parameters, corresponding to the change of the electrical parameters of a HED e from their nominal values of "on line" by means of setting in the blocks "Slider Gain"and "Slider Gain4".

The static (resonance) characteristics of the power oscillation circuit shown in figure 1a correspond to the serial connection of the capacitor *C* and automatically stabilized supply voltage U_s = 380 *V* presented in figure 3.

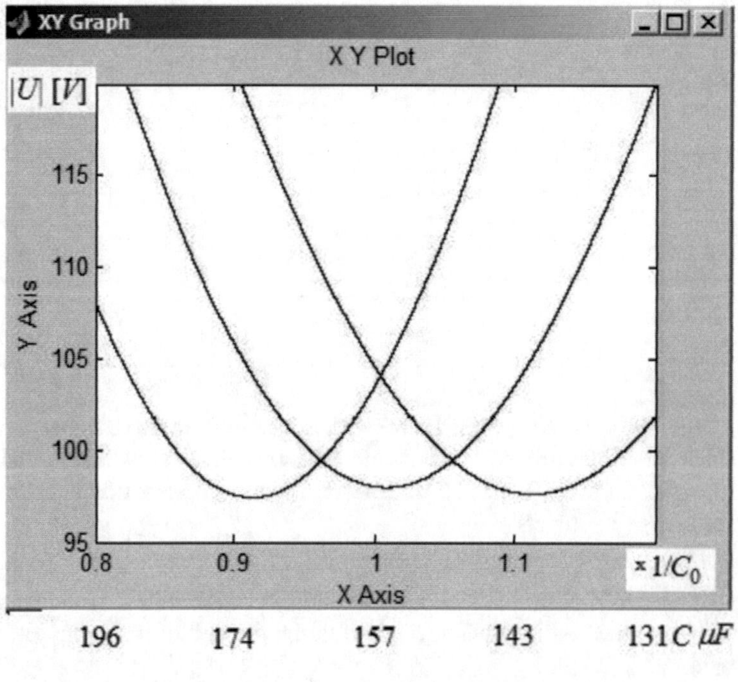

Figure 3. The resonance (static) characteristics of *U*(1/C) and *U*(C) for the aggregate of GCs when changing the number of GCs at ± 10%. The nominal number of the GCs in an aggregate corresponds to the curve in the center of the figure. The minimum of $\frac{dU}{dC} = 0$ corresponds to Cosφ = 1. The workspace for the values of Cosφ < 1 is to the right of the local extrema.

Conclusion

1. The model of the electric-drive of high-speed separation device for isotope separation with the serial capacitive compensation of the reactive component of the total power in terms of the Simulink Toolbox of MATLAB as the extreme dependence of the output voltage U at the cleats terminals of SFC from the capacitance C is developed.
2. It is shown that the energy efficiency $\mathrm{Cos}\varphi = 1$ for the power system "HD-FSC" is achieved in the state of the resonance for the power oscillating circuit included the compensating capacitance C with the inductance – active parameters of the electric-drive. The nominal voltage U_s for the power supply of the electric-drives of GCs is automatically stabilized when the output voltage FSC is equal to $U = U_s/Q$, where Q is the quality factor power resonant circuit.
3. The model allows to reflect the technological changes of the electrical parameters of the local power system "HED-FSC" with sequential compensation of the reactive component of the total power in the restructuring of the assembly of GCs and the drift of their parameters of the electric drives.
4. The build model is based on the methodology of the structural modeling of dynamic processes in electric circuits.

Acknowledgement

The author would like to thank Professor V D Borisevich for discussions, which enhanced the quality of the paper.

References

[1] Crossland Ed I 2012 *Nuclear Fuel Cycle Science and Engineering* (Woodhead Publishing).
[2] Juromskiy V M 2015 *Phys. Proc.* **72** 146-9.
[3] Borisevich V D, Juromskiy V M and Trushkin N S 2013 *Nucl. Phys. Eng.* **4** 434–5 (in Russian).
[4] Aleksakov G N, Gavrilov V V and Fedorov V A 1991 *Structural models of dynamic processes* (Moscow: Engineering center MEPhI) (in Russian).
[5] Dabney D B and Harman T L 2001 *Mastering Simulink 4* (Prentice Hall Upper Saddle River New Jersey).

Journal of Physics: Conference Series **751** (2016) 012011 doi:10.1088/1742-6596/751/1/012011

Adaptive system for automatic stabilization of the power factor for electric drives of separation device by means of serially connected capacitors bank

V D Borisevich and V M Juromskiy

National Research Nuclear University MEPhI (Moscow Engineering Physics Institute), Kashirskoe highway 31, Moscow, 115409, Russia

E-mail: VDBorisevich@mephi.ru

Abstract. A method for designing adaptive systems for automatic extremum search to stabilize the power factor of local electric power system of electric is considered. It consists in application of the serially connected capacitors compensating the reactive component of the total electric power of in parallel connected centrifugal machines usually called as an aggregate. Operation of the system just demands measuring voltage at the output of the static frequency converter for electric drives. The proposed control system is designed to stabilize the power factor close to unity in a case of alteration of parameters of a separation cascade or a single separation device in an aggregate. Such system can be operated continuously or connected occasionally depending on a technological situation. In addition, it totally excludes the phenomenon of overcompensation.

1. Introduction

As is known, the ability of the AC drive to produce the mechanical work is determined by the active power $P=UI\text{Cos}\varphi$, where φ is the phase shift between the voltage V and the current Iin the AC electric circuit. The reactive power $Q=U\text{Sin}\varphi$, creating the electromagnetic fields in the windings of the electric drive, does not perform the useful work, but loads additionally the power circuit, circulating in the circuit "generator – electric drive". The total (apparent) power of the electric drive calculates as $S=\sqrt{P^2+Q^2}$ and measures by the current-voltage method. The energy efficiency of an electric drive is estimated as the ratio of the active power to the apparent one and calls as the power factor (PF) or $\text{Cos}\varphi= P/\sqrt{P^2+Q^2}$. Also known that the power factor for an electric drive has the maximum value at full mechanical load ($\text{Cos}\varphi$=0.7–0.8) and reduces idling to $\text{Cos}\varphi$=0.2–0.3. A fraction of the reactive power varies from about 50% P to (4 – 5)P and even more. According to the general-purpose industrial standard in many countries for PF equal to 0.95 the reactive component should be around 30% of the active power P. Thus, the designers of electric drives face with the following problems: 1 - current I in electric circuits is determined in addition to the active component by the reactive one of the total power S, which increases the Ohmic losses in the power electric circuits; 2 - similarly, the reactive current additionally load the output circuits of a generator what results to necessity of increase in the power characteristics (dimensions, cost) of a generator or reduction of the active power amount which can be obtained from a generator. Therefore, designers are well aware of the law: more power, less PF that connected with loss of energy in distribution circuits, generators, and other equipment of power systems. All statements above are totally applied to a drive of the separation device for separation of isotopes (GCs). It is quite obvious that minimizing the consumption of electric power at the separation plants of GCs is the direct way to reduce the cost of electricity supplied by nuclear power plants [1].

2. Problem statement

The hysteresis synchronous motors (herewith after HM) with PF of (0.2–0.4), which are powered by the static frequency converter (SFC) forming the local power system "HM-SFC" with the non-standard frequency, are used as the drivers of the GCs [2]. Recently it has been demonstrated that increase of the PF in such power systems is possible either by traditional parallel inclusion of the compensating capacitors relatively to an unit of HMs in an aggregate of GCs, what provides the value of Cosφ about 0.95–1.0 [2] or by connection in the series of such capacitors [3].The features of the series compensation are based on the fundamental non-dependence of the resonance frequency $\omega_0 = 1/\sqrt{LC}$ of the serial power oscillating circuit *L-R-C* on the active resistance *R*. In the case under investigation it means independence on all kinds of the active electrical resistances in the power system and reduction of the output voltage in a SFC providing the staffing voltage for a HM at the resonance frequency ω_0 proportionally to the quality factor $Q = \omega L/R$. At a constant value of the capacitance *C* of the compensating capacitor corresponding to the standard requirement for Cosφ= 0.90 – 0.95, the most unfavorable for the operation of the power system is the inductive load reduction due to the destruction of GCs or the planned reconfiguration of the separation cascade. In this case, the power system can pass through the state when Cosφ = 1 and reach the undesirable mode known as "overcompensation" with a change in the nature of the power supply from the inductive to capacitive one. Restoring the given PF is based on changing the value of the compensating capacitors *C* manually or automatically according to the principle of the negative feedback. In the latter case, the displacement of the power system to the overcompensation mode leads to the changing a sign of the negative feedback in the automatic control circuit of the capacity *C*, which leads to loss of its function. In this paper the HM-FSC complex is considered as the power system stabilized in the state of the resonance at the frequency power of the adaptive automatic system. It is based on the automatic search of the minimum of the voltage *U* at the output cleats of a FSC as the function of modification of the compensating capacitance *C* which is connected in the series with a HM.

The extremum (minimum) of the *U*(C) function under the condition of the resonance corresponds to Cosφ =1. The use of this algorithm demonstrates that it totally eliminates the phenomenon of overcompensation. The problem is considered in the framework of the engineering approach. To be specific, we set that a HM has the frequency of the supply f=1800*Hz*, the supply voltage U_s =380 *V*, and its other parameters are as follows: $L = 50\ \mu H$, R=0.15 Ω (the quality factor Q~3.8; Cosφ ~0.25).

From the position of the resonance phenomenon in the local power circuit HM-FSC with the serial capacitor compensation of the reactive component of the apparent power and the constant value of U_s the dependence of *U*[*V*] at the exit of a FSC on the value of the compensating capacitor *CμF* takes the form of the extreme characteristic presented as the curve 2 in figure 1 for which dU/dC=0 what corresponds to the equality Cosφ=1. Alteration of the drive parameters, when for some reason the number of HM is changed in an aggregate leads to modification of the *U*(*C*) dependence in the coordinate space {*U*; *C*} shown in figure 1. It may move to the left or right sides simultaneously displacing the voltage extremum corresponding to the dU/dC=0 condition (see the curves 1, 2, and 3 in figure1).

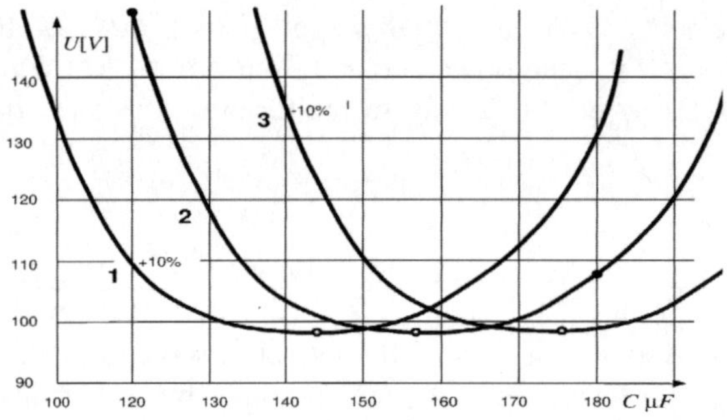

Figure 1. The qualitative dependence of the voltage U [V] on the output cleats of a FSC on the value of the compensating capacitance $C\mu F$ in the power system HM-FSC.

The curve 2 in figure 1 corresponds to the initial number of HMs, as well as the curves 1 and 3 correspond to the decreased and increased by 10% number of HMs, respectively. After the resonance tuning the value of C_0=157 μF for the initial number of HMs corresponding to the curve 2 in figure 1, the load increase leads to deterioration of the PF value (the curve 3) whereas the load decrease results to the undesirable phenomenon of "overcompensation" (the curve 1).

Because of the frequency resonance independence of the value of the active resistance R of a HMat the constant voltage $U_s$and the constant value of the Q-factor for the altering number of HMs, the minimum voltage U at the output of a FSC does not displace in a vertical direction in the coordinate space $\{U;\ C\}$ unlike a similar situation in the case of the parallel resonance [2].

3. Object of the extremum control

The attribute of the maximum energy efficiency for the local energy system HM-FSC (at the condition of U_s=const) is adopted the minimum extreme dependence of the voltage $U(C)$ at the output cleats of a FSC the corresponding dU/dC=0 (min) which means for the specific process conditions of the resonance in the series R-L-C power circuit the absence of the reactive component in the apparent power consumption and Cosφ=1.

The control problem consists in identification and track the value of dU/dC=0 (min) of the extremum characteristics of $U(C)$ in the coordinate space $\{U;\ C\}$ when the simultaneous automatic stabilization of the U_s voltage takes place.The supply voltage $U_s$of an electric drive is stabilized by a control loop with the negative feedback, including the components as follows. These are the sensor for the U_s value; the element comparing $U_s(t)$ with a given value of U_{spec}=380 V; an automatic controller with an integrator as of the control law; a FSC, an output signal U from which changes by the output of a regulator; and a HM is represented by the L-R parameters (see figure 2).

4. Algorithm for the automatic search of the extremum

The work of the step-by-step automatic algorithm to search the extremum is based on determining the direction of alteration in the search coordinates with $C(t)$ relatively to the operating point $\{U;\ C\}$ in the extreme dependence $U(C)$ in which dU/dC=0. In the finite differences it looks like $\Delta U_n/\Delta C_n=0$. Here $n = 1, 2,...\infty;\Delta C_n=C_n-C_{n-1}$ is an amplitude of a search step (the capacitance value for one tuning compensating capacitor) in the range of duration T for the nth search step. While C_n, $C_{n-1}$are the values of a capacitor in the circuit L-R-C in the moments of time T_n, T_{n-1}, and $\Delta U_n=U_n-U_{n-1}$is the increment of the U voltage at the output cleats of a FSC in the logic as follows:

$$(\text{Sign}\ [\Delta C_n \Delta U_n] < 0) \rightarrow (\Delta C_{n+1} > 0)$$

$$(\text{Sign}\ [\Delta C_n \Delta U_n] > 0) \rightarrow (\Delta C_{n+1} < 0)$$

$$\text{and } \Delta C_{n+1} = \sum_{n=1}^{n=\infty} \Delta C_n .$$

The algorithmis characterized by the self-oscillating process relatively to the value $\Delta U_n/\Delta C_n$=0.

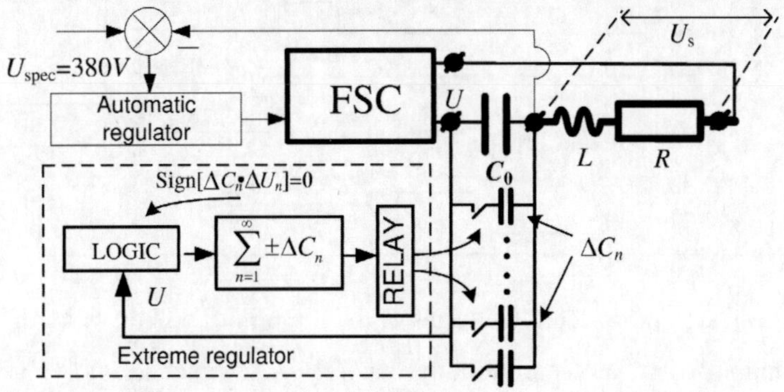

Figure 2. The system of the automatic optimization of the PF in the HM-FSC complex with the regulatory impact by a change of the capacitance with respect to the value of the nominal C_0 capacitance.

The notations used in figure 2 are as follows: L is the inductance; R is the active resistance of the HM windings; C_0 is the principal capacitor of the serial resonant contour L-R-(C_0 +ΔC_n); ΔC_n are the switchable tuning capacitors. The searching impact $\sum_{n=0}^{\infty} \pm\Delta C_n$ is generated by the solid relays, switching the tuning capacitors ΔC_n. Since the work of the step-by-step searching algorithm is based on the assessment of the partial derivatives of the static extreme characteristic, the duration T of the searching "step" must be greater than the settling time for the transient processes in the system of stabilization of U_s.

5. System model and its tuning

In this case the control object is the resonant system corresponding to the curve 2 in figure 1, in which the minimum value of $U(C)$ corresponds to the ratio U_s/Q=380/3.8=100 V.

The control algorithm is implemented as a digital-based discrete functional blocks at the Simulink Toolbox of MATLAB [4].The sampling time for modeling is adopted as the duration of the searching step T of the search algorithm. The model $U(C) = \alpha C^2 + C_0$ that corresponds to the curve 2 in figure 1, is based on the multiplication block "Prodakt 1", in which the parameter "Gain 1" is customized as α= 0.034 $V/\mu F$ and C_0= 157 μF (''Constant 4").

The deviations $U(C)$ the extreme values of which areshown in figure 1 are carried out in the element "Sine Wave", by displacement of the C coordinates according to the harmonic low. The change in the number of HMs is carried out in the block "Step 1".The first difference at the entrance ΔC_n and the exit ΔU_n of the characteristics $U(C)$ are calculated using the time delay T (in one "step") with $W(z)$ = $1/z$, where z is the argument of the discrete Z-transform analysis of the digital control systems [4]. The terms of reverse $[\Delta C_n \, \Delta U_n] = 0$ are implemented the block "Product 2", and by the operator "Sign". The search impact is formed by the discrete integrator $W(z) = T/(z-1)$. The increment output of the integrator ΔC_n per unit "step" is configured by the transmission factor of the amplifier "Gain 4". The indicators "Scope" and "XY Graf" are intended for monitoring of the system parameters.

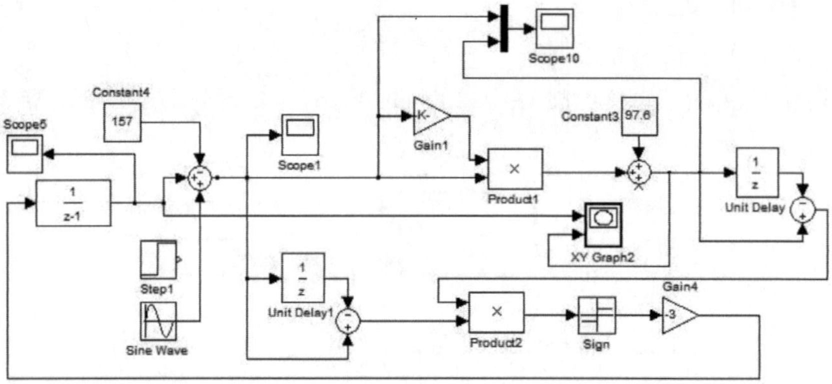

Figure3. The system model in the terms of the Simulink Toolbox of MATLAB.

During the simulation it is taken conditionally that T = 1. In practice, the value of T should be longer that the time of the forced discharge of the special industrial capacitors for compensation of the reactive power (in dozens of seconds) and more that the regulation time in the system of automatic stabilization U_s (in seconds). The amplitude of the search step ΔC_n is selected from the conditions of the confident distinction increment of the ΔU_n voltage from the exit of a SFC and following to the results of simulation is approximately (1-3)μF.

6. Results of simulation and discussion

Figure4 illustrates the situation when a part of GCs in an aggregate is fallen out what results to sharp displacement of the characteristic $U(C)$. In the model it means the transition from the curve2 to the curve1 in figure 1 with appearance of the “overcompensation” phenomenon or the jump from the curve 1 to the curve 2 in the same figure. As can be seen, the search algorithm takes out the power system in a new state of resonance with $\mathrm{Cos}\varphi = 1$ after a few “steps” $\Delta C_n = 3\mu F$.

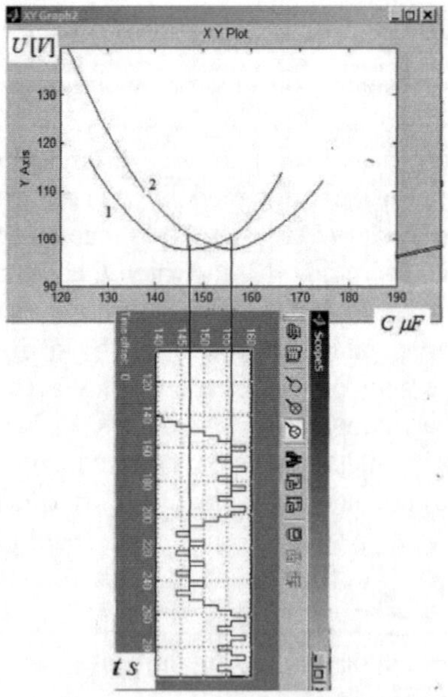

Figure 4. Working off the resonance in the power circuit in the case of a failure with GCs. ΔC= 3μF.

After tuning of the system by means of the searching algorithm the resonance of the energy system at C=157μF (the curve 2 in figure 4) at the time moment t=200s the total number of GCs in a separation cascade was changed (the curve1 in figure 4) and at the time t=250s the number of GCs was restored (the curve2 in the same figure 1). Figure5 demonstrates the auto-search of the mode Cosφ=1when takes place the slow shift of $U(C)$ following to the harmonic low with the frequency 0.001 Hz between the curves1 and 3in figure 1corresponding to the drift of the electricparameters of the HMs in time. The centers of the self-oscillation mode on the search coordinate C_n, where $\Delta U_n/\Delta C_n$ = 0 (upper part of the figure 4) corresponds to moving over the coordinate C with the extrema caused by the drift of $U(C)$. The lower part of figure 5 illustrates a change of the search coordinate C_n in tracking the resonance of the "drifting" $U(C)$.

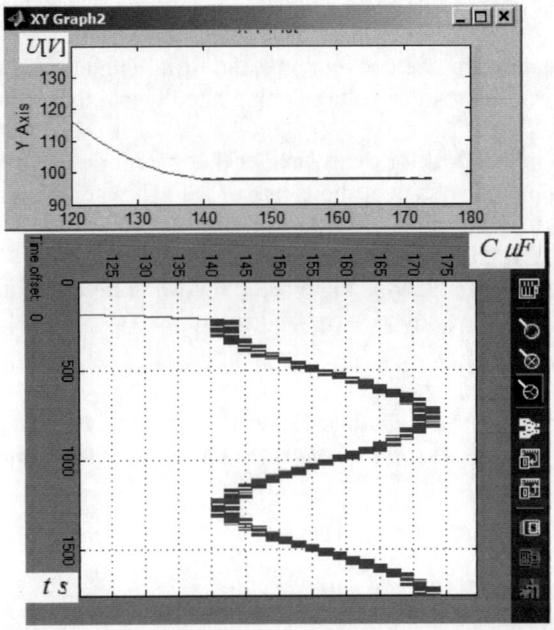

Figure5. Automatic search of Cosφ=1 with a slow drift of $U(C)$.

In the process of fixing the steady-state regime of search, the deviation on both sides of the minimum $U(C)$ ("Scope 10") does not exceed one Volt, and the errors of deviation from the specified value PF = 1 of an energy system are insignificant. Note that the regime of "overcompensation" in the power system is also absent. The results of simulation showed in figures 4 and 5, allow us to make a conclusion on the possibility of realization of the working mode for the energy system with Cosφ = 1.

7. The practical implementation of the system

The control algorithm is implemented with the standard functional resources of the general industrial automatic controllers with freely programmable structure. The electrical industry produces a wide range of the specialized capacitors for the reactive power compensation in the power circuits and the high-current thyristors for their switching. In the mode of resonance and the stable supplying voltage U_s=380V for a GC, the voltage U at the cleats of a FSC is reduced in proportion to the Q-factor of the power oscillating contour and is equal approximately to 100V, what reduces the performance requirements for a FSC.

In case of the industrial conditions the control system of the power factor can be operated continuously or switched periodically as well as be fulfilled interactively by an operator depending on the technological situation.

Journal of Physics: Conference Series **751** (2016) 012011 doi:10.1088/1742-6596/751/1/012011

8. Alternative applications

The developed system can be applied to stabilize Cos φ=1 not only in the case under investigation, but in numerous industrial technological systems and objects with variable inductive/resistive loads. As the illustrative examples one may review the induction melting furnaces, the electrolysis process, the high-power electric motors for various applications, i.e. practically at any industrial enterprise or in stand-alone (on-board) power systems with limited resources, including the case when the power or on-board voltage in a circuit is less than in the staffing or in unstable conditions.

Conclusions

1. It is presented the engineering approach to design the adaptive system for automatic stabilization of the PF for the electric drives of GC based on an automatic search for the minimum of the output voltage U at the cleats of a FSC on the value of the compensating capacitance C connected in series with the electric drive.
2. The control system is realized by the conventional industrial engineering means of controls. In the working mode it just requires the regular voltage sensors to control the output voltage for a FSC and a HM.
3. The output voltage for a in the Q-factor times smaller than that for a HM in the optimum regime.
4.The instability in the supply of voltage at the cleats of a FSC, when the parameters of GCs or HM are changed, do not effect to the nominal value of the supply voltage of a HM and the energy efficiency of the power system "HM-SFC".
5. In this working mode of the adaptive systemthe mode of"overcompensation" is totally excluded.

References

[1] Crossland Ed I 2012 *Nuclear Fuel Cycle Science and Engineering* (Woodhead Publ).
[2] Juromskiy V M 2015 *Phys. Proc.* **72** 146-9.
[3] Borisevich V D, Juromskiy V M and Trushkin N S 2013 *Nucl. Phys. Eng*. **4** 434–5 (in Russian).
[4] Dabney D B, Harman T L 2001 *Mastering Simulink 4* (Prentice Hall Upper Saddle River New Jersey).

Journal of Physics: Conference Series **751** (2016) 012012 doi:10.1088/1742-6596/751/1/012012

Reduction of uranium hexafluoride to tetrafluoride by using the hydrogen atoms

B P Aleksandrov[1], E B Gordon[2], A V Ivanov[3], A A Kotov[4] and V E Smirnov[1]

[1] NPO Energomash , Khimki, Moscow region, 141400 Russia
[2] Institute of Problems of Chemical Physics, RAS, Chernogolovka, Moscow region, 142432 Russia
[3] Scientific Research Institute of Chemical Technology (JSC "VNIIHT"), Moscow, 115409 Russia
[4] State Research Center of Russian Federation Troitsk Institute for Innovation & Fusion Research (SRC RF TRINITI), Troitsk, Moscow, 142190 Russia

Abstract. We consider the reduction of UF_6 to UF_4 by chemical reaction with hydrogen atoms originated in the powerful chemical generator. The principal design of such a chemical convertor is described. The results of the mathematical modeling of the thermodynamics and kinetics of the UF_6 to UF_4 reduction process are analyzed. The few options for the hydrogen atom generator design are proposed. A layout of the experimental setup with the chemical reactor is presented. The high efficiency together with the ability of the process scaling without loss of its efficiency makes this approach to the uranium hexafluoride depletion into tetrafluoride promising for its application in the industry.

1. Introduction

For the long period of nuclear materials production the more than 1.5 million tons of depleted uranium hexafluoride (DUH) has been accumulated in the world, with an annual increase of ~ 70 thousand tons. Storage of such DUH quantities is a serious and sophisticated problem, which encourages the search for the processes for its conversion into more safe for storage compounds (e.g., U_3O_8 or UF_4) being commercially demanded products [1,2].

One of well-known and investigated in Russia methods of the DUH conversion into more safe uranium tetrafluoride (DUT) is its reduction in hydrofluoric flame [1-3].

The reduction reaction of $UF_6 + H_2 \Rightarrow UF_4 + 2HF$ due to its high activation energy requires the reaction temperature of 1500-2000 K [4]. It is assumed that the reagent can be heated in hydrofluoric flame: $H_2 + F_2 \Rightarrow 2HF$. The final products should be uranium tetrafluoride and dehydrated hydrogen fluoride. The main advantage of this method is the minimum energy expenses because the energy for heating the mixture released in chemical reaction.

The UF_6 reduction reaction with hydrogen becomes exothermic only after the UF_4 gas solidification, which can occur only at temperatures below 1600 K [4]. Thus, for successful conversion of UF_6 to UF_4 it is necessary to satisfy the contradictory requirements for the temperature to be sufficiently high for the reduction of UF_6 in gas phase, and, simultaneously, enough low for the further realization of UF_4 solidification. The solution of this contradiction can be contributed with the heterogeneous catalytic reduction of UF_6 on the reactor walls and/or on the surface of just formed solid UF_4 dust. The whole reduction process becomes in this case exothermic and self-sustaining [5,6].

It was proposed in [5, 7] to carry out the UF_6 reduction with pre-atomized hydrogen produced by a hydrogen atom generator, which makes the process more manageable and easy realized. It was proposed for that to use as the powerful chemical generator the system developed in the "NPO Energomash named after Academician VP Glushko " for CW chemical HF lasers (HFL), where the H atoms were the waste product. This system produces as much as 1 mole of hydrogen atoms per second and it was designed to work at high temperatures in corrosive environment (HF).
An important difference of this method for the uranium hexafluoride reduction is the absence in the initial mixture of molecular fluorine, which leads to fast competing $H + F_2 \Rightarrow HF + F$ reaction of H atoms decay, and the presence of supersonic jet, preventing the heat propagation upstream and allowingsuch a way to work at high temperatures without perishing the mixing chamber [5].

Application of the hydrogen atom generator expands the abilities of the chemical reactor, provides high efficiency during process scaling-up, and open the possibility to study various schemes of the reduction process organizing.

National technologies of DUH reduction in hydrofluoric flame developed in JSC "ARRICT" and allowing to have DUT as the product are attractive because of the partial compensation of energy expenses for conversion process by the implementation of commercial products having a national economic significance. Indeed, in addition to dehydrated hydrogen fluoride as well as commodity products for microelectronics such as silicomethane, silicon tetrafluoride and others may be produced.

2. Scheme of the chemical reactor with an atomic hydrogen generator

The principal design of the reactor for UF_6 reduction with using the generator of atomic hydrogen where UF_6 supply into the nozzle arrays shown in figure 1 [6].

Thermochemical generators of hydrogen atoms having the efficiency required for an industrial application which are the main unit of the chemical reactor were created for CW HF chemical lasers (HFL). The typical generator is presented in figure 2 [7]. It consists of several basic units: array of burners, combustion chamber, nozzle array and reactor.

It operates as follows. In the combustion chamber the reaction of H_2 molecules with excess amount of F_2 takes place. The excessive weakly-bound F_2 molecules (dissociation energy of F_2 molecule on F atoms is only 37 kcal/mole) entirely dissociated at high temperature to fluorine atoms. These atoms are then tempered passing through the supersonic nozzle array and are in-flow mixed with molecular hydrogen. The subsequent fast chemical reaction $F + H_2 \Rightarrow HF(v) + H$ results in H atoms formation which enter in the aimed chemical reaction with DUH.

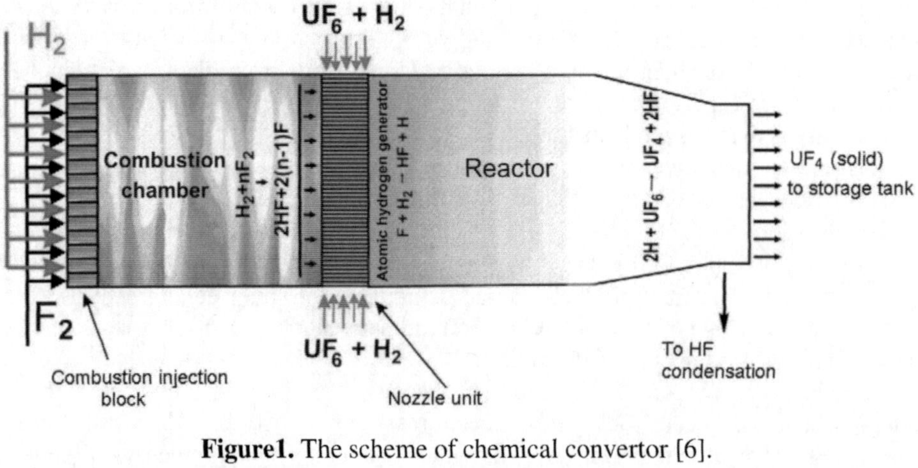

Figure1. The scheme of chemical convertor [6].

The proposed technology for uranium hexafluoride to tetrafluoride reduction in the chemical reactor with an atomic hydrogen generator is close ideologically to the method for UF_6 to UF_4 reduction with hydrogen in the hydrofluoric flame. This technology combines the advantages of the UF_6 reduction method using molecular hydrogen in the hydrofluoric flame (having minimal energy

Journal of Physics: Conference Series **751** (2016) 012012 doi:10.1088/1742-6596/751/1/012012

expenses) with removing the problems caused by endothermicity of the process at the stage preceding UF_4 condensation. This is due to the fact that both reactions of the UF_6 to UF_4 reduction with atomic hydrogen

$$UF_6 + H \Rightarrow UF_5 + HF + 69{,}5 \text{ kcal/mole and } UF_5 + H \Rightarrow UF_4 + HF + 34.6 \text{ kcal/mole}$$

are highly exothermic and rather fast at temperatures of several hundred K. That is why no outer energy is needed to organize stationary burning and keep it stable at sufficiently low temperature.

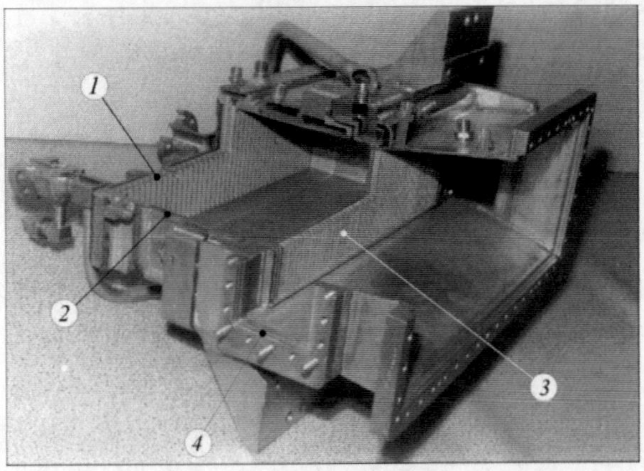

Figure 2. Hydrogen atoms generator. 1-burner array, 2-combustion chamber, 3-nozzle array and 4 - reactor [8].

3. Numerical modeling of reduction processes

The chemical reactor with hydrogen atoms generator allows organize various schemes of the reduction processes. Consider some of them.

3.1. Reduction with atomic hydrogen

Let us consider the work of a chemical reactor in the mode of uranium hexafluoride reduction with atomic hydrogen. A mixture of fluorine and hydrogen molecules is supplied to the combustor, while uranium hexafluoride with secondary hydrogen – to the nozzle unit. The mixture of weakly coupled molecules of fluorine and hydrogen molecules in excess of fluorine ($[F_2] / [H_2] = n > 1$) is ignited at the inlet to the combustion chamber. The reaction $H_2 + nF_2 \Rightarrow 2HF + 2(n-1)F$ is highly exothermic and the heat released is sufficient for dissociation of F_2 molecules excess. Combustion products arrive into the reaction zone through supersonic oxidizer nozzles where mix with secondary hydrogen supplied through interleave neighbor fuel nozzles. As a result of the reaction of $2(n-1)F+2(n-1)H_2 \Rightarrow 2(n-1)HF+2(n-1)H$, hydrogen atoms H are formed and mixed with UF_6 in the reactor (and if necessary with a buffer gas - diluent). If the process is organized in such a manner that the UF_6 is reduced to uranium tetrafluoride according to formula: $2(n-1)H+(n-1)UF_6 \Rightarrow (n-1)UF_4+2(n-1)HF$, then for every mole of the recovered UF_6 2 moles of hydrogen atoms are needed.

Kinetic calculations of the reduction process for this mode of reduction are made in [5, 6]. The ultimate goal of the one-dimentional calculations in [5] was to determine the degree of UF_6 reduction depending on the stoichiometry and temperature of the mixture, reagent flow rate, and the selectivity of the target product - UF_4. In the mathematical description of chemical reductions in the reactor the following simplifying assumptions were made in this study: the reagent mixing was assumed to be instant, the linear flow rate of the reactants - constant and equals 200 m / c; due to large areas of the reactor the heat-removing was neglected and the temperature of nozzle unit was assigned to constant value, while the temperature of the reagents was adjusted by the amount of the diluent gas N_2 added to the flow and the processes of diffusion and thermal conductivity were not taken into account. The coordinate along the axis of the reactor serves as an independent variable and the concentration and temperature at the reactor inlet as the initial conditions.

In figure 3, the results of calculations of the main characteristics for the reduction process are demonstrated [5] for the initial $[UF_6]_0$ concentration of $2{\cdot}10^{17}$ cm^{-3}. From this figure, which shows the concentration profiles of the reactants and products it can be seen that the initial reaction product is UF_5, its subsequent reduction and formation of solid particles occurs after the moment when almost no UF_6 left.

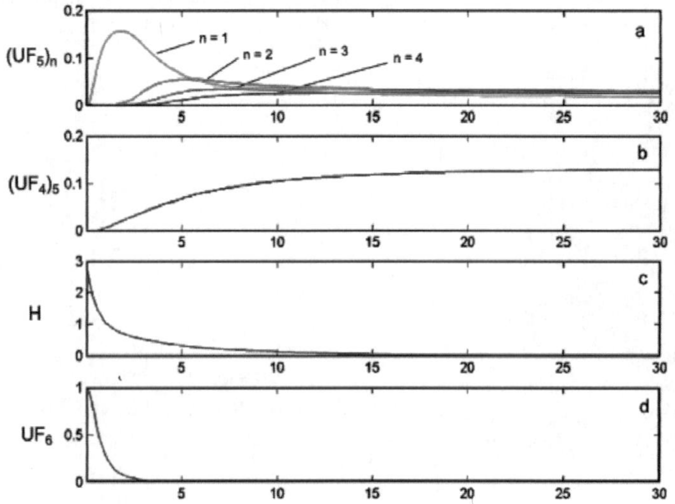

Figure 3. The concentrations of $(UF_5)n$, $(UF_4)_5$, H, and UF_6, as the functions of the distance X along the reactor. Flow velocity is $2{\cdot}10^4$cm/sec, $[UF_6]_0 = 1$ ($2{\cdot}10^{17}$ cm^{-3}), $[H]_0 = 3$, with no dilution with nitrogen [5].

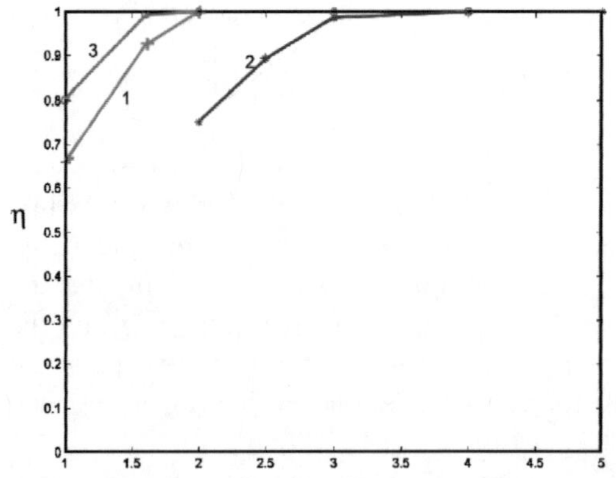

Figure 4. The yield of UF_6 conversion ($\eta = 1 - [UF_6]_c / [UF_6]_0$), *vs* the content of hydrogen atoms (1, 2) and fluorine (3) in the working mixtures; $[UF_6]_0 = 1$ ($2{\cdot}10^{17}cm^{-3}$). 1, 3 - without diluting the mixture (only HF_0), 2 -with dilution with nitrogen to $[M] = [N_2] + [HF]_0 = 10$ [5].

The depth of UF_6 reduction, $\eta = 1 - [UF_6]_c/[UF_6]_0$ becomes full immediately after meeting the stoichiometric requirement $[H]_0$ ($[F]_0$)/$[UF_6]_0 = 2$, and only a ten-fold dilution of the mixture causes the need to increase the concentration of atoms in half (figure 4).

Numerical simulation of the UF_6 reduction by H atoms regarding the geometry of the nozzle array is considered in [6], figure5. The composition of the initial mixture is defined by conventional formula: $[H_2]^{(1)} + \alpha\ [F_2]^{(1)} + 2(\alpha - 1)K[H_2]^{(2)} + 2(\alpha - 1)\beta\ [UF_6]^{(2)}$). Symbols (1), (2) denote, the combustion chamber and the reactor zone respectively. At the description of chemical reactions and vibrational kinetics in the mixing zone a kinetic model of HF-laser was used supplemented by the reactions involving uranium components. The maximum concentration of atomic fluorine is achieved at $\alpha \equiv [F_2]: [H_2] \sim 3$. In this case, the reduction of one molecule of UF_6 to UF_4 requires at least 2.5 molecules of H_2 and 1.5 molecule of F_2, forming five molecules of HF [6].

Figure 5 shows the results of calculations for the nozzle array spacing of 7 mm for the composition $\alpha = 3$, $K = 2.2$, $\beta = 0.3$. The molecular weights of the first and second flows are
$\mu^{(1)} = 19.5$ and $\mu^{(2)} = 43.2$.

The ratio of the fuel (H_2) nozzle's exit area to oxidizer (F_2) one is chosen to be equal to 2.5: 1. Calculation was performed for the pressure at nozzle's exit plane p_0 = 10 Torr.

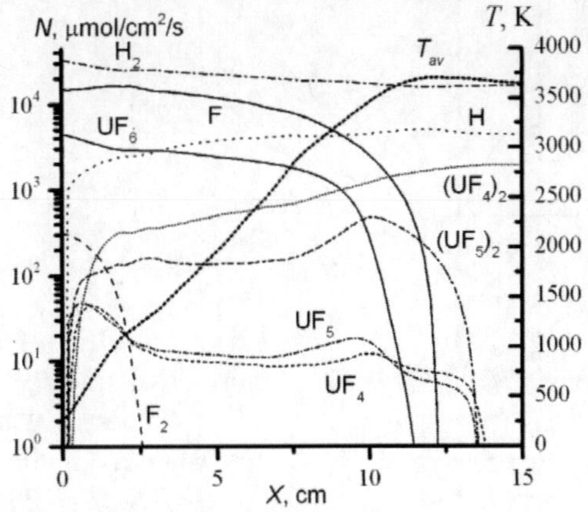

Figure 5. Distribution along the stream of the molar flows of components and of the temperature averaged over the cross-section. P_0 = 10 Torr and nozzle array spacing is h= 7 mm [6].

The total hydrogen flow rate per nozzle array unit area for this variant is 0.08 g / cm^2 / sec, the fluorine flow rate is 0.44 g / cm^2 / sec, and uranium hexafluoride flow rate is 1.58 g / cm^2 / sec. It is shown that a full reduction of hexafluoride and creation of the targeted product occurs in this case, at a length of about 15 cm, figure 5. A feature of this scheme is the need to realize the mixing of supersonic flows, one of which contains F atoms, and the second - H_2 and UF_6 mixture. The mixing is mainly due to the diffusion of the most lightweight component. At first, H_2 diffuses into the stream containing F atoms, and then the hydrogen atoms diffuse back into the flow of H_2 with UF_6, providing the reactions of hexafluoride and pentafluoride reduction [6].

3.2.Termochemical reduction.

In this mode, fluorine molecules and a mixture of hydrogen and uranium hexafluoride is fed into combustor, (figure 2), i.e., only one chamber (combustion chamber) is used. Calculation of the equilibrium composition of a stoichiometric mixture of F_2 molecules with molecules of the hydrogen and uranium hexafluoride is made in [4]. The problem of finding the final thermodynamic equilibrium composition of the mixture of substances involved in the reactions is determined not by the type of chemical reactions, but only the composition of the initial mixture and the process of transition to the final state (at constant pressure, volume, temperature, etc.). This greatly simplifies the calculation task and allows to reduce it to a system of ordinary nonlinear equations.

Dependences of thermodynamic equilibrium concentrations and their temperature on the initial mixture composition concentrations were obtained at solving the system of equations describing the equilibrium state in the system UF_n - H_2 (H) - F_2 (F) - HF. The initial mixture is represented in a general form: $H_2+(\alpha+\beta)F_2+(1-\alpha)UF_6$. One can obtain any composition of interest by varying the coefficients α and β. A complete partial pressure and the temperature of the individual components in thermodynamically equilibrium mixture were obtained depending on the composition of the initial mixture of $H_2+\alpha F_2+(1-\alpha)UF_6$ at the initial temperature T_0 = 373 K and p_0= 1.5 atm.

In figure 6, the calculated thermodynamic equilibrium concentration ratio for UF_5 / UF_4 with taking into account gas and condensed phases for the starting composition $H_2+\alpha F_2+(1-\alpha)UF_6$ at T_0 = 373 K and p_0 = 1 ... 10 atm is shown at a linear scale.

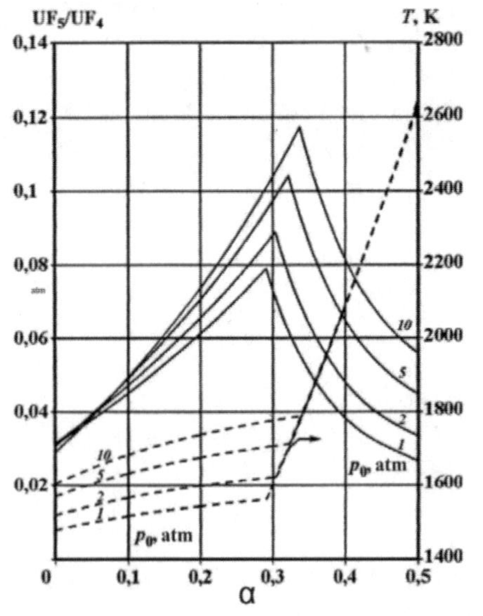

Figure 6. UF_5 / UF_4 concentration ratio in the thermodynamically equilibrium mixture, taking into account gas and condensed phases coexistanse, for the initial composition $H_2+\alpha F_2+(1-\alpha)UF_6$ at $T_0 = 373$ K and $p_0 = 1 \ldots 10$ atm. The values of the pressure are shown above the appropriate curves, [4].

The figure clearly shows that the content of UF_5 in the equilibrium mixture is small and varies within the range of $3 \div 12\%$ of the total UF_4 depending on α and pressure p_0. A sharp curve bend of the dependency $[UF_5] / [UF_4]$ on α is due to the start of the formation of a condensed phase.

The formation of the condensed phase, on the one hand, removes the condensed component of the individual component concentration from the process of formation of an equilibrium gas composition; on the other hand, this is accompanied by the release of additional heat to the mixture due to the heat of gas-liquid phase transition. The formation of the liquid phase causes a drop of the UF_4 concentration in the gas phase, so the temperature dramatically slows down its decline due to heat release during the phase transition. Therefore, despite the temperature decreasing at α reduction, the relative concentration of UF_5 starts to fall.

3.3. Combined reduction (with molecular and atomic hydrogen together).

In this mode, the uranium hexafluoride is supplied to the first zone (combustion chamber) together with fluorine and hydrogen, the combustion products flow into the second zone (reactor) through the oxidizer nozzles of the generator, and the hydrogen is fed there through the fuel nozzles, figure 2. This method provides a two-stage reduction - the thermal conversion in the first zone and an additional reduction by H atoms in the second zone. Numerical simulation of the reduction process for two different geometries of the nozzle array is given in [6]. The fields of pressure and temperature in the area of mixing the oxidizer and the fuel streams, as well as the distribution along the stream of the component's mole flow rates and of the temperature averaged over the cross section, are obtained , figure 7.

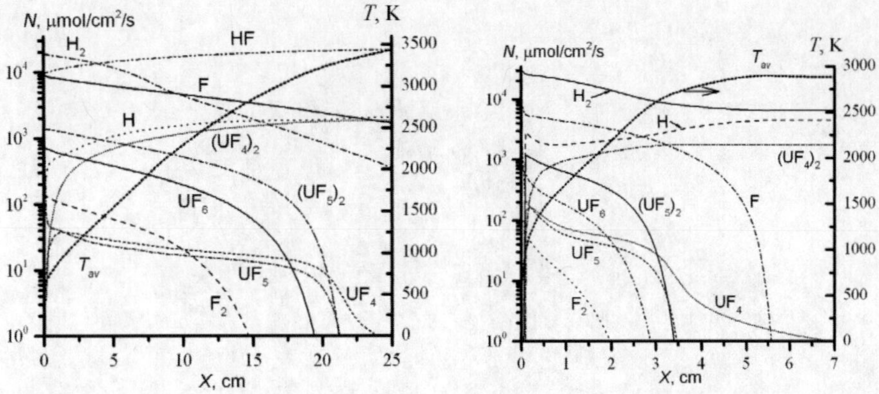

Figure 7. Distribution of the mole flow rates of components and the temperature averaged over the cross section of reactor. P_0 = 10 Torr. The nozzle array spacing is supposed to be 14 mm *(left)* and 7 mm *(right)* [6].

Figure 7*(left)* displays the downward distributions of the mole flow rates of components and of the static temperature averaged over cross-section with taking into account the heat capacities of components. As seen, the uranium pentafluoride UF_5 near the outlet of the nozzle array rapidly and almost completely dimerizes increasing the average temperature of the flow by ~ 130K. At a distance of a few millimeteres from the nozzle array the processes of uranium fluoride reduction come into effect, as evidenced by the decrease of H concentration. The decrease of the UF_6 concentration by two orders of magnitude and the creation of the target product UF_4 and $(UF_4)_2$ almost completely ends at distances of ~20 cm from the nozzle array exit plane. The average static pressure increases rather quickly with temperature increasing: at the distance of 1 cm to ~ 20 Torr, at 6 cm to 32 Torr, and at 30 cm – to 83 Torr. The length of the reduction for this kinetics is of the order of 15-20 cm.

Calculation was also made at halving the mixing step of the nozzle array h =7 mm, the results are shown in figure 7 *(right)*. At a more intense mixing the reduction zone shortens, in this case - to ~ 3.5 cm, that corresponds to the length of hydrogen diffusion into the oxidizer stream.

The pressure at the nozzle unit exit was also varied. The higher is the pressure, the more easy to organize the pumping system. Fourfold increasing of the pressure at the nozzle unit exit – up to p_0 = 40 Torr at an array spacing of 7 mm, naturally, slows the mixing of the secondary oxidizer and fuel flows, and lengthens the reduction zone to ~ 17 cm. This pressure increasing (and hence, flow rate)) leads to an increase of the pressure at the end of the reduction zone to ~ 220 Torr, that makes it possible to relatively easy to perfume the exhaust of the combustion products at atmospheric pressure [6]. Performed calculations show the fundamental possibility of UF_6 to UF_4 reduction using atomic hydrogen generators of supersonic cw chemical lasers of high performance.

However, too much variation in literature for values of the constants of elementary chemical processes, in particular, of the reactions H + UF_6 and H + UF_5, creates certain difficulties in formulating the requirements for the design of a supersonic nozzle unit for variants of the reactor for DUH reduction using atomic hydrogen. For-more accurate calculations it is necessary to clarify the experimental rate constants of certain chemical reactions (direct and inverse).

4. Experimental setup with a hydrogen atoms generator.

Figure 8 schematically illustrates the layout of the experimental chemical reactor with a hydrogen atoms generator. There are a few basic units of the experimental setup: gas supply system (gases H_2, F_2, N_2, etc.), system of preparation of UF_6, a chemical reactor, diagnostics, device of particulate collection UF_4 (cyclone unit, metal ceramic filters), vacuum receiver, ventilation systems and special ventilation. The entire installation is inside gas safety container.

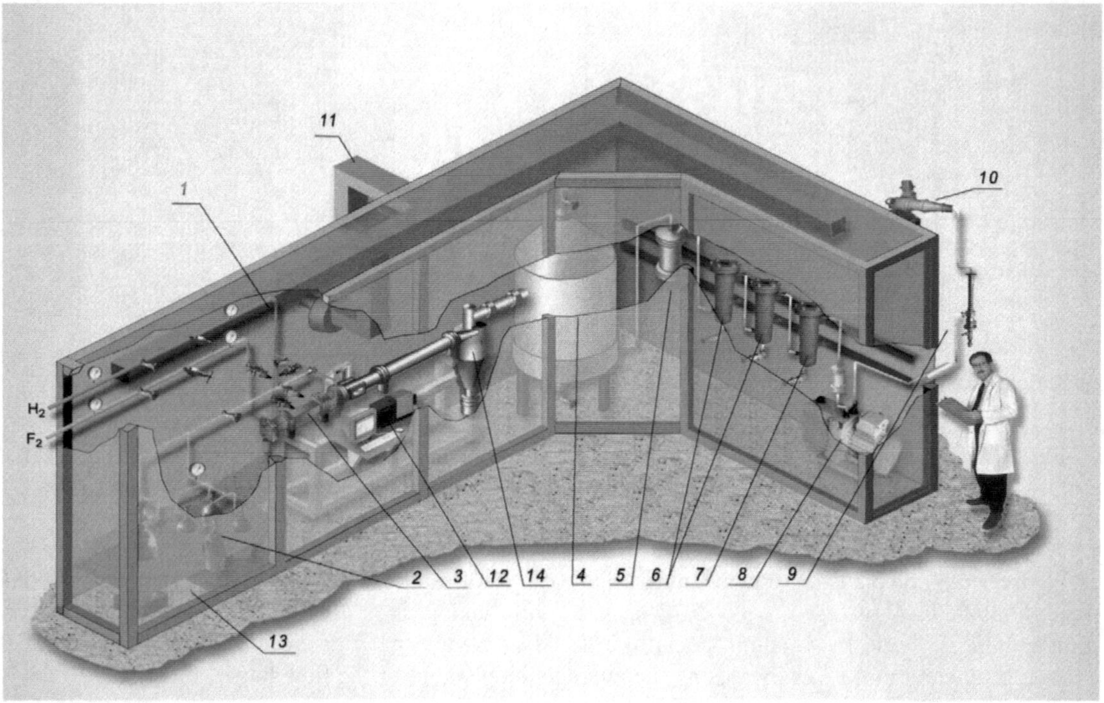

Figure 8. A general view of the experimental setup with the hydrogen atoms generator. 1 - system of gas supply, 2 - cylinders for UF_6, 3 - chemical reactor, 4 - vacuum receiver, 5 - cryogenic trap, 6,7 - columns with NaF and filters, 8 - vacuum pump, 9 - hydrogen afterburner, 10 - special ventilation system, 11 - ventilation system, 12 - Fourier spectrometer, 13 - gas safety container, 14 - cyclone unit for UF_4.

The system of UF_6 preparation should provide the necessary flow rate of UF_6. The diagnostic system includes sensors of temperature, pressure, and flow rate; optical measurements (Fourier spectrometer) are needed to determine the concentrations and temperature of chemical mixture reagents. The vacuum receiver provides the coordination of the gas-dynamic channel of the chemical reactor. The system is equipped with a special ventilation set of filters and cryogenic traps to neutralize the exhaust.

The above-considered options of using the hydrogen atoms generator designed by the «NPO Energomash», allow a defluorination of uranium hexafluoride to the reduced consumption of up to
5 g / cm^2 / s, which is quite enough to clarify the calculation of experimental models and processing of the setup components. With this performance the setup has quite acceptable dimensions, figure 8. Theoretically, the continuous operation of the hydrogen atoms generator with its area of 400 cm^2 is capable of reducing with a capacity of up to 7 tons of depleted uranium hexafluoride per hour, provided that the gasification system and supply of the required flow rate of DUH exist. [6]
Mathematical modeling of thermodynamics and kinetics of the UF_6 to UF_4 reduction demonstrates the feasibility of the proposed method at quite reasonable parameters of the experimental device. Nevertheless, there is a certain variation in the data for some of the reaction rate constants, their refinement could lead to adjustments in the working parameters of the experimental chemical reactor.

5. Discussion

To be self-sustaining and providing a high yield the process of the UF_6 to UF_4 reduction must be exothermic. The analysis performed shows that the reaction of $UF_6 + F_2 + H_2 \Rightarrow UF_4 + HF$ at any ratio of reagent contents may be exothermic only provided the UF_4 solidification during the reaction course. But in accordance with the general considerations and numerical calculation presented in this paper, the UF_4 condensation occurs only if the reaction temperature is less than some specified value T *. Therefore, in principle, there are three possible approaches to the the UF_6 to UF_4 reduction when using $F_2 + H_2$ reaction.

1. The reduction reaction in a mixture of $UF_6 + F_2 + H_2$, in a burner when, either by selecting the relative concentrations of the reactants, or by diluting the active mixture with an inert gas, the combustion temperature is less than T *. In this case it is possible to arrange a self-sustaining combustion in the form of a "smoky" torch, therein the smoke is particles of UF_4, which are carried away by the flow and can land on the cooled panels downstream. Wherein, of course, there are no difficulties in the transition to large-scale reactor. Unfortunately, the level of knowledge of the rate constants for chemical reactions occurring in the reactor does not allow us to be sure that this mode can be implemented because a relatively low temperature of the reaction means small rates of elementary processes. Furthermore, in this mode, it is difficult to maintain a stable combustion for a long time due to the nonlinear dependence of the reaction rates on the rate of UF_4 condensation. Experience of managing the combustion of hydrocarbons on conditions with heavy soot supports this conclusion.

2. Carrying out the reaction $UF_6 + F_2 + H_2$ in a burner at temperatures higher than T *, without any dilution of the active mixture with inert gas, but introducing into the reaction zone the panels cooled from inside to the temperature less than T *. Then, the product will condense on panels surfaces and the reaction rate will be high, including quite probable catalytic effect of surfaces coated with UF_4. In this regard, of interest are the experiments that have been already performed in JSC "ARRICT" as well as those that can be performed in the future. The disadvantages should include the complexity of the introduction of panels into the reaction zone, especially at large flows of reagents.

3. Using the chemical generator of hydrogen atoms. Since the $UF_6 + H \rightarrow UF_4 + HF$ reaction is exothermic even in the absence of UF_4 condensation, it will occur under all conditions and at any scale of the burner. Positive is the fact that in this case the product is formed downstream and beyond the reaction zone. Negative factor in this approach is that the consumption/ of the reagents and, above all, fluorine per one molecule of the converted uranium fluoride is higher than in the first two approaches, because of the additional fluorine should be spent in a hydrogen atoms chemical generator of. This approach is the most technically complex, but its complexity is mainly associated with the generator of atoms, and it should be taken into account that such generators producing flows of atoms required for a full-scale setup already exist.

In general, one should take into account that these approaches are not entirely independent. For example, in the frames of the third approach, one can try to reduce the initial concentration of hydrogen atoms by adding a certain amount of molecular hydrogen and fluorine into the mixer, thereby reducing the necessary flow of fluorine.

6.Conclusion

Taking into account the extensive experience of this paper authors in work with fluorine, hydrogen, fluorides and uranium, one should be argued that the problem of the of UF_6 to UF_4 reduction by using fluorine with hydrogen at high reactant flows will be successfully solved. When using the more technically complex and more costly approach with a chemical generator of hydrogen atoms, the combined schemes can be applied simultaneously to minimize the expense of fluorine.

The proposed method for reduction of uranium hexafluoride by use of a generator of hydrogen atoms is protected by RF patent №2204529 from 20.05.2003, [7].

References

[1] Vlasov A A et al 2000 *Proceedings of the 5-th Russian(International) Scientific Conference "Physical and chemical processes in the selection of atoms and molecules"* 25.

[2] Gordon E B et al. 2000 *Proceedings of the 5-th Russia (International) Scientific Conference "Physical-Chemical Processes in the Selection of atoms and molecules"* 35.

[3] Boot L A et al 2016 *Chem. Tech.* **1** 28.

[4] Beznozdrev V N, Fedorov A M, Kolesnikov Yu A, Malyuta D D 2005 *Proceedings of NPO Energomash* **23** 382.

[5] Gordon E B, Dubovitsky V A, Kolesnikov Yu A et al 2002 *Proceedings of the 7th All - Russia(International) Scientific Conference "Physical and chemical processes in the selection of atoms and molecules"* 100.

[6] Alexandrov B P, Smirnov V E, Beznozdrev V N 2014 *Proceedings of the 3rd International Scientific Conference "Innovative projects and technologies of nuclear energy"* **2** 35.

[7] Gordon E B et al 2003. *Russian patent number 2204529, registered in the State Register of Inventions of the Russian Federation* with a priority from 28.03.2001.

[8] Aleksandrov B P, Bashkin A S, Beznozdrev V N, Parfenov M V, Pirogov N A, Semenov S N 2003 *Quantum Electron.* **33** 25.

Magnetron sputtering of copper on thermosensitive polymer materials of the gas centrifuge rotors

V Borisevich[1], S Senchenkov[2] and D Titov[1]

[1] National Research Nuclear University MEPhI (Moscow Engineering Physics Institute), Kashirskoe highway 31, Moscow, 115409, Russia
[2] National Research Centre "Kurchatov Institute", 1, ac. Kurchatov square, 123098, Moscow, Russia

E-mail: vdborisevich@mephi.ru

Abstract. Magnetron sputtering is the well-known and widely-used deposition technique for coating versatile high-quality and well-adhered films. However, the technology has some limitations, caused by high temperatures on the coating surface. The paper is devoted to the experimental development of a process of magnetron sputtering of copper on a surface coated with a thermosensitive polymer made of carbon fiber with epoxide binder. This process is applied for balancing a rotor of a gas centrifuge for isotope separation. The optimum operating parameters of the process are found and discussed. They were in quantitative agreement with data obtained by means of non-stationary modeling based on a global description of plasma in the typical geometry of the magnetron discharges obtained in independent research. The structure of the resulting layer is investigated.

1. Introduction

The development of magnetron sputtering paved the way for the coating of various substrates with materials that previously were impossible by thermal and ion sputtering. The main advantage of this technique is that bombarding the substrate with high-energy electrons becomes unnecessary for that leads to uncontrolled heating of the substrate and damages structures.

According to expert evaluations, high-power pulsed magnetron sputtering (HPPMS) has not exhausted its potential in that it enables dense plasma with a high content of a target component to be generated [1-2]. To date, many polymer materials are greatly restricted in applications in HPPMS because of sensitivity to thermal loading. As is known, the general value of the substrate temperature during the sputtering process is above 500°C, which is not suitable for many kinds of thermosensitive materials. No doubt, that the substrate temperature can be lowered by cooling and lowering the deposition rate but it greatly increases the cost of the sputtering process. [3]. It is also important in our case that a reduction in the process temperature would avoid surface tensions in the substrate that result in damage and flaking [4–5].

We deal with balancing of the gas centrifuge rotors, which is a classical problem of a turbomachinery. Being unbalanced these machines start to vibrate and it can lead to catastrophic failure. The balancing of three-dimensional rotating bodies is the most difficult case because it demands to place the balancing masses in a few places on a rotating body surface simultaneously. Referring those who are interested in the details of the rotor balancing process to the paper [6], we will focus on the development of the magnetron sputtering system and the search of its optimum parameters. Being intended for mass production, this technique should be simple and cheap enough. The additional limiting condition for it is a restriction on a surface temperature of a rotor caused by its surface coated with a thermosensitive polymer. In our case it is carbon fiber with epoxide bounder.

Recently, a model analysis of HPPMS of copper by means of the non-stationary two-zone model on the effects of voltage characteristics on HPPMS of copper was done by Kozaket. al. [7-8], where the conclusion was made that the average target power density in a pulse is, in addition to the target voltage, a fundamental quantity determining the average ionized fraction of copper atoms in the flux onto the substrate and the normalized deposition rate. Another conclusion was that a compromise between a high degree of ionization of the sputtered target material atoms and an acceptable deposition rate should be found by tuning the average target power density delivered in a pulse. It was demonstrated that sufficiently long pulses (at least 100 μs) are needed to achieve high target power densities in a pulse and, consequently, high degrees of ionization of the sputtered atoms. The results of our experiments are compared with that model analysis below in the section Results and Discussion.

The feature of our problem consists in balancing of the spinning at high speed rotor covered with thermosensitive polymer. The idea was born to adapt HPPMS of copper for that. We found a paper of Kozak [6], but decided to verify the results of this theoretical research because the temperature limitation on the HPPMS process was added as the used carbon fiber with epoxide binder is destroyed by heating over 70°C. This limitation led us to question of limiting the deposition rate to keep our material from harm but keeping it rather high to be industry efficient.

As the possible means to solve this problem, we propose a cheap and easy method to operate a balanced magnetron with a low current through the substrate that would enable magnetron sputtering to be used on thermosensitive polymer material without significant substrate heating (t_{sub}< 70 °C).

As is well known, oxides and nitrides of titanium and aluminum are the most frequently used materials in magnetron sputtering. The choice of usage copper instead of the above-mentioned materials was dictated by the fact that copper has one of the highest sputtering coefficients, and, of course, it is much cheaper than they are. In addition, we realize that magnetron sputtering of copper may be used in various applications for creation of electrical and thermo-conductive layers on thermo-sensitive plastics [9-11].

2. Description of stand and method

The experimental stand we propose for the magnetron sputtering with copper of polymer materials consists of a vacuum camera with flanges for the magnetron modules. The modules are situated at five balancing planes to provide the rotor to come through resonant frequencies without destruction (see figure 1). Carbon fiber covered with epoxide is used as the substrate because it is widely used and easily obtained. The major limitation imposed on the process is a substrate temperature that should not exceed 70°C.

The material for the vacuum camera walls is stainless steel. Two pumps are used in the stand: a backing pump of pumping speed 5.5 *l*/s and a turbomolecular pump of pumping speed 150 *l*/s(for N_2). The vacuum camera is pumped down to a pressure of 1.3×10^{-3} Pa. The balanced magnetron is used with a cathode made of copper M1 with surface area of approx. 7570 mm^2. The magnetron sputtering modules are connected to a power supply, which generates powerful pulses of negative polarity with maximum voltage of 3 kV for durations of 50–300 μs. A thermocouple sensor and a vacuum gauge trigger shutting of the valve and to measure pressure. Pure argon (of 99.998% purity) is used as the process gas. The argon flow rate of 4.72×10−5 *l*/s is controlled by an analog flow meter. The total volume of the vacuum camera is 100 *l*. The weight scales have an accuracy of ±0.5 mg giving a measurement limit of 420 g and ±10 mg; a measurement limit of 4.200 g is used for the sputtered mass measurements.

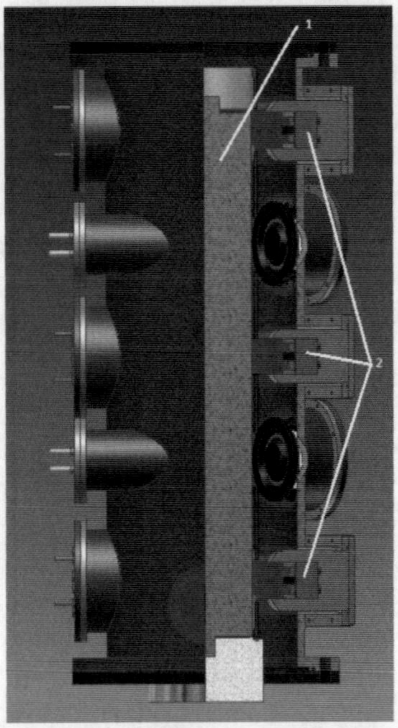

Figure 1. Design of the stand. 1 – polymer material of the rotor, 2 –magnetron modules

For sustained work for a magnetron, a constant pressure needs to be maintained. Indeed, this is not an easy task because the pressure to supply a constant argon flow to the installation must not fall below 10^{-3} Pa, otherwise the magnetron operation mode changes. Moreover, a good vacuum without a trace of oxygen is needed for magnetron sputtering with copper. Non-compliance with this condition leads to reactive deposition of copper oxide, for which adhesion is much worse. Additionally, the results from copper oxide sputtering are much worse compared with that for pure copper.

3. Special Requirements of the magnetron design

The optimal settings which are described below were found in a series of experiments yielding a sputtering speed of about 8.82 μm/mm^2·h that produced a sustained and predictable mode of operation for the magnetron. If the magnetron design (figure 2) does not follow these requirements, negative effects take place, the most unpleasant being the discharge ignition behind instead of in front of the target. A "rear" discharge glows brightly and large currents are realized in this mode, although gradually front discharge pulls over the magnetic field to the region above the target. Concurrently, the current drops by one order of magnitude and subsequently sputtering speed decreases. Simultaneously, the vacuum camera pollution is very fast; the maximum deposition rate achieved in this case does not exceed 6.05 μm/mm^2·h.

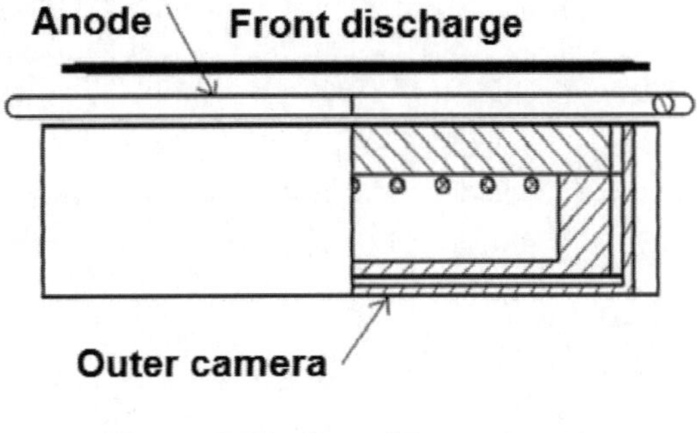

Figure 2. Design of the magnetron.

4. The optimum regime of the magnetron

During testing of the magnetron sputtering system the discharge voltage vs applied voltage graph was obtained at various pulse durations (100, 200, and 300 μs) at a frequency of 200 Hz (see figure 3).

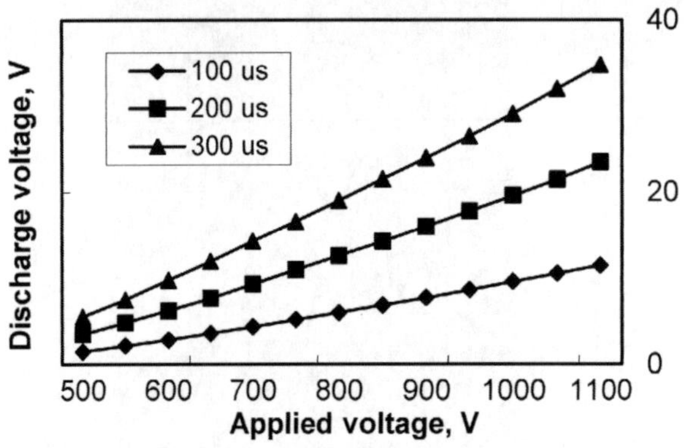

Figure 3. Discharge voltage vs applied voltage for various pulse durations.

The voltage dependence of the discharge ignition delay is shown in figure 4. The delay time is inversely proportional to the supplied voltage and is nearly 30 μs at low voltage. Hence, we concluded that the pulse duration of 100 μs is not optimal because a considerable amount of that time (up to 1/3) is spent on discharge ignition. Note that the value obtained for the delay time from these experiments is consistent with that in the work of Anders et al. [12]. The pulse duration of 300 μs is also not optimal because of heating of the polymer substrate and dissipating resistor. The experiments were therefore performed with the 200 μs pulse duration.

Additionally, discharge voltage vs applied voltage dependence at different pulse frequencies with pulse duration of 200 μs was obtained in experiments. These results are presented in figure 5. Clearly, from the data obtained, a higher pulse frequency leads to greater sputtering of the target and to greater heating of the substrate. As one can see the pulse frequency of 400 Hz and 600 Hz leads to great deposition rate, but these sputtering conditions totally destroy the substrate.

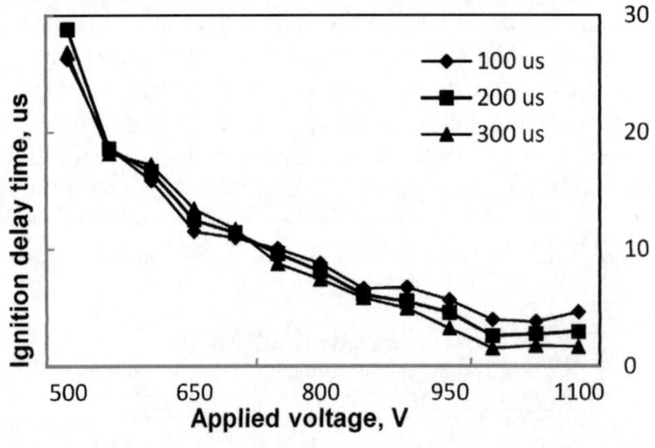

Figure 4. Voltage dependence of discharge ignition delay.

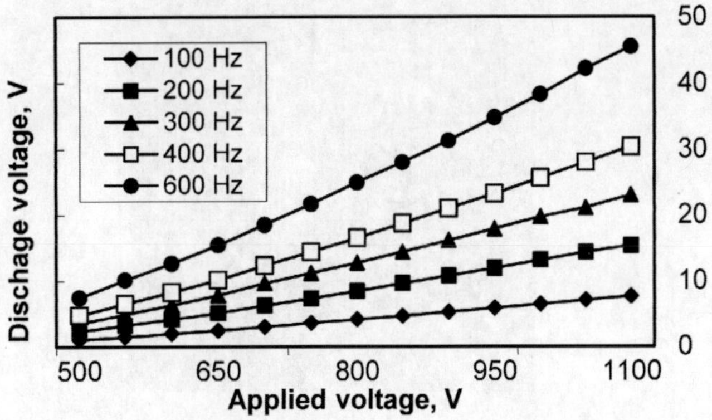

Figure 5. Discharge voltage vs applied voltage for various pulse frequencies.

The dependence of the substrate temperature T_{sub} on sputtering time t_{sput} under optimal operating settings (V_{imp}= 1 kV, t_{imp}=200 μs, f_{imp}=300 Hz) is exhibited in figure 6. One sees substrate temperatures being restricted to about 63°C; short breaks in magnetron use (up to 30 s) leads to a sharp decrease in substrate temperature. The temperature was occasionally measured by a thermocouple which was situated next to the center of the sputtered spot. It was protected from the copper ions to prevent errors.

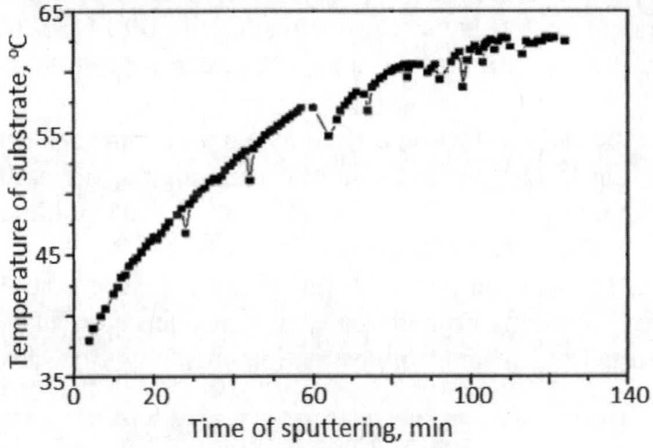

Figure 6. Dependence of a substrate temperature on sputtering time.

We believe that it differs from the regular mode of the conventional DC magnetron, using which we would be unable to operate with such low temperatures of the substrate.

According to [7], the high degree of ionization of the sputtered copper atoms leads necessarily to lower deposition rates when compared with DC magnetron sputtering at the same target power density. Due to this fact, we were able to carry out sputtering at a temperature limited by the properties of a polymeric substrate.

5. Results and discussion

The copper film was studied by TESCAN scanning electron microscope VEGA3 at NRNU MEPhI. As one can see in figure 7, the surface consists of four layers: the bottom is the layer of carbon fiber; next layer is epoxide binder; then goes the layer of dense copper and the upper layer is copper of the columnar morphology. The total thickness of a copper film is. approx.15 μm.

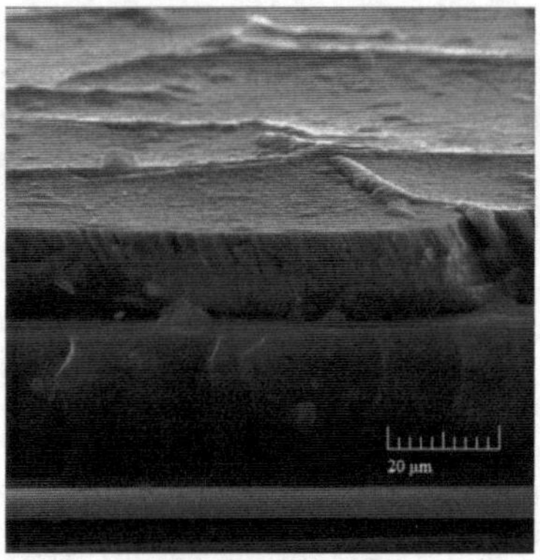

Figure 7. SEM photo of the surface.

The columnar structure is clearly visible on the surface layer of the film. The observed phenomenon of a column-like structure in sputtered deposited films has been written in a review of Windischmann [13]. In [14] it is explained by the presence of oxygen in the discharge area. The presence of oxygen on the surface and inside the copper layer in an amount of maximum 8% was detected by the conducted qualitative spectral analysis of the film and substrate. We attribute the reason of it by a weak leakage to our vacuum system.

They predict that when the amount of oxygen is reduced, the densification of the copper layer will take place. Bombarding ions will erase the peaks and fill the valleys further densifying the film, which in turn will lead to the upgraded mechanical properties of the film [15]. It will be one of the future goals in our further work on the problem to achieve more planar structure of a copper layer on a polymer substrate.

The interface between the carbon fiber, epoxy and sputtered copper is demonstrated in figure 8. At the lower side of the picture the yellow dots show carbon atoms; the carbon fiber at the bottom is clearly visible (compare with figure 7). The epoxide occupies the middle part of the picture. The amount of the carbon there is obviously lower. As for the upper side of the picture, where green dots represent copper, one can see that the bottom layer of the copper surface is very dense. There is a mixture of copper and carbon atoms near the outer side of the layer. The adhesion of the sputtered layer was tested by the stress test. No signs of destruction or flaking off were detected as a result of the test.

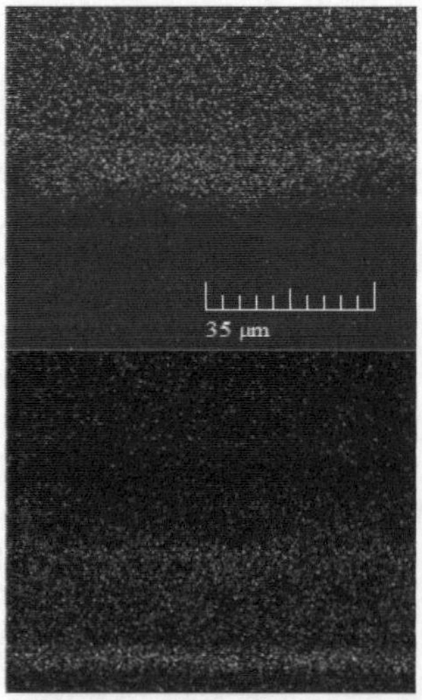

Figure 8. The distribution of the atoms on the surface cut. Green dots are cooper (upper) and yellow are carbon (lower).

Note that all physical regularities observed in our experiments are in qualitative agreement with that for the large currents (up to 80 A) applied in modeling [7-8]. As it was shown in the research of Kozak, the pulse length of 200 μs is very significant time mark as the average target power densities in a pulse, the average ionized fraction of the sputtered copper atoms in the flux onto the substrate as well as the normalized deposition rate reach their saturation by this time. The discharge parameters in [7-8] and in our research were obtained at different experimental stands and at various working conditions, nevertheless, they are in a good enough agreement. It allows us to assume that these parameters are general for this type of the discharge. This fact creates good and stable operational conditions for wide application, which is one of the purposes of this research.

6. Conclusions

The effective HPPMS of copper on the thermosensitive polymer material to balance the rotor of centrifuges for isotope separation is demonstrated at the specially designed experimental stand. Carbon fiber coated with epoxide was used as a substrate. The balanced magnetron, operating with pulse duration of 200 μs and pulse frequency of 300 Hz, enabled a sputtering speed of about 8.82 $\mu m/mm^2 \cdot h$ to be achieved without significant heating of the substrate (T_{sub}<70 °C, max T_{sub}=63 °C). The dependence of substrate temperature on sputtering time was obtained. The dependence of magnetron discharge on various pulse durations and pulse frequencies were determined.

The experimental results obtained confirm the theoretical findings by Kozak et al. on the fundamental character of three process parameters namely, the average target power density in a pulse, the average ionized fraction of sputtered copper atoms in the flux onto the substrate as well as the normalized deposition rate. These parameters reach their saturation in 200μs what also in good agreement with the model results for one of the considered values of the magnetic field. It seems that these parameters are general for the discharge regardless of the operating conditions.

The proposed method of non-destructive mass sputtering on thermosensitive polymer surfaces opens the possibility of a significant expansion of the range of materials used for HPPMS as well as application of the latter. The usage of HPPMS lead to the development of the gas centrifuge rotor balancing method which is cheaper and faster compared with the convenient one.

Acknowledgments

This work was performed at the Center of Physical and Chemical Technologies, National Research Centre "Kurchatov Institute". Authors would like to thank Mr. A.G.Antipov and Mr. N.I.Timofeev for their help in designing the experimental stand.

References

[1] Christie D J 2005 *J. Vac. Sci. Technol. A* **23** 330.
[2] Sarakinos K, Alami J and Konstantinidis S 2010 *Surf. Coat. Technol.* **204** 1661.
[3] Shaginyan L R, Han J G, Shaginyan V R and Musil J *J. Vac. Sci. Technol. A* **24** 4.
[4] Alami J, Bolz S and Sarakinos K 2009 *J. Alloy. Compd.* **483** 530.
[5] Helmersson U, Lattemann M, Bohlmark J, Ehiasarian A P and Gudmundsson J T 2006 *Thin Solid Films* **513** 1.
[6] Whitely S 1984 *Rev. Mod. Phys.***56** 67.
[7] Kozak T and Vlcek J 2013 *Plasma Sources Sci. Technol.* **22** 015009.
[8] Kozak T and Pajdarova A D 2011 *J. Appl. Phys.* **110** 103303.
[9] Kelly P J and Arnell R D 2000 *Vacuum* **56** 159.
[10] Lundin D and Sarakinos K 2012 *J. Mat. Res.* **27** 780.
[11] Yu X, Wang C, Liu Y, Yu D and Xing T 2006 *Plasma Sci. Techol.* **1** 337.
[12] Anders A, Andersson J and Ehiasarian A 2007 *J. Appl. Phys.* **102** 113303.
[13] Windischmann H 1992 *Crit. Rev. Solid State Mat. Sci.* **17** 547
[14] Doerner M and Nix W 1988 *Crit. Rev. Solid State Mat. Sci.* **14** 225.
[15] Thornton J 1977 *Thin Solid Films* **40** 335.

3D numerical study of a feed jet in a rotating flow-field

D J Jiang and S Zeng

Department of Engineering Physics, Tsinghua University, Beijing 100084, China

E-mail: jiangdj@tsinghua.edu.cn

Abstract. A contribution of a feed drive in the total counter-current flow in a gas centrifuge for isotope separation is an important problem in optimization of its separation performance. A 3D model is used to simulate flow structures of a feed jet in a rotating flow-field. By using a CFD code, the details of a feed jet are obtained under the axial feed jet boundary condition. It is demonstrated that because of the vacuum regime in the region near the axis of rotation, the results of numerical simulation of a CFD code bring errors. The 3D DSMC simulation is a feasible method to overcome this problem in the future.

1. Introduction

A gas centrifuge (GC) is a rapidly rotating hollow cylinder, in which an isotope mixture is separated by the centrifugal force in a radial direction. Further separation in a GC is carried out in the axial direction by means of the axial counter-current flow. Usually, it excites in a working chamber of a GC to get higher separation efficiency by three possible ways: the thermal drive by a temperature gradient on a sidewall of a centrifuge machine, the mechanical drive takes place by the mechanical braking of a rotating flow by a stationary scoop, and the feed drive induces by retarding influence of the ingoing flow to the GC [1].

A process gas is introduced into the spinning rotor from a stationary central post and removes from stationary pipes called scoops located at either end of the rotor. Note that originally the supersonic feed jet flow locates in the rarefied flow region near the axis of rotation (a vacuum core). Fu et. al [2] applied the eigenvector method to study the radial feed jet. The conclusion was that the feed flow speed becomes the subsonic one on the boundary between a vacuum core and a viscous gas. Roblin et al. [3] used the direct simulation by the Monte-Carlo method (DSMC) to simulate the flow field of the radial feed jet, and the feed conditions on the viscous boundary were modified according to the results of the simulation.

As it turned out, DSMC is a feasible method to make a numerical simulation of the rarefied flow in the vacuum core, but very time consuming one. Based on our previous work of the 3D radial feed jet in a GC [4], in this paper the 3D flow structure of the axial feed jet in the rapidly rotating flow is investigated by the CFD method.

2. Calculation procedure

The control 3D Navier-Stokes equations of the feed jet flow are as follows,

$$\frac{\partial U}{\partial t}+\frac{\partial(F-F_v)}{\partial x}+\frac{\partial(G-G_v)}{\partial y}+\frac{\partial(H-H_v)}{\partial z}=0$$

where the footnote "v" means a viscous term,

$$\boldsymbol{U}=\begin{bmatrix}\rho\\ \rho u\\ \rho v\\ \rho w\\ e\end{bmatrix},\quad \boldsymbol{F}=\begin{bmatrix}\rho u\\ p+\rho u^2\\ \rho uv\\ \rho uw\\ (e+p)u\end{bmatrix},\quad \mathbf{G}=\begin{bmatrix}\rho v\\ \rho uv\\ \rho v^2+p\\ \rho vw\\ (e+p)v\end{bmatrix},\quad \mathbf{H}=\begin{bmatrix}\rho w\\ \rho uw\\ \rho vw\\ \rho w^2+p\\ (e+p)w\end{bmatrix},$$

$$\boldsymbol{F}_v=\begin{bmatrix}0\\ \tau_{xx}\\ \tau_{xy}\\ \tau_{xz}\\ k\dfrac{\partial T}{\partial x}+u\tau_{xx}+v\tau_{xy}+w\tau_{xz}\end{bmatrix},\quad \boldsymbol{G}_v=\begin{bmatrix}0\\ \tau_{xy}\\ \tau_{yy}\\ \tau_{yz}\\ k\dfrac{\partial T}{\partial y}+u\tau_{xy}+v\tau_{yy}+w\tau_{yz}\end{bmatrix},\quad \boldsymbol{H}_v=\begin{bmatrix}0\\ \tau_{xz}\\ \tau_{yz}\\ \tau_{zz}\\ k\dfrac{\partial T}{\partial z}+u\tau_{xz}+v\tau_{yz}+w\tau_{zz}\end{bmatrix},$$

$$\tau_{xx}=2\mu\frac{\partial u}{\partial x}-\frac{2}{3}\mu\left(\frac{\partial u}{\partial x}+\frac{\partial v}{\partial y}+\frac{\partial w}{\partial z}\right),\quad \tau_{xy}=\mu\left(\frac{\partial u}{\partial y}+\frac{\partial v}{\partial x}\right),$$

$$\tau_{yy}=2\mu\frac{\partial v}{\partial y}-\frac{2}{3}\mu\left(\frac{\partial u}{\partial x}+\frac{\partial v}{\partial y}+\frac{\partial w}{\partial z}\right),\quad \tau_{yz}=\mu\left(\frac{\partial v}{\partial z}+\frac{\partial w}{\partial y}\right),$$

$$\tau_{zz}=2\mu\frac{\partial w}{\partial z}-\frac{2}{3}\mu\left(\frac{\partial u}{\partial x}+\frac{\partial v}{\partial y}+\frac{\partial w}{\partial z}\right),\quad \tau_{xz}=\mu\left(\frac{\partial u}{\partial z}+\frac{\partial w}{\partial x}\right).$$

Here *u, v, w* are the Cartesian velocity vector components, ρ is the density of a process gas, *p* is the pressure of the process gas, and e is the total energy per unit volume.

The Finite Volume Method (FVM) was adopted to discretize the control equations and the 2-order implicit upwind scheme was used to capture the shock waves. The realizable turbulence model was applied for simulation the turbulence caused by a supersonic feed jet.

3. Analysis model and calculation conditions

The analysis model and coordinate system are shown in figure 1. The process gas flows into a GC is passing through the feed hole in a stationary stand. In order to evaluate the interaction between a feed flow and a rotating gas, a rotating rotor wall is in considered in the 3D approach. Only the flow region below the feed inlet boundary plane is calculated to induce the computation workload. Also the flow in a waste chamber is fully not considered.

Using the Iguassu model centrifuge [5], the calculation domain is the central area near a feed hole. All other parameters used in calculation are listed in table 1.

The hexahedral mesh system is generated in the simulation procedure. The typical grids are demonstrated in figure 2. We refined the grid near the rotor and the feed inlets to employ the parameters variation of this area. 2,016,000 hexahedral cells were used in this simulation.

Table 1 Parameters of the Iguassu Centrifuge

Parameter	Value
Radius of rotor	$r_a = 6$ cm
Height of rotor	$8r_a$
Angular velocity	$\omega=10^4$ rad/s
Peripheral speed	600 m/s
Wall pressure	13 300 Pa
Average temperature	$T_0 = 300$ K

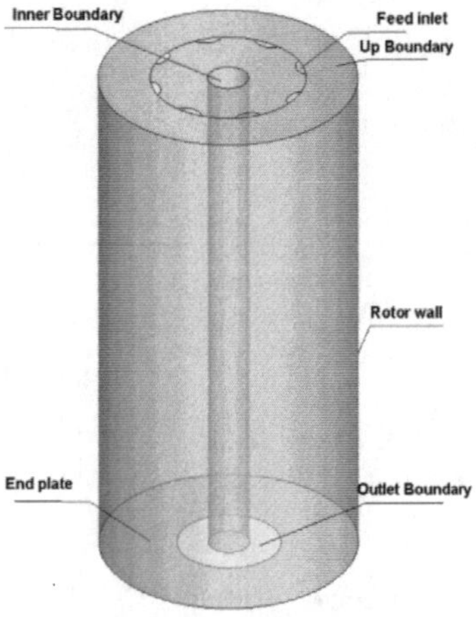

Figure 1. Analysis model for the feed inlet, stationary feed pipe, rotating rotor wall, and end plate in absence of a withdraw chamber.

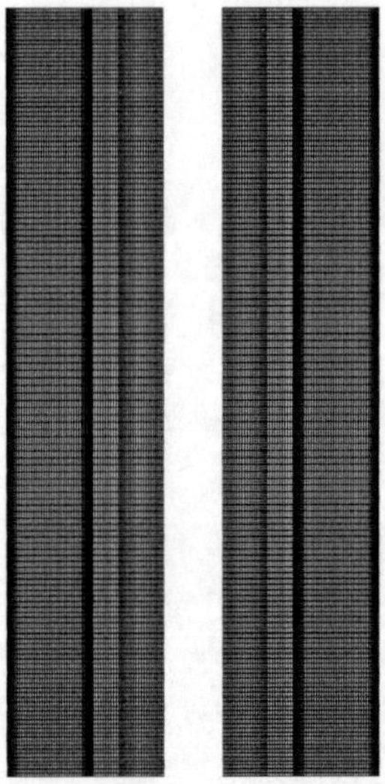

Figure 2. Generated meshes in an angular section

Figure 3. Meshes in an axial section

The boundary condition settings used in calculation are shown in Table 2.

Table 2 Boundary conditions

Boundary	Boundary Type	Conditions setting
Feed inlet	Pressure inlet	Pressure, velocity, temperature
Up boundary	Rotating wall	Rotating speed; temperature; No-slip
Inner Boundary	Station wall	temperature; No-slip
Rotor wall	Rotating wall	Rotating speed; temperature; No-slip
Outlet Boundary	$T_0 = 300$ K	Pressure, temperature

The initial pressure was given according to the rotating rigid body assumption as following:

$$p = p_w \exp\left(\frac{M_r \Omega^2 r_a^2}{RT_0} (\frac{r^2}{r_a^2} - 1) \right) .$$

The initial temperature is fixed as the uniform distribution of 300 K. Also, the velocity in the radial and axial directions is set as zero. The tangential velocity is established according to the formula below

$$v_\theta = \omega r \, .$$

4. Results analysis and discussion

Expand waves were captured in simulation, which can be seen from the flow distribution in the section x=0 in Figure 4. The following phenomena can be observed:

(1) In axial direction, the expansion waves appear near the feed inlet, but disappear rapidly.

(2) The gas flowing into the gas centrifuge affects with the rotating gas, so the tangential velocity decreases near the feed jet flow.

(3) The stream lines composited by the radial and axial velocity note that the feed gas flows into the centre region when it enters into gas centrifuge.

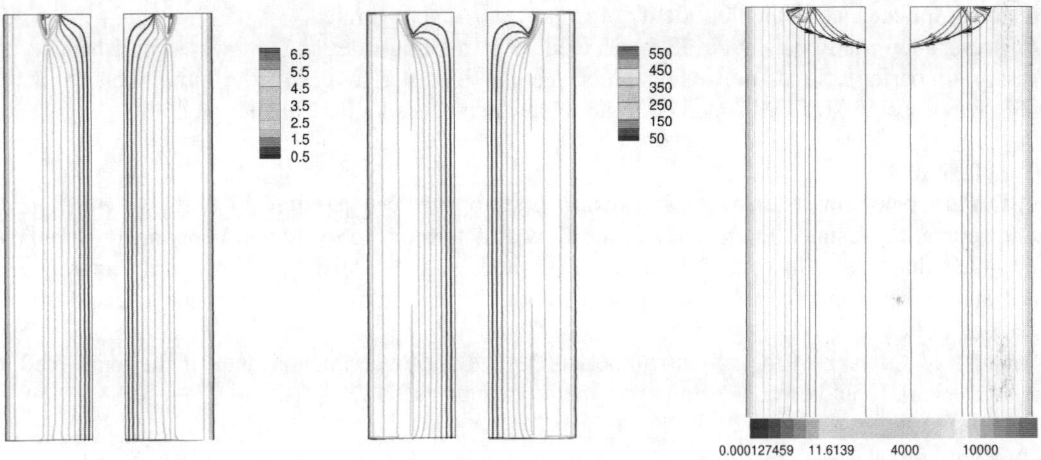

(a) Mach number (b) Tangential velocity (c) Pressure

Figure 4. The parameters of the flow distribution in the x=0 section.

The obvious deformation of the feed jet in the tangential direction is clearly visible. Bethink that only the axial feed jet (without the radial and tangential velocities) was simulated in this research. The affected range in axial range by feed jet is only about 1 cm, which is smaller than expected. It means that there is no much effect on the counter-current by feed jet.

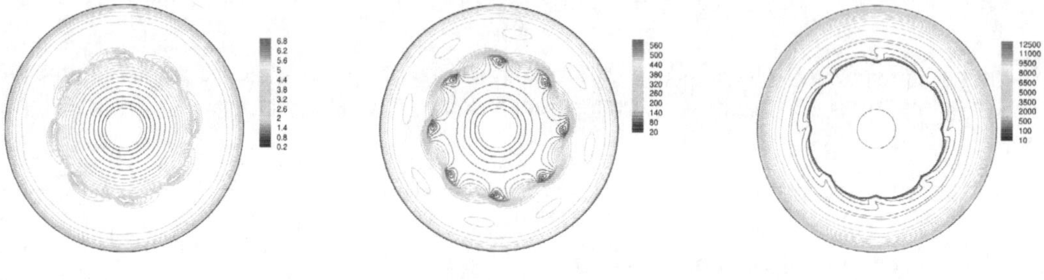

(a) Mach number (b) Tangential velocity (c) Pressure

Figure 5. The parameters of the flow distribution in the axial cross section near the feed jet inlet

Two phenomena need to be discussed:

(1) The affected flow region is quite small, especially in axial direction and the edge of the explosion wave is clear. In fact, the influenced area should be larger due to the low pressure in centre region.

(2) The gas flows into the centre area as soon as it enters the rotating flow field. The reason is the loss of the axial momentum of the gas and the radial gradient of the pressure. It means that the influence of the feed jet on the flow field in the gas centrifuge is limited.

At present few study papers on 3D axial feed jet in rotating rarefied flow were published, so it is necessary to perform simulation using other method in order to verify the simulation results. A feasible way is using 3D DSMC method to do this, which needs higher calculation ability.

5. Conclusions

The 3D numerical simulation was carried out to study the feed jet flow in the gas centrifuge for isotope separation. Using CFD technique, the flow structure of the feed jet was obtained. Simulation result notes that the influence of feed jet is limited, and the feed gas flows into centre region immediately when it entered GC. The expand waves decrease in axial direction; the affected range is only about 1 cm.

Because of the lack of the similar numerical results and experimental data of the axial feed jet, more study work should be done using other methods, such as 3D DSMC.

Acknowledgement

This research was carried out in the framework of the project No 1151101155 supported by NSFC.

References

[1] Doneddu F, Roblin F and Wood H G 2000 *Sep. Sci. Technol.* **35** 1207-21.
[2] Fu R et al 1995 *Tsinghua Sci. Technol.* **1** 61-5.
[3] Roblin P et al 2000 *22nd International Symposium on Rarefied Gas Dynamics* (Sydney Australia).
[4] Jiang D J, Zeng S 2008 *Int. Workshop on Separation Phenomena in Liquids and Gases* (Brazil).
[5] Kai T et al 2000 *J. Nuc. Sci. Technol* **37** 153-65.

Homotopic method applied to solving the flow field in a gas centrifuge

Y N Zhang and S Zeng

Department of Engineering Physics, Tsinghua University, Beijing 100084, China

E-mail: zhangyunan@outlook.com

Abstract. To investigate the flow field in a hyper-speed gas centrifuge, a hybrid difference scheme is used to discretize the axisymmetric Navier-Stokes equation. Source terms are included to simulate the injection and extraction of gas, also the mechanical drive of the scoops. The nonlinearity is obvious as the drive is strong. A Newton iteration used to solve the equation system becomes sensitive to the initial guess of the solution, which makes it difficult to converge. A homotopic method with self-adaptive steps is adopted to cope with this problem and to accelerate the iteration process. Numerical experiments simulating different strengths of scoop drive prove the effectiveness of the algorithm.

1. Introduction

Hyper-speed gas centrifuges are elementary devices to separate isotope mixtures. The injection and extraction of working gas, the temperature distribution on the rotor wall of a centrifuge and the drag effect exerted by the extraction scoops stimulate a complex secondary flow inside the rotor, which has very important influences on the separation power of the centrifuge. Due to the high speed of present centrifuges (over 600m/s), it is very difficult to study the flow field experimentally. As a result, numerical study is more preferable.

A traditional way of numerically simulating a centrifuge flow field is to use linearized N-S equations, e.g., work by Cloutman [1], Dickinson [2], Kozlov [3]. However, only when the effects stimulating the secondary flow (drive effects) are not strong, the linearized models can give reasonable predictions. More precise models use the nonlinear N-S equations. Typical works include Lahargue [4], Harada [5], Kirkpatrick [6], Kai [7][8][9], and Omnes [10], etc. To solve the resulting nonlinear algebraic equation system, usually an iterative method, for example a Newton iteration, is applied. In the case of strong drives, the flow field presents strong nonlinearity and the iteration process becomes sensitive to initial solution, making it difficult to converge or divergent. Therefore, the stability and speed of an iteration method is very important for successful numerical simulation of a gas centrifuge flow field.

Here we present a predictor-corrector homotopic method to accelerate Newton iteration for solving centrifuge flow field. A self-adaptive step strategy is used to obtain fast and robust convergence. To show the effectiveness of the method, simulation of the flow field in an Iguassu model centrifuge is performed. Different drive effects are considered, and the mechanical drive of scoops is adjusted to simulate the disturbances of the flow field with different strengths. Numerical experiment results prove the robustness of the method. A simple separation analysis is also carried out to show the flexibility and success of the method in dealing with different combinations of drive effects.

2. Computational Model

Here the axisymmetric N-S equations in cylindrical coordinates are used to describe the flow field inside a gas centrifuge rotor. The characteristic quantities a (radius of the rotor), Ωa (rotor peripheral speed), p_w (pressure on rotor wall at rigid body rotation) and T_0 (average temperature along rotor wall) are used to nondimensionalize the equation system:

$$\frac{1}{r}\frac{\partial}{\partial r}(r\rho\rho_0 u)+\frac{\partial}{\partial z}(\rho\rho_0 w)=\dot{m} \tag{1}$$

$$\rho\rho_0\left[u\frac{\partial u}{\partial r}+w\frac{\partial u}{\partial z}-\frac{(r+v)^2}{r}\right]+\frac{1}{2A}\frac{\partial}{\partial r}(\rho_0 p)=F_r+EG_r+\dot{m}(u_m-u) \tag{2}$$

$$\rho\rho_0\left[u\frac{\partial v}{\partial r}+w\frac{\partial v}{\partial z}+u\left(2+\frac{v}{r}\right)\right]=F_\theta+EG_\theta+\dot{m}(v_m-v) \tag{3}$$

$$\rho\rho_0\left(u\frac{\partial w}{\partial r}+w\frac{\partial w}{\partial z}\right)+\frac{1}{2A}\rho_0\frac{\partial p}{\partial z}=F_z+EG_z+\dot{m}(w_m-w) \tag{4}$$

$$\begin{aligned}\rho\rho_0\left(u\frac{\partial T}{\partial r}+w\frac{\partial T}{\partial z}\right)+(\gamma-1)\rho\rho_0 T\left(\frac{\partial u}{\partial r}+\frac{u}{r}+\frac{\partial w}{\partial z}\right)=(\gamma-1)(2AE\Phi+\dot{q})+\frac{\gamma E}{Pr}G_T\\+\dot{m}\left\{(T_m-T)+(\gamma-1)A\left[(u_m-u)^2+(v_m-v)^2+(w_m-w)^2\right]\right\}\end{aligned} \tag{5}$$

where γ, A, E and Pr are the specific heat ratio, speed factor, Ekman number and Prandtl number respectively. The source terms $\dot{m}$, $\mathbf{F}=(F_r,F_\theta,F_z)^T$ and $\dot{q}$ are included to simulate the gas injection (or extraction), mechanical drives and heat production in the rotor.

Variables ρ, T and $\mathbf{V}=(u,v,w)^T$ in the equations represent the deviations of the gas density, temperature and velocity from the rigid body rotation state, whose density variation is exponential:

$$\rho_0=\exp\left[A(r^2-1)\right] \tag{6}$$

so the dimensionless density and temperature of the flow field is $\rho\rho_0$ and T, and the dimensionless velocity field is $(u,r+v,w)^T$.

For simplicity, we use symbols G_r, G_θ, G_z and Φ to represent the dissipative terms in the momentum equations and energy equation, G_T to represent the thermal diffusion term in the energy equation, which have the following forms, respectively:

$$G_r=\frac{4}{3}\left(\frac{\partial^2 u}{\partial r^2}+\frac{1}{r}\frac{\partial u}{\partial r}-\frac{u}{r^2}\right)+\frac{\partial^2 u}{\partial z^2}+\frac{1}{3}\frac{\partial^2 w}{\partial r\partial z} \tag{7}$$

$$G_\theta=\frac{\partial^2 v}{\partial r^2}+\frac{1}{r}\frac{\partial v}{\partial r}-\frac{v}{r^2}+\frac{\partial^2 v}{\partial z^2} \tag{8}$$

$$G_z=\frac{\partial^2 w}{\partial r^2}+\frac{1}{r}\frac{\partial w}{\partial r}+\frac{4}{3}\frac{\partial^2 w}{\partial z^2}+\frac{1}{3r}\frac{\partial u}{\partial z}+\frac{1}{3}\frac{\partial^2 u}{\partial r\partial z} \tag{9}$$

$$G_{\mathrm{T}} = \frac{\partial^2 T}{\partial r^2} + \frac{1}{r}\frac{\partial T}{\partial r} + \frac{\partial^2 T}{\partial z^2} \tag{10}$$

$$\begin{aligned}\Phi = & -\frac{2}{3}\left(\frac{\partial u}{\partial r} + \frac{u}{r} + \frac{\partial w}{\partial z}\right)^2 + 2\left[\left(\frac{\partial u}{\partial r}\right)^2 + \left(\frac{\partial w}{\partial z}\right)^2 + \left(\frac{u}{r}\right)^2\right] \\ & + \left[\left(\frac{\partial u}{\partial z} + \frac{\partial w}{\partial r}\right)^2 + \left(\frac{\partial v}{\partial z}\right)^2 + \left(\frac{\partial v}{\partial r} - \frac{v}{r}\right)^2\right]\end{aligned} \tag{11}$$

In addition, the perfect gas assumption is used and the dimensionless state equation is given below:

$$p = \rho T \tag{12}$$

To determine the solution to this system, we adopt the following boundary conditions:
On solid walls:

$$\mathbf{V}_{\mathrm{b}} = \mathbf{0}, \left(\frac{\partial \rho}{\partial n}\right)_{\mathrm{b}} = 0, T_{\mathrm{b}} = T_{\mathrm{wall}} \tag{13}$$

On the vacuum core boundary (at the radius position r_{in}):

$$\left(\frac{\partial \rho}{\partial r}\right)_{r_{\mathrm{in}}} = \left(\frac{\partial T}{\partial r}\right)_{r_{\mathrm{in}}} = \left(\frac{\partial w}{\partial r}\right)_{r_{\mathrm{in}}} = 0 \tag{14}$$

$$u = 0 \tag{15}$$

$$\left(\frac{\partial (v/r)}{\partial r}\right)_{r_{\mathrm{in}}} = 0 \tag{16}$$

As the above boundary conditions impose no restriction on the gas content in the rotor, a subsidiary condition restricting the holdup H of the centrifuge should be provided:

$$\iiint_{\mathrm{V}} \rho \rho_0 \mathrm{d}\tau = H \tag{17}$$

The integration here is carried out over the volume of the centrifuge considered.

3. Numerical Method

The N-S equations are discretized on a staggered grid [7][10]. To suppress non-physical oscillations in advection dominating grid cells, a hybrid scheme is used to discretize the convection terms. If the grid Reynolds number R_{Δ} is less than 2, a second-order central difference scheme is used, otherwise a first-order upwind scheme is adopted.

To solve the resulting discretized nonlinear equation system, usually iterative methods are applied. A widely used method is the Newton iteration. However, achieving convergence is not always an easy task if the system has a strong nonlinearity. In this circumstance, the iteration process can be sensitive to the initial guess: if the initial guess is not close enough to the real solution, the iteration may converge very slowly or even diverge.

In computations of various areas, employing the continuation technique or homotopic method is a strategy to tackle such a problem [11]. We will demonstrate in the following how to use it to accelerate and stabilize the iteration process in solving the N-S equations.

3.1. Concept and implementation of homotopic method

Here we briefly introduce the basic concept of homotopic method. As our target is to apply the technique to the problem of solving centrifuge flow field, no detailed mathematical discussion is given. Readers with interest are referred to reference [11].

Suppose the nonlinear equation system to be solved is denoted as a $\mathbf{R}^n \mapsto \mathbf{R}^n$ mapping:

$$F(X)=0 \tag{18}$$

Let us define another $\mathbf{R}^{n+1} \mapsto \mathbf{R}^n$ mapping called $H(X,\alpha)$, which satisfies the following conditions:

$$\begin{cases} H(X,0)=G(X) \\ H(X,1)=F(X) \end{cases} \tag{19}$$

where $G(X)$ is a function with known zeros.

The solution set of system $H(X,\alpha)=0$ forms a curve $c(\alpha)$ in $\mathbf{R}^{n+1}$ with α as its parameter. Curve c connects X_0, the solution of $G(X)=0$, with the final solution X^*. If we can trace this curve, the final solution can be achieved. Since c is called a homotopic curve, this solution technique is known as homotopic method.

According to the definition of homotopic curve:

$$H(c(\alpha),\alpha)=0 \tag{20}$$

Differentiating the equation with respect to α:

$$H'(c)c'(\alpha)=0 \tag{21}$$

where $H'(c)=(\partial H/\partial X)_c$ and $c'(\alpha)=\mathrm{d}c/\mathrm{d}\alpha$, then we have an initial value problem as follows:

$$H'(c)c'(\alpha)=0 \tag{22}$$

$$c(0)=X_0 \tag{23}$$

This problem can be solved by many numerical methods, such as Euler method, Runge-Kutta method, etc. Here we present a predictor-corrector homotopic method:

Algorithm A
(1) Assign the starting point $c(0)=X_0$ and step length h. Let $\alpha_0=0$ and $n=0$.
(2) Solve the linear equation system (22) and obtain the solution $c'(\alpha_n)$.
(3) Calculate the predictor $\tilde{c}(\alpha_n)=c(\alpha_n)+hc'(\alpha_n)$. Let $\alpha_{n+1}=\alpha_n+h$.
(4) Solve the nonlinear equation system $H(c(\alpha_{n+1}),\alpha_{n+1})=0$ with $\tilde{c}(\alpha_n)$ as the initial guess.
(5) If $\alpha_{n+1}=1$, then $X^*=c(\alpha_{n+1})$ is the final solution, or otherwise let $n:=n+1$ and go to (2).

In Algorithm A the predictor step (3) uses Euler method to trace the next point on the curve. The corrector step (4) uses a nonlinear solver to correct the prediction. As our aim is to get the final solution X^*, i.e., the end of curve c, the precision requirement for the intermediate tracing points can be relaxed to save computation cost. Therefore, a Newton method as the corrector step's nonlinear solver is sufficient. A convergence requirement ε_1 controls the solver precision.

Note that Algorithm A uses a constant step length, which may be costly because h should be small for the first few steps and larger for the rest steps. So Algorithm A is modified by using a self-adaptive step strategy as follows:

Algorithm B

---

(1) Assign the starting point $c(0) = X_0$ and step length h. Let $\alpha_0 = 0$ and $n = 0$.

(2) Solve the linear equation system (22) and get the solution $c'(\alpha_n)$.

(3) Calculate the predictor $\tilde{c}(\alpha_n) = c(\alpha_n) + hc'(\alpha_n)$. Let $\alpha_{n+1} = \alpha_n + h$.

(4) Solve the nonlinear equation system $H(c(\alpha_{n+1}), \alpha_{n+1}) = 0$ with $\tilde{c}(\alpha_n)$ as the initial guess. The convergence requirement is $\left\| H\left(c_{n+1}, \alpha_{n+1}\right)\right\| < \varepsilon_1$. If convergent within N_1 steps, go to (5), or otherwise, let $h := h/2$ and go to (3).

(5) If $\alpha_{n+1} = 1$, solve the nonlinear system $H(X^*, 1) = 0$ with $c(\alpha_{n+1})$ as the initial guess, or otherwise let $h := 2h$, $n := n+1$ and go to (2).

3.2. Homotopic mapping for GC N-S equations

In this section we will illustrate the way to construct the homotopic mapping H for the N-S equations in a gas centrifuge. Previous study shows the main factors to cause large disturbances in the rotor are the gas injection of the feed flow and mechanical drive exerted by the extraction scoops. In our model, these effects are simulated by the N-S equation system with source terms:

$$F(X) = G(X) + S(X) = 0 \tag{24}$$

where $G(X)$ is the discretized N-S equation system without source terms and $S(X)$ is an $\mathbf{R}^n$ vector for the source terms.

We suggest constructing H as the following:

$$H(X, \alpha) = G(X) + \alpha S(X) \tag{25}$$

The equation system $G(X) = 0$ is relatively easy to solve because there are no large disturbances introduced in the flow field, as in the case of the rigid body rotation. With this solution as the initial guess X_0, the homotopic method in above sections can be performed.

4. Numerical experiments and discussion

In this part we present some results of numerical experiments to show the effectiveness of the homotopic method. The simulated model centrifuge is an Iguassu model. Table 1 presents the main working parameters of the centrifuge.

To simulate the flow field with different strengths of disturbances, we adjust the strength of the momentum source terms $\mathbf{F}$. In the previous studies, different methods simulated the influence of scoops, such as those in [5][6][9][10]. An intuitive model is to use a disk rotating at a different speed from the rotor [5][6]. We imitate this model to set the strength of $\mathbf{F}$ as in equation (26).

$$F_\theta = \frac{4\pi}{3\delta_E} E\left(\frac{\Delta\Omega}{\Omega}\right)\left(\frac{r_2^3 - r_1^3}{\Delta V}\right) \tag{26}$$

where $[r_1, r_2]$ is the radial range of the source; ΔV is the volume of the source; δ_E and $\Delta\Omega/\Omega$ are, respectively, an estimate of the thickness and the relative difference of the rotation speed of the disk

Journal of Physics: Conference Series **751** (2016) 012015 doi:10.1088/1742-6596/751/1/012015

model. In this way, the retardation effect of the angular momentum source term is approximately the same as a disk model with a $\Delta\Omega$ difference in the rotation speed.

Table 1. Main working parameters for the Iguassu model centrifuge

parameter	value
angular speed Ω (rad/s)	1.0×10^4
mean temperature T_0 (K)	300
waste end temperature T_{bot} (K)	300.5
product end temperature T_{top} (K)	299.5
wall pressure p_w (kPa)	13.33
feed flow rate F (mg/s)	12
flow cut θ	0.5

A 116×256 mesh is used to discretize the equations. Five cases with different strengths of scoop drive are considered. A direct Newton method, Algorithms A and B are tested on these cases. The Newton method fails in all cases, and the comparison between the two homotopic methods are shown in table 2, where CT is the number of total P-C steps used and IT is the total Newton iteration count. The maximum allowed IT is 150.

From table 2 we see the maximum $\Delta\Omega/\Omega$ reaches 100%, which means that the retardation effect of the angular momentum sink approximates a still disk in the gas. In these cases, the flow field definitely deviates from the rigid body rotation state, especially in the extraction chambers. Consequently, the direct Newton method fails to converge. Algorithm A fails in the last 2 cases because the step length must keep small enough, resulting in too large a total number of iterations. In contrast, Algorithm B adjusts its step lengths whenever needed and the computation cost can be reduced significantly.

Table 2. Comparison between two homotopic algorithms

case	$\frac{\Delta\Omega}{\Omega}$ (%)	Algorithm A		Algorithm B	
		CT	IT	CT	IT
1	10	15	38	7	49
2	20	28	66	8	54
3	40	37	96	10	72
4	80	—	—	11	84
5	100	—	—	11	100

In figure 1 stream lines with $\Delta\Omega/\Omega=0$ and 10% are shown. The stream lines are derived from the stream function defined as below [12]:

$$\frac{\partial\psi}{\partial r}=r\rho\rho_0 w,\ \frac{\partial\psi}{\partial z}=-r\rho\rho_0 u-\int_r^1 r'\dot{m}^* dr' \tag{27}$$

After the introduction of mechanical drive, the stream lines change a lot. Without any mechanical drive, the thermal drive along the rotor wall forms a circulation near the wall. The feed flow drives two reverse small circulations, which have an adverse effect on the separation process. In the case with a 10% mechanical drive, the thermally driven circulation has a larger range: the gas flows from the inner hole of the waste baffle, reflects on the product baffle, and returns to the waste chamber through the outer hole of the waste baffle. However, there is still a strong reverse circulation stimulated by the product scoop. This is probably because the product mechanical drive is too strong. In order to achieve a satisfactory separation efficiency the strength of the mechanical drive should be optimized.

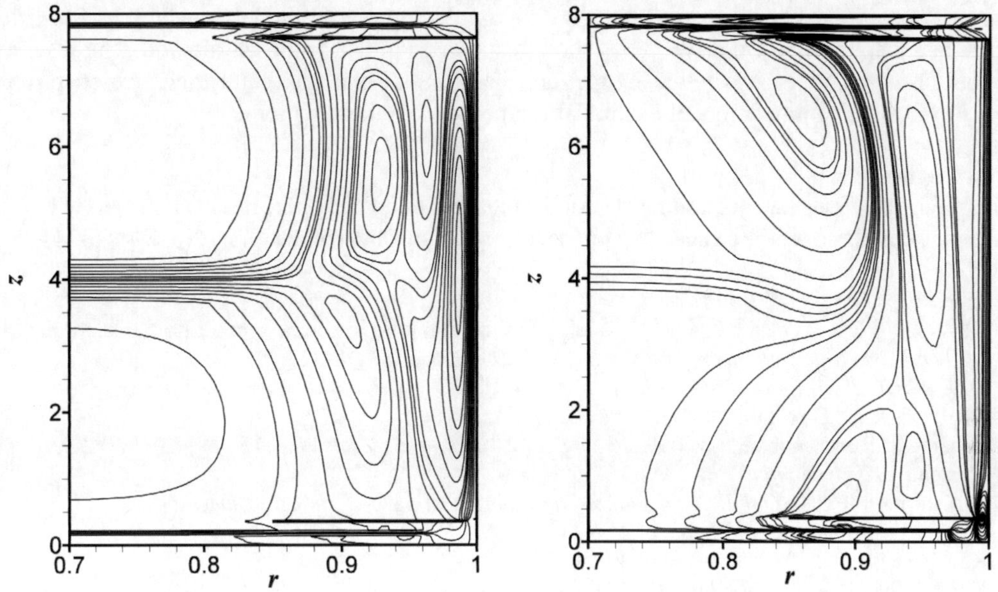

(a) $\Delta\Omega/\Omega=0$ (b) $\Delta\Omega/\Omega=10\%$

Figure 1. Stream lines with different strengths of mechanical drives

Figure 2 shows the angular velocity perturbation in the radial direction at $z=0.2$ in the waste chamber. It is clear that in some areas of the rotor, the gas slows down, exceeding over 50% of the peripheral speed. This is by any means not a small perturbation, and, therefore, a linear model cannot and should not be used to simulate flow in these regions.

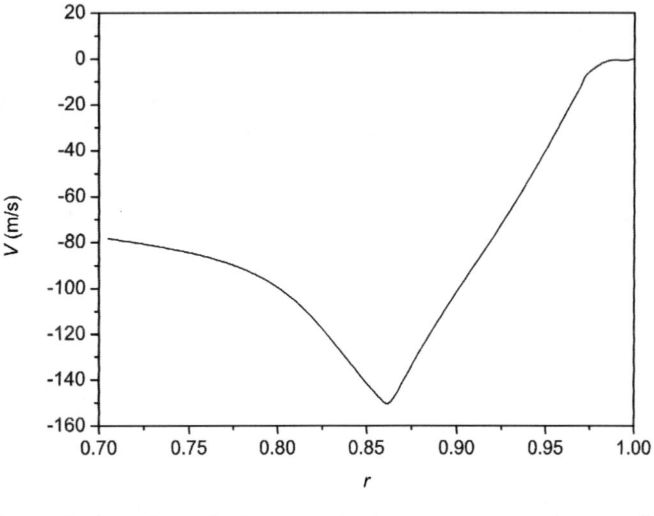

Figure 2. Angular velocity perturbation on cross section $z=0.2$

5. Conclusion

A hybrid difference scheme is used to discretize N-S equations in a gas centrifuge. In cases with large perturbations, solving the resulting nonlinear equation system is difficult because of the sensitivity to initial solution by using a traditional iterative method. A predictor-corrector homotopic method with adaptive steps is applied to solve this problem. Basic concept and also implementation procedure are presented.

Numerical experiments computing the flow fields of the Iguassu model centrifuge with different strengths of scoop mechanical drive verify the effectiveness of the proposed method. The results also show the strong nonlinearity in the extraction chambers of the gas centrifuge, which prove the necessity of using a nonlinear model in simulating the flows in these regions.

Acknowledgment

This research is supported by National Natural Science Foundation of China (Grant No. 11575097) and National Key Basic Research and Development Program of China under Grant No.2014CB744100.

References

[1] Cloutman L D and Gentry R A 1983 Numerical simulation of the countercurrent flow in a gas centrifuge *Los Almos National Laboratory* LA-8972-MC(UC22) DE83-013323
[2] Dickinson G J and Jones I P 1981 *J. Fluid. Mech.* **107** 89–107
[3] Kozlov S A and Eriklintsev I V 2013 *Phys. Scr.* **T155**
[4] Lahargue J P and Soubbaramyer 1977 *Comput. Method. Appl. Mech. Eng.* **15** 259–73
[5] Harada I 1980 *J. Comput. Phys.* **38** 335–56
[6] Kirkpatrick J R 1986 *Oak Ridge National Laboratory* TN(USA) Rep. K/CSD/TM-63
[7] Kai T 1977 *J. Nucl. Sci. Technol.* **14** 267–81
[8] Kai T 1983 *J. Nucl. Sci. Technol.* **20** 49–502
[9] Kai T 1999 *J. Nucl. Sci. Technol.* **37** 153–65
[10] Omnes P 2007 *Computer Fluids* **36** 1028–39
[11] Allgower E L and Georg K 2003 *Introduction to Numerical Continuation Method* (Philadelphia: SIAM)
[12] Wood H G and Morton J B 1980 *J. Fluid. Mech.* **101** 1–31

Waves in a gas centrifuge

S V Bogovalov, V A Kislov and I V Tronin

National Research Nuclear University MEPhI (Moscow Engineering Physics Institute), Kashirskoe highway 31, Moscow, 115409, Russia

E-mail: svbogovalov@mephi.ru

Abstract.
Impact of the pulsed braking force on the axial gas circulation and gas content in centrifuges for uranium isotope separation was investigated by the method of numerical simulation. Pulsed brake of the rotating gas by the momentum source results into generation of the waves which propagate along the rotor of the centrifuge. The waves almost doubles the axial circulation flux in the working camera in compare with the case of the steady state breaking force with the same average power in the model under the consideration. Flux through the hole in the bottom baffle on 15% exceeds the flux in the stationary case for the same pressure and temperature in the model. We argue that the waves reduce the pressure in the GC on the same 15%.

1. Introduction

Isotope separation in gas centrifuges (hereafter GC) are explored for production of enriched uranium for nuclear power stations. Rotors of contemporary gas centrifuges rotate with linear velocities 600-700 m/s and have radius 6-8 cm [2]. Scoops located at the bottom and top ends of the rotor are used to eliminate product and waste UF6 the GC. Simultaneously the scoops brake the gas. The braking force results into axial circulation of the gas. Additionally, the interaction of the supersonically rotating gas with the scoops is accompanied by formation of waves which propagate in the working camera of the rotor forming spiral waves. The waves propagate in the axial direction and are quickly transformed into linear waves. Scheme of the propagation of the shock wave in the axial direction is shown in 1.

The acoustic waves are able to produce a specific gas flow due to absorption of the waves and transfer of the momentum from the waves to gas. These are so called acoustic flows [14], [15]. This mechanism of excitation of the axial circulation essentially differs from the traditional methods of the circulation excitation. Therefore, it is important to understand the physics of the waves in the GC and their impact on the characteristics of the GC.

2. Properties of the linear waves in the GC

Shock waves are damped quickly. They are transformed into linear waves propagating in the main part of the GC. The characteristics of the linear waves were investigated in [3]. The dispersion law for linear waves in strong centrifugal field is shown in fig. 2. It can be seen that three families of the linear waves are in the strong centrifugal field: upper, lower and sound.

Upper and lower families of the waves are damped quickly because energy of these waves is concentrated near the axis of rotation. The energy of the sound waves is concentrated near the rotor wall. According to our estimates the damping length of these waves always exceeds the

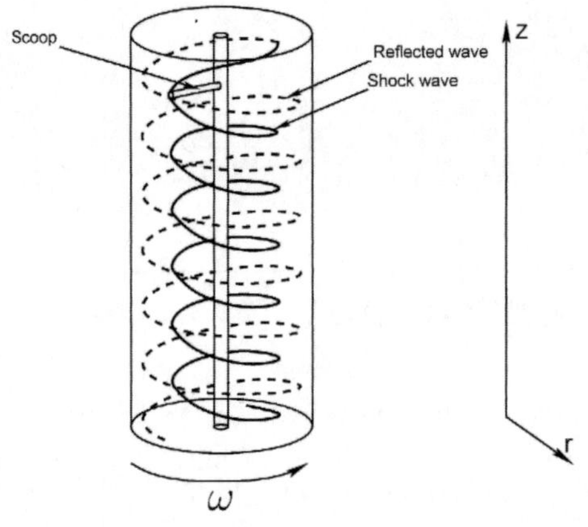

Figure 1. Scheme of the shock wave in the gas centrifuge

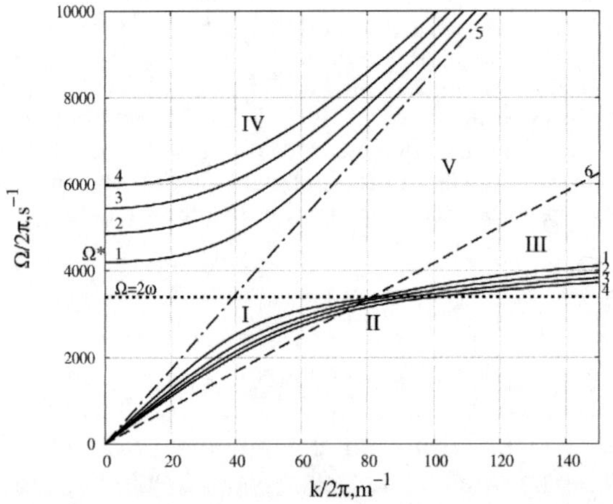

Figure 2. Dispersion law for the linear waves in strong centrifugal field. Lines 1-4 upper and lower families, 5 sound wave

length of the rotor. This prediction was confirmed in [4] by the numerical simulation. Only one type of the waves which corresponds to the dispersion law $\omega = kc$ survives in the GC.

3. The impact of the waves on the axial circulation

3.1. Formulation of the problem

The waves produced by the scoops are stationary in the laboratory frame system. Computer modeling of the gas flow in the GC is performed in rotating frame system [5, 6, 7, 8, 9]. In this frame system the waves are not stationary. They propagate along the GC as running spiral waves. 2D axisymmetric model of source-sinks is explored for modeling of the gas flow in the GC [7]. We use axisymmetric model of the sources and sinks in the rotating frame system as well. The sources of the momentum and energy are not stationary in our model. Therefore, they generate running axisymmetric waves.

Special computer code has been developed in MEPhI for numerical simulation of the non-stationary gas dynamics in strong centrifugal fields. The results of verification of the code were presented in [10, 11].

We use a model which contains two cameras as it is shown in fig. 3. The GC is closed. We do not consider here the problems connected with feed, product and waste fluxes. This allows us to investigate the role of the waves in the simplest model without mixing with other processes which has no direct relation to our problem. A small region is selected In the upper camera where the source of the momentum and energy is introduced into the flow. We consider two types of the source. The first one is the conventional steady state source. The second one is the pulsed source of energy and momentum with the same as in the first case integral of the breaking force over the period of rotation. The shape of the pulse of the momentum source is shown in fig. 4. The largest part of the breaking force affects the gas during 1/10 period of rotation of the rotor. The working camera is located under the upper camera. They are separated by the baffle having two concentric holes: near the axis and at the wall of the rotor. These holes make possible free circulation of the gas in the rotor.

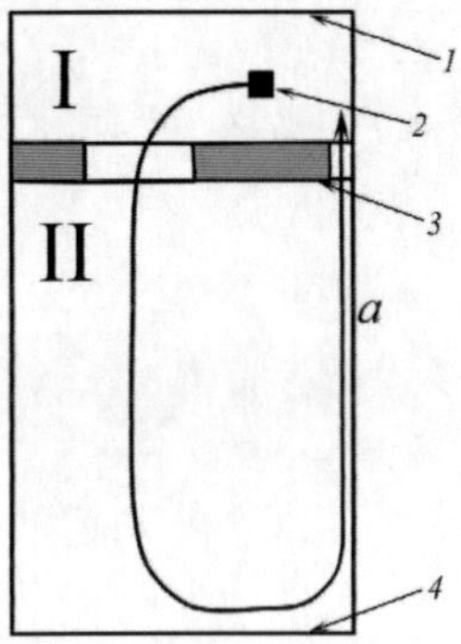

Figure 3. Scheme of the computational domain. 1 top cap, 2 source of the momentum and energy, 3 top baffle, 4 end cap

The pressure in the GC falls down to the axis of rotation with the rate 3-4 orders of magnitude per 1 cm [1]. There is a vacuum near the axis, where hydrodynamic equations are not applicable. This region is not interesting for us because there is no gas here. To exclude this region from the computational domain an additional boundary is introduced at some radius where the density of the gas is already very small but equations of hydrodynamics is still possible to apply [9, 13]. Boundary conditions on this wall correspond to the rigid adiabatic free slip wall. No slip isothermal walls where assumed at all other walls and the baffle.

3.2. Solution

Parameters of GC are typical for so called Iguasu centrifuge [16, 12]. They are given in table 1.

The problem has been solved as a time dependent with time step equal to $\frac{1}{30}\tau_0$ to resolve with an acceptable accuracy the dependence of the braking force on time, where τ_0 is the period of the GC rotation. The simulation has been performed for a sufficiently long time to obtain sustained time dependent flow. The pressure at the lower end cap has been used to control the regime of the flow.

3.3. Axial circulation

The braking force produces unbalanced radial gradient of pressure. The gas moves to the axis of rotation after the slowdown increasing the pressure near the axis. This results into expansion of the gas and its motion through the hole of the baffle. This way the perturbation generated in the upper camera propagates into the working camera. Here the perturbation takes a form of the wave which can propagate to the bottom end cap of the GC. The wave propagates back after

Table 1. Parameters of the model Iguasu GC.

Parameter	Value
Molar mass, M	352 g/mol
Radius of the rotor, a	0.065 m
Angular frequency, ω	$2\pi \cdot 1700\ s^{-1}$
Rotor temperature, T_0	300 K
Pressure near the rotor wall, p_w	80 mmHg
Specific heat at constant pressure, c_p	385 $J \cdot K/kg$
Viscosity, η	$1.83 \cdot 10^{-5}\ Pa \cdot s$
Thermal conductivity, κ	0.0061 $W/m \cdot K$
Length of the working camera, L	0.224 m
The outer radius the hole of the baffle, r_h	59.1 mm

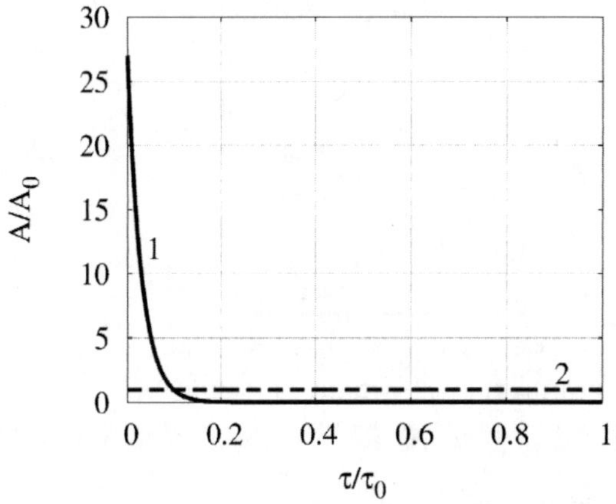

Figure 4. The dimensionless braking force in the top source versus dimensionless period of rotation. 1 nonstationary case, 2 stationary case

reflection from the bottom end cap. Multiple reflection of the wave from the end caps results into rather complicated time dependent flow. Snapshot of this flow is shown in fig. 6. It follows from this figure that the wave satisfies to the dispersion equation of the form $\omega = kc$, where c is the conventional sound velocity, k is the wave vector and is the angular frequency of the wave. The existence of this type of waves in strong centrifugal field has been predicted recently in the work [3]. According to this work the waves with another dispersion law are possible. However, these waves are damped due to the molecular viscosity very quickly. On this reason they are not found in our simulations.

Comparison of the velocity field obtained with the stationary breaking force (see fig. 5) with the snapshot of the velocity field with the pulsating breaking force, fig. 6, shows that they are different. In the non-stationary case there are velocity perturbations which exceed the velocity in the stationary case. For the practical applications it is important to estimate the impact of the waves on the separation and efficiency of the GC. However, the snapshot of the flow field does not allow us to do that. It is necessary to consider the averaged flow field.

The axial circulation flow in the rotor can be described by the integral functions showing the mass flux m_- moving down the rotor and mass flux m_+ moving up the rotor. In the stationary case they should coincide due to the equation of discontinuity. In the non-stationary case

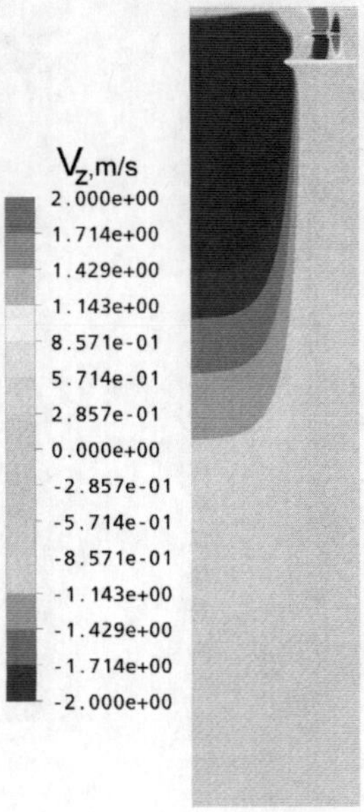

Figure 5. Axial velocity in the working camera oft he GC in the stationary case

the same functions after averaging also should coincide with each other. This coincidence is an additional test of the correctness of the simulation. These mass fluxes are shown in fig. 7. It is the most important that the averaged mass flux twice exceeds the mass flux of the axial circulation produced by the steady state breaking force. The nature of this effect can be understood in terms of the acoustic flow [14, 15]. The scoop generates all possible types of waves including ones with fast damping due to the molecular viscosity. These waves transfer their momentum to the gas increasing the mass flux.

Thus, the waves generated by the scoops essentially change the integral characteristics of the axial circulation. This can result into difference between the calculated in the conventional (stationary) model parameters of the GC and real parameters.

4. The impact of the waves on the gas content

4.1. Formulation of the problem

The scheme of the computational domain for this case is presented in fig. 8. The domain consists of three cameras: top, working and bottom. Product (enriched) and waste (depleted) fluxes are removed in the bottom and top cameras, respectively by the sinks shown by solid points 6 and 2. Feed flow (F) is injected at the middle of the working camera. Circulation flow (C) is induced due to two mechanisms: temperature gradient ΔT at the rotor wall and brake of the gas in the top camera. The braking force is located in the point source 2. Parameters used for the computer simulation of the Iguasu centrifuge are presented in tab. 1 with some changes. The length of the rotor is equal 0.526 m and ΔT is equal 10 K. We used the same boundary conditions as in previous section and compared the cases with steady state and pulsed sources. The dependence on time of the pulsed braking force is shown in fig. 4.

4.2. Gas content

Initially the solution has been performed for the steady state case. The waste flux has been specified in the point 2 equal to 0.55 of the feed flux. The pressure in the working camera is

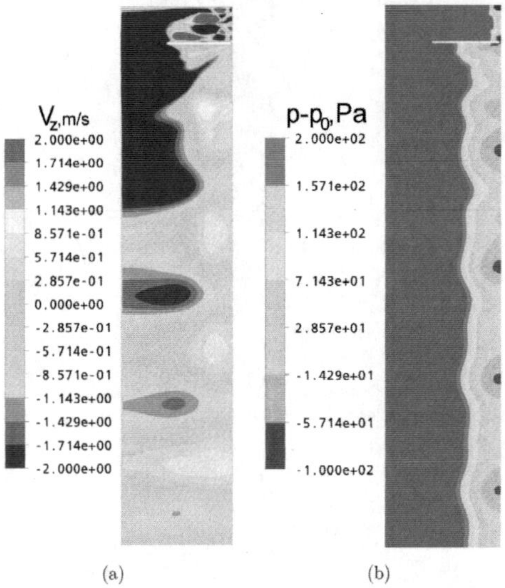

Figure 6. The waves in the working camera of the GC: (a) axial velocity; (b) $(p - p_0)\,e^{-\frac{M\omega^2 r^2}{2RT_0}}$, where $p = p_w e^{\frac{M\omega^2\left(r^2-a^2\right)}{2RT_0}}$

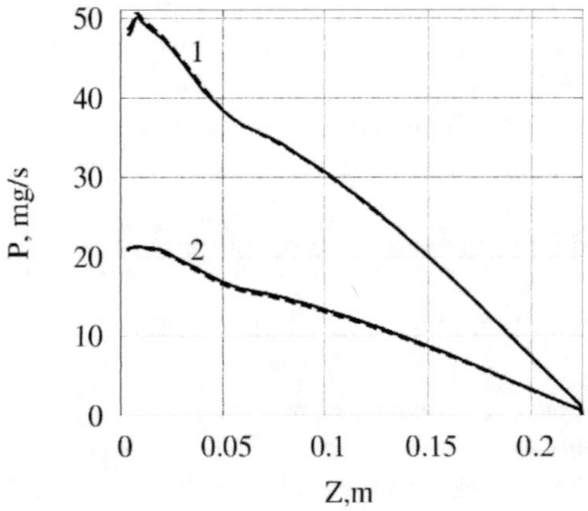

Figure 7. Dependence of m_+ (solid lines) and m_- (dashed lines) on axial coordinate z. Zero of this coordinate corresponds to the point right under the upper baffle. 1 averaged nonstationary case, 2- stationary case

obtained in the process of the solution. It equals to 79 mm Hg. This pressure provides the product flux satisfying to the mass conservation. It equals to 7.2 mg/s. The stream lines of the stationary flow are shown in fig. 9.

In the transient case the sink fluxes and pressure were taken initially like in the stationary one. The simulations show that in the transient flow the flux through the hole in the bottom baffle appears 15% higher than the product flux. This means that the mass is not conserved in the transient case. The mass conservation will take place when the pressure in the working

Journal of Physics: Conference Series **751** (2016) 012016 doi:10.1088/1742-6596/751/1/012016

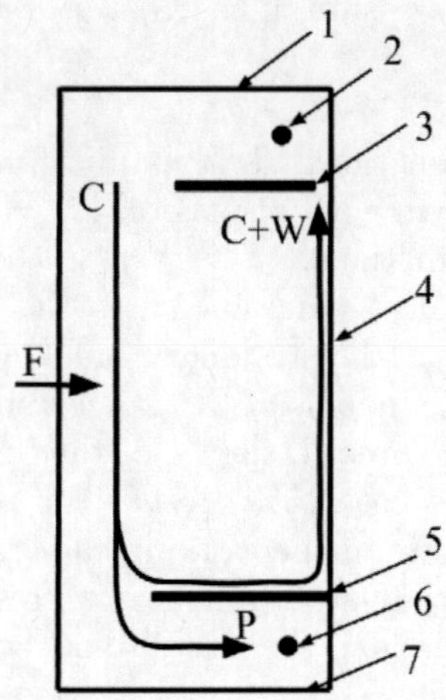

Figure 8. Scheme of the computational domain. F, P, W feed, product and waste fluxes, respectively, and C circulation flow. 1 top cap, 2 source of the momentum and energy, sink of the waste flux, 3 the top baffle, 4 rotor wall, 5 the bottom baffle, 6 sink of the product flux, 7 the bottom cap.

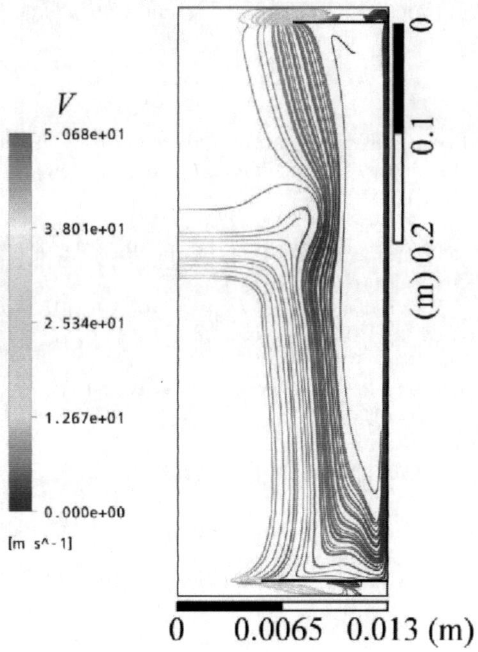

Figure 9. The stream line in the case of the stationary braking force. Scale in radial direction multiplied 15 times.

camera falls down to reduce the mass flux through the hole in the bottom baffle to the value equal to the product flux. But this takes a lot of computational time. Duration of such simulation exceeds a reasonable value. Nevertheless, we can conclude from this simulation that the waves can essentially change the working pressure in the GC.

5. Conclusion

The influence of the linear waves on the hydrodynamic flow in the GC is investigated. We obtained that the axisymmetric waves with dispersion law =kc are excited due to this force in accordance with the theoretical predictions.

The pulsating breaking force almost twice increases the mass flux in the axial circulation flow in compare with the mass flux at the stationary breaking force in the closed GC with two cameras. Conventional axisymmetric models of the gas flow in the GC do not take into account this effect. This could result into incorrect estimate of the optimal breaking force.

The product flux in the transient case 15% exceeds the product flux of the stationary case for the same pressure and temperature in the working camera in the case of the GC with three cameras. This means that in the steady-state regime the pressure in the working camera of the transient case is expected to be lower than the pressure in the stationary case approximately on the same 15%.

Acknowledgements

The work has been performed under the support of the Ministry of education and science of Russia, grant no. 3.726.2014/K. The computer simulations have been partially performed on the computer cluster of National research nuclear university(MEPhI), Basov farm. This work was also performed within the framework of the Center "Physics of nonequilibrium atomic systems and composites" supported by MEPhI Academic Excellence Project (contracts No. 02.a03.21.0005, 27.08.2013).

References

[1] Borisevich V D et al. 2011 *Fizicheskie osnovy razdeleniya isotopov v gazovoi centrifuge (Physical basics of the isotope separation in gas centrifuge)* (Izdatel'skii dom MEI, Moscow), in russian
[2] Glaser A 2008 *Sci. Glob. Secur.* **16(12)** 126
[3] Bogovalov S V, Kislov V A and Tronin I V 2015 *Theor. Comput. Fluid Dyn.* **29 (1-2)** 111
[4] Bogovalov S V, Kislov V A and Tronin I V 2015 *AIP Conf. Proc.* **1648** 030008
[5] Wood H G, Morton J B 1980 *J. Fluid Mech.* **101 (1)** 1
[6] Doneddu F, Roblin P, Wood H G 2000 *Sep. Sci. Technol.* **35 (8)** 1207
[7] Matsuda T, Sakurai T and Takeda H 1975 *J. Fluid Mech.* **69 (1)** 197
[8] Aisen E M et al. 1997 *Mathem. Mod.* **9 (4)** 27-38
[9] Von Halle E, Wood H G and Lowry R A 1983 *Nucl. Technol.* **62 (6)** 325
[10] Abramov V I et al. 2013 *Comput. Math. Math. Phys.* **53(6)** 789
[11] Bogovalov S V et al. 2013 *Computers and Fluids* **86** 177
[12] Bogovalov S V et al. 2014 *Sep. Sci. Technol.* **49 (3)** 329
[13] Borisevich V D, Levin E V and Naumochkin V V 1991 *Sov. At. Energy* **70 (1)** 3642
[14] Rayleigh, Lord 1883 *London Ser. A* **175** 121
[15] Rayleigh, Lord 1896 *MacMillan London*
[16] Rodrigues N A S, Schwab C and Wood H G 1996 *Proceeding of the Fifth Workshop on Separation Phenomena in Liquids and Gases*

The computer simulation of 3d gas dynamics in a gas centrifuge

V D Borman, S V Bogovalov, V D Borisevich, I V Tronin and V N Tronin

National Research Nuclear University MEPhI (Moscow Engineering Physics Institute), Kashirskoe highway 31, Moscow, 115409, Russia

E-mail: svbogovalov@mephi.ru

Abstract. We argue on the basis of the results of 2D analysis of the gas flow in gas centrifuges that a reliable calculation of the circulation of the gas and gas content in the gas centrifuge is possible only in frameworks of 3D numerical simulation of gas dynamics in the gas centrifuge (hereafter GC). The group from National research nuclear university, MEPhI, has created a computer code for 3D simulation of the gas flow in GC. The results of the computer simulations of the gas flows in GC are presented. A model Iguassu centrifuge is explored for the simulations. A nonaxisymmetric gas flow is produced due to interaction of the hypersonic rotating flow with the scoops for extraction of the product and waste flows from the GC. The scoops produce shock waves penetrating into a working camera of the GC and form spiral waves there.

1. Introduction

Numerical simulation of the gas flow and diffusion of the mixture of isotopes is the most reliable tool to obtain information about processes in the gas centrifuges. Starting with the work [1], axisymmetric 2D models are widely explored for the numerical simulations. This model appeares very useful as for academic study of the gas flow in the extremal conditions of hypersonic gas flow in GC as for engenering design of new GC's. A range of works have been performed on the basis of this model [2, 3, 4, 5]. No doubts that this model will provide us more new information about the gas dynamics in GC in future. Nevertheless, there are processes in GC that have 3D geometry and it becomes more and more clear that essentially they effect on the gas dynamics in the GC. Engineering calculations of the gas dynamics in the GC are impossible without taking into account these processes.

In this work, we outline the basic results obtained in National Research Nuclear University, MEPhI on the numerical simulation of the gas dynamics in 2D and 3D models of GC.

2. The role of 3D flows in gas centrifuges

Strong centrifugal field is not the only factor,which provides an efficient isotope separation in GC. A range of physical processes is explored to produce an additional axial circulation, which dramatically increases the efficiency of the isotope separation in the GC. This secondary circulation is one of the key features of the industrial GC.

Pair of scoops, located near the end caps of the GC, are used to extract enriched and depleted gas mixture from the GC. Simultaneously, they provide the additional axial circulation due to the mechanical brake of the gas. Mach number of the gas in GC is of the order ~ 7. Interaction of the gas with the scoops is accompanied by formation of strong shock wave, which propagates along the rotational axis. The shock wave forms a spiral wave propagating from one end of the GC to another. It can reflect from the end caps of the GC forming waves running in both directions along the rotational axis. The amplitude of the shock waves is damped rather quickly. After the damping, they propagate in the working chamber of the GC in the form of small amplitude waves. Therefore, to understand the impact of the waves on the dynamics of the gas it is reasonable to start from study of the physics of the waves in the gas compressed by the strong centrifugal fields typical for the GC.

It is well known that the waves in an uniform gas can produce a flow due to their absorption [6]. These are so-called acoustic flows described by Lord Rayleigh [7, 8]. The acoustic flows are produced at the transfer of the energy and momentum from the waves to the gas due to the molecular viscosity.

The waves generated by the scoops and propagating along the axis of GC can also provide an additional mechanism of the secondary axial circulation in the GC due to their absorption. This mechanism can essentially differ from the conventional mechanism of the circulation generation. The waves can propagate and therefore transfer the breaking torque at larger distance from the scoops than it happens at the axisymmetric brake. Our recent numerical experiments confirm that the waves essentially changed the axial circulation [9].

There is another important motivation for study the role of the spiral waves on the gas flow in GC. It appears that the waves can essentially change the gas content in the GC due to modification of the gas flow through a hole connecting the working and waste camera of the GC. Numerical simulations show that the waves can change the gas content in the GC on 15-20% [10].

We are at the very beginning of the study of the role of the waves in the physics of GC. We already found some new physical phenomena due to the spiral waves and apparently, more effects will be found in future. Nevertheless, the impact of the waves on the axial circulation and gas content in GC give us convincing reasons to develop numerical technology for fully 3D simulation of the gas flow in GC. Comparison of the gas flow with and without waves is possible only in the 2D model. Nevertheless, the real waves in the GC, has 3D spiral shape. Therefore, the engineering calculation of the gas dynamics in the GC is possible only in 3D geometry.

3. The basic requests to the software for 3D simulation

The numerical simulation of the gas flow in the conditions of the GC by the conventional numerical codes is impossible because of extremal conditions in the GC. There are at least a few reasons requesting developing of specialized codes for simulation of the gas dynamics in the gas centrifuges. The most important among them are:

(i) Strong variation of density and pressure along radius. All the schemes based on the control volume method use interpolation schemes for calculation of the fluxes of mass, momentum and energy at the surfaces of the control volumes. Strong gradients of pressure give unrealistic fluxes, which produce unphysical solutions. Therefore, a specialized numerical scheme should be developed.

(ii) Velocities of the axial circulation are of the order of a few centimeters per second. They are small compared with the rotational velocity of the gas, which can achieve 600-700 m/s. Accurate calculation of the axial circulation of the axial circulation demands exploration of the rotational frame system. In this system, the rigid body rotation of the gas equals to zero. This is a conventional approach in the 2D axisymmetric models. The axial symmetry of the model is violated by the scoops in 3D models. The scoops rotate in the rotating frame. Therefore, one of the problems of the 3D numerical simulation is development of the codes, which allow us to reveal axial circulation with accuracy better that $2 \cdot 10^{-5}$ at moderate mesh resolution in the laboratory frame system.

(iii) The scoops have rather complicated geometrical shape. Reproduction of the geometry of the scoops and other geometrical details of the GC is important for accuracy of the calculations. At present time, the universal technology for mesh generation around of the bodies with an arbitrary geometry is to use tetrahedral mesh. At the same time, this is not reasonable to use the tetrahedral mesh in the entire computational domain. This type of mesh creates too strong numerical diffusion. It is better to use hexahedral mesh in the largest part of the computational domain. Therefore, the numerical technology should include the mesh generation containing tetrahedral and hexahedral cells.
In addition, the 3D simulations demand huge memory and CPU time. They should be performed in multiprocessor regime. On these reasons the 3D simulations of the gas dynamics in GC has been performed in the very limited cases. As a rule, only fragments of the total flow have been modeled in 3D simulations before [11].

4. Computational Technology of National Research Nuclear University, MEPhI

For development of our original specialized numerical code, we used two schemes of interpolation. First one is the Rhie - Chow interpolation scheme proposed in [12]. Some details of modification of the scheme are presented in [13]. Second one is a second order Godunov interpolation scheme [14]. Our experience shows that the realization of the Godunov scheme is simpler and gives more stable code.

The problem of the gas flow in 3D geometry is solved as a steady state problem by the relaxation method starting from some initial state of gas. Implicit scheme is explored in all cases. The simulations have been performed on a computer claster containing 525 nodes. Verification of the code has been performed with the special tasks developed by us specially for the verification of the numerical codes for simulation of the gas flows in strong centrifugal fields [15, 16].

The mesh generator exploring combined tetrahedral and hexahedral meshes has been developed. This generator allows us to reproduce the geometry of the scoops with high accuracy. The example of the mesh around on of the scoops is shown in fig.1.

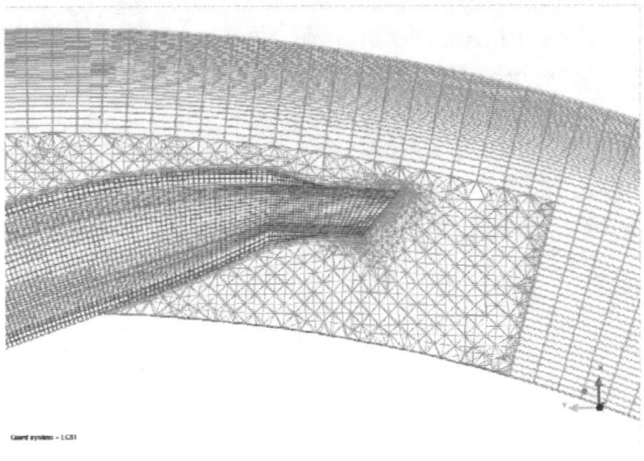

Figure 1. Mesh around a scoop. The mesh is combined of tetrahedral and hexahedral cells.

We imposed the following boundary conditions at the boundaries of the computational domain:

(i) No slip walls were used at the walls of the rotor, walls of the scoops and baffles.

(ii) A temperature distribution is imposed at the walls of the rotor.

(iii) Zero thermal resistance is imposed at the walls of the scoops. The problem of the scoop heating is solved simultaneously with the hydro dynamical problem.
(iv) The computational domain is limited by an artificial inner wall located at the radius where the Knudsen number is close to 1. The hydro dynamical approximation is not valid at smaller radius. Free slip condition is imposed at this wall. The feed flux is also imposed on this wall as a boundary condition.
(v) A specified waste mass flow is imposed at the output from the upper (waste) scoop.
(vi) A small pressure is specified at the output from the lower (product) scoop.

Figure 2. The geometry of the computational model. The volume of the computational domain is located between the wall of the rotor and the inner wall located at the radius where the Knudsen number is close to 1. The location where the feed flux is imposed is shown in green at the inner wall. The volume of the computational domain is separated by the baffles (shown in blue) on 3 chambers connected by concentric holes. The scoops are located in the waste and product chambers at the right part of the figure.

Geometry of the computational domain is shown in fig. 2. The computational domain includes the volume of the rotor filled by the working gas UF_6, baffles, scoops and the artificial inner wall. The baffles divide the volume of the rotoron three chambers: working, waste and product chambers. Parameters of the rotor correspond to the parameters of Fthe Iguasu centrifuge introduced in [17]. However, the length of the rotor is intentionally reduced to 20 cm to reduce the size of the mesh and computational time. The mesh generator produced the mesh containing $1.2 \cdot 10^7$ nodes.

5. The basic results of the 3D simulations

As it was expected, the numerical simulation has shown that the scoops produce strong shock waves. Fig. 3 demonstrates the velocity distribution of the gas on some cylindrical surface. The shock waves freely penetrate from the waste and product chambers into the working chamber

Journal of Physics: Conference Series **751** (2016) 012017 doi:10.1088/1742-6596/751/1/012017

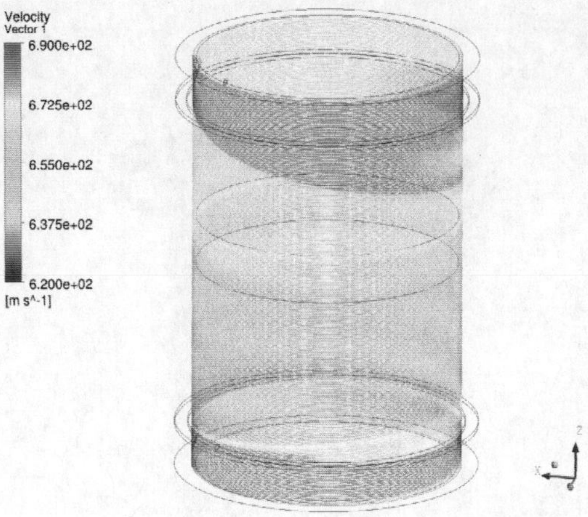

Figure 3. Velocity distribution on a cylindrical surface located in the computational domain. The scoops are located on the left part of the figure. All the geometry is shown by the contour lines. The scoops produce the spiral shock waves penetrating into the working chamber.

forming the spiral wave there. The amplitude of the spiral shock waves is reduced rather quickly. They transform into acoustic spiral waves propagating along the rotor.

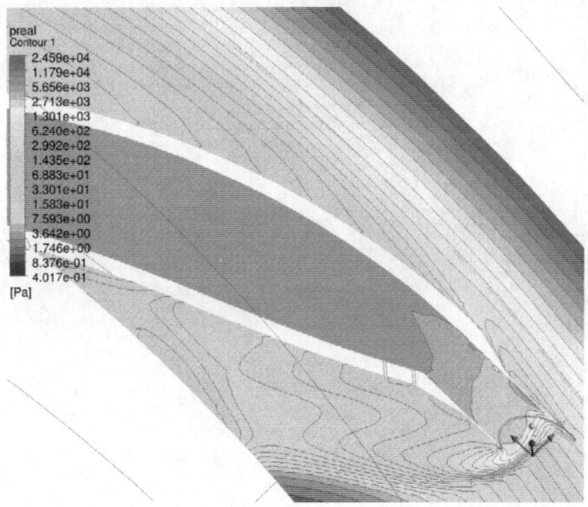

Figure 4. Pressure distribution around the waste scoop in the plane cutting the scoop symmetrically perpendicular to the rotational axis.

The flow near the scoops has rather complicated structure. For example, the pressure distribution around the waste scoop is shown in fig. 4 in the plane perpendicular to the rotational axis cutting the scoop symmetrically. A bow shock is formed at the very nose of the scoop.

The flow through the concentric holes in the baffles separating the rotor on the chambers has essentially 3D character. Fig. 5 shows the flow through the hole in the waste baffle separating upper (waste) chamber from the working chamber. There is the flow from the waste chamber to the working chamber (down) and in the opposite direction (up). This strongly disagrees with the conventional 2D models, which demonstrate only the flow in one direction from the waste chamber. Nevertheless, the integral mass flow is directed from the waste chamber to the working

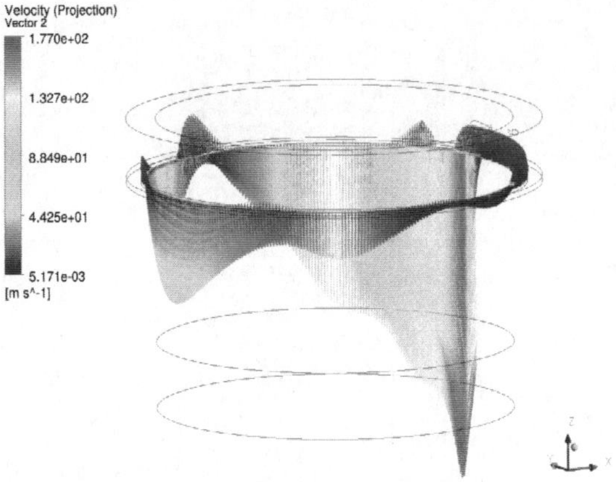

Figure 5. The flow of the gas through the hole in the waste baffle. The waves generated by the scoop produce rather complicated flow from the waste chamber to the working chamber and in the opposite direction. The integral flow is directed down.

one like in the 2D model.

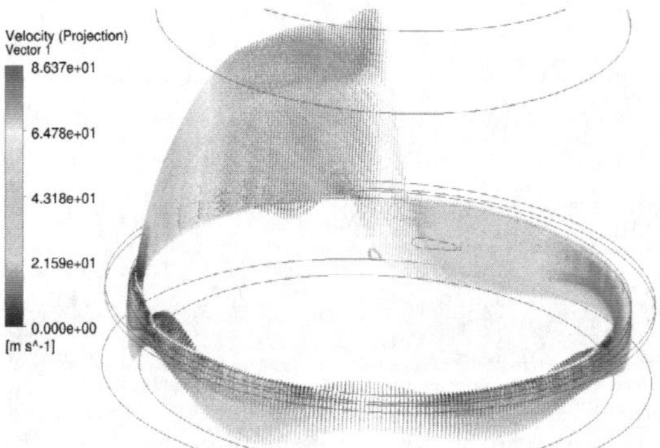

Figure 6. The flow of the gas through the hole in the product baffle. The waves generated by the scoop produce the flow from the working chamber to the product chamber and in the opposite direction. The integral flow is directed down.

Almost the same picture we observe in fig. 6 for the flow through the hole in the lower (product) baffle. There are flows in opposite direction. Moreover, the largest velocity takes place for the flow directed from the product to the working camera (up). Nevertheless, the integral flow is directed from the working to the product chamber.

Fig. 3 demonstrates the shock wave propagation along the rotational axis.The shock waves are damped rather quickly and transform into the acoustic waves propagating in the gas subjected to the strong compression due to the centrifugal force. Fig. 7 shows that these waves can propagate along the entire working chamber without any visible damping. The waste and product chambers are removed from this figure because of strong pressure variation there.

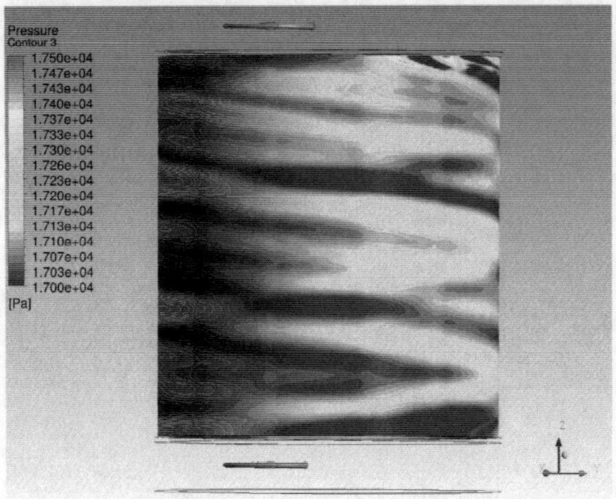

Figure 7. Variation of the pressure in the working chamber produced by the waves..

6. Conclusions

The scoops of GC produce spiral waves, which can propagate in the axial direction along the rotor. These waves can be modeled numerically only in 3D model of GC. Our estimates on the basis of nonstationary flows in 2D models show that the waves can affect the axial circulation and gas content in the GC. Therefore, the engineering calculation of the gas flow in the GC with accuracy of the order of a few percents is possible only in 3D model. The conventional 2D models can give deviation from real optimized parameters on the level 20 – 30% due to the processes having essentially 3D character. Our experience shows that fully 3D simulation is possible with the existing software and hardware and is limited only by the memory and CPU time.

Acknowledgements

The work has been performed under the support of the Ministry of education and science of Russia, grant no. 3.726.2014/K. The computer simulations have been partially performed on the computer cluster of National research nuclear university(MEPhI), Basov farm. This work was also performed within the framework of the Center "Physics of nonequilibrium atomic systems and composites" supported by MEPhI Academic Excellence Project (contracts No. 02.a03.21.0005, 27.08.2013).

References

[1] Kai T 1977 *J Nucl. Sci. Technol* **14** 267
[2] Matsuda T, Sakurai T and Takeda H 1975 *J. Fluid Mech.* **69** 197
[3] Park J K, Hyun J M 1990 *Fluid Dyn. Res.* **6** 139
[4] Pradhan S, Kumaran V 2011 *J. Fluid Mech.* **686** 109
[5] Zhang W C, Jiang D J and Zeng S 2014 *At. Energy Sci. Technol.* **48(2)** 331
[6] Landau L D, Lifshitz E M 1987 *Theoretical Physics* vol. 5 (Butterworth-Heinemann: Oxford)
[7] Rayleigh, Lord 1883 *London Ser. A* **175** 121
[8] Rayleigh, Lord 1896 *MacMillan London*
[9] Bogovalov S V, Kislov V A and Tronin I V 2015 *Theor.Comput. Fluid Dyn.* **29** 111
[10] Bogovalov S V, Kislov V A and Tronin I V 2015 *AIP Conf. Proc.* **1648** 030008
[11] Jiang D, Zeng S 2006 *14th Intl. Conf. on Nucl. Engn (ICONE14)* (Miami, Florida, USA, ASME Conf. Proc.) ICONE14-89783 402
[12] Rhie C M, Chow W L 1983 *AIAA* **21** 1525
[13] Bogovalov S V, Tronin I V 2015 *Comput. Math. Math. Phys.* **55** 1727

[14] Godunov S K, Zabrodin A V and Ivanov M Ya 1976 *Nauka* (in Russian)
[15] Abramov V I et al. 2013 *Comput. Math. Math. Phys.* **53** 789
[16] Bogovalov S V et al. 2013 *Computers and fluids* **86** 177
[17] Rodrigues N A S, Schwab C and Wood H G 1996 *Proceeding of the Fifth Workshop on Separation Phenomena in Liquids and Gases*

Thermodynamic modelling of detonation H-N-O high explosives

Yu A Bogdanova[1], S A Gubin[1], A A Anikeev[1] and S B Victorov[2]

[1] National Research Nuclear University MEPhI (Moscow Engineering Physics Institute) Kashirskoye shosse 31, Moscow 115409, Russia
[2] OpenSearchServer, Paris, France

E-mail: bogdanova.youlia@bk.ru

Abstract. The multiphase model of a detonation products based on the equations of state (EOS) of chemically reacting H-N-O systems which is used in the thermodynamic TDS code is presented. This model is applicable over a wide range of temperatures and density. This model consists of theoretically reasonable EOS for a multicomponent gas (fluid) phase. The calculations of detonation based on the presented model to be in good agreement with experimental data.

1. Introduction

The knowledge of thermodynamic properties of the chemical reactive systems consisting of elements H, N, O is necessary in solving of various practical problems in many fields of science and equipment. In particular, thermodynamic data of equilibrium conditions of such multicomponent systems at high pressures and temperatures are necessary in chemistry and physics of detonation and shocks [1-3]. Therefore development of reliable equation of state (EOS) models for such systems is of a great theoretical and practical interest.

In this work the multiphase EOS model for the chemical reacting H-N-O - systems which was realized into the thermodynamic TDS code [4] is presented. The model is applicable over a wide range of temperatures and density covering the area from normal pressure up to tens of gigapascals. This model includes theoretically reasonable EOS for a multicomponent gaseous (fluid) phase. The thermodynamic calculations based on the presented model allow to obtain the data of thermodynamic properties and chemical composition of the studied system. The examples presented in this work show that the offered EOS provide the exact thermodynamic description of conditions of H-N-O systems reached in static, dynamic (shock wave) and detonation experiments.

2. Equations of state

It is known that all thermodynamic properties of system, including product composition, can be found by differentiation of free Helmholtz energy.

In the present work it is supposed that the multicomponent gas (fluid) phase of products contains any combination of the following molecules and atoms: N_2, O_2, O, N, NO, N_2O, NO_2, H_2O, H_2, H and NH_3. It is assumed that molecules i and j of a multicomponent fluid phase interact via a spherically symmetric Exp-6 potential (modified Buckingham potential):

$$\varphi_{ij}(r)=\frac{\varepsilon_{ij}}{\alpha_{ij}-6}\left(6\exp\left[\alpha_{ij}\left(1-\frac{r}{r_{m,ij}}\right)\right]-\alpha_{ij}\left(\frac{r_{m,ij}}{r}\right)^{6}\right), r\geq c_{ij} \quad (1)$$

$$\varphi_{ij}(r)=+\infty, r\leq c_{ij}, \quad (2)$$

where $\varepsilon_{ij} > 0$ and corresponds to the potential well depth ($\min[\varphi_{ij}(r)] = -\varepsilon_{ij}$), $r_{m,ij}$ is the distance between the centres of molecules at which the potential energy takes the minimum value ($\varphi_{ij}(r_{m,ij}) = -\varepsilon_{ij}$), and α_{ij} is the parameter determining the repulsive stiffness.
For polar molecules in [5, 6] the modified form of potential of Exp-6 considering electrostatic interactions of molecules in which well-depth depends on temperature was offered:

$$\varepsilon_{ii}(T)=\varepsilon_{0,ii}\left(1+\frac{\lambda_{ii}}{T}\right) \quad (2)$$

where λ_{ii} –parameter which is responsible for the accounting of electrostatic effects.

In [7] it was shown that thermodynamic calculation of isotherms of ammonia and its mixtures with hydrogen with EOS on the basis of potentials (1)–(3) have better agreement with experimental data, than when using standard three-parametrical potential of Exp-6 (1)-(2).

Such potential (1)-(3) provides the realistic description of intermolecular forces in dense gases at high pressures and temperatures relevant to problems of detonation and shock waves.

In this work the improved version [8] of the perturbation theory KLRR is used. This version has the higher accuracy of calculation of thermodynamic parameters of a pure fluid at the high density, and especially for stiff potentials ($\alpha > 14$), in comparison with the original version [9] and the BBH version [10] of the KLRR theory is used.

The potential parameters have been determined mostly by matching experimental Hugoniot data and available results of static experiments for the region of moderate pressures and temperatures. Potential parameters are defined in [8] and for the substances investigated in this work are presented in table 1.

Table1. Potential parameters of Exp-6 (1)–(2)

Molecule	$\varepsilon_{0,ii}/k_B$ K	$r_{m,ii}$, m^{-10}	α_{ii}	λ_{ii}, K
N_2	100.6	4.25	12.3	0
N	120.0	2.65	10.4	0
O_2	96.2	3.79	14.7	0
O	277.0	2.57	11.5	0
NO	140.2	3.70	13.9	0
NO_2	326.2	4.27	13.8	0
N_2O	234.3	4.27	13.8	0
H_2	36.9	3.67	10.6	0
H	110.0	1.55	10.0	0
H_2O	195.0	3.22	13.4	640.0
NH_3	207.0	3.69	12.8	199.0

In multicomponent mixtures except like-pair interactions it is necessary to consider unlike-pair interactions. Potential parameters for unlike interactions ($i \neq j$) can be set in an obvious form (if these parameters are known), or formally expressed via like-pair potential parameters for the corresponding pairs of molecules:

$$r_{m,ij}=k_{ij}\frac{r_{m,ii}+r_{m,jj}}{2};\ \varepsilon_{ij}=l_{ij}\sqrt{\varepsilon_{ii}\varepsilon_{jj}}\ ;\ \alpha_{ij}=m_{ij}\sqrt{\alpha_{ii}\alpha_{jj}} \quad (3)$$

In expressions (4) k_{ij}, l_{ij} and m_{ij} – the correcting multipliers which values are usually close to unit. These multipliers can be defined if there is suitable theoretical information or experimental data.

Otherwise $k_{ij} = l_{ij} = m_{ij} = 1$that reduces formulas (3) to classical Lorentz–Berthelot combination rules [11].

The unlike-pair Exp-6 parameters are additive in present work, i.e. they follow the Lorentz–Berthelot combination rules ($k_{ij} = l_{ij} = m_{ij} = 1$), for all pairs except for H_2O–N_2, for which we found the best-fitting value of $k_{H2O-N2} = 1.02$.

To calculate the thermodynamic properties of a multicomponent fluid phase in [12] an improved van der Waals one fluid (vdW1f) model is proposed, which assumes a mixture of chemical species, interacting via potentials (1), to be a hypothetical one-component fluid with an effective Exp-6 potential. The composition-dependent parameters of the effective potential are given by

$$r_m = \left[\sum_{i=1}^{n}\sum_{j=1}^{n} x_i x_j r_{m,ij}^3\right]^{1/3}; \varepsilon = \sum_{i=1}^{n}\sum_{j=1}^{n} x_i x_j \varepsilon_{ij} r_{m,ij}^3 \Big/ \left(r_m^3\right); \alpha = \sum_{i=1}^{n}\sum_{j=1}^{n} x_i x_j \alpha_{ij} \varepsilon_{ij} r_{m,ij}^3 \Big/ \left(\varepsilon \cdot r_m^3\right) \quad (4)$$

where the summation extends over all ij pairs, n is the number of the components, and x_i is the mole fraction of species i. The reliability of vdW1f in predicting the excess thermodynamic parameters of a multicomponent fluid has been shown by Ree [12], who is the author of the α-mixing rule.

3. Results

The presented multicomponent model of detonation products based on EOS of dense gas was applied to calculation of thermodynamic properties and composition of detonation products of some H-N-O systems in the wide range of pressure and temperatures. Results of calculations are in good agreement with experimental data of shock Hugoniots (figure 1), static measurements (figure 2) and detonation experiments (table 2).

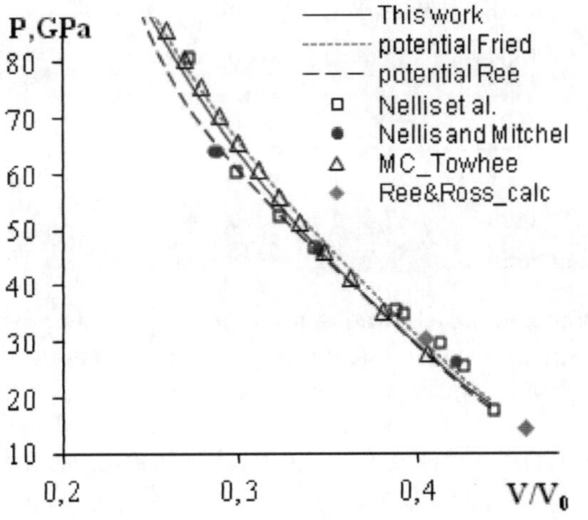

Figure 1. Shock Hugoniots for liquid nitrogen in *P-V*. Symbols - experiments [16, 17] and MC simulation [18], lines – this work. Results of our calculations with potentials [13, 14] are showed also.

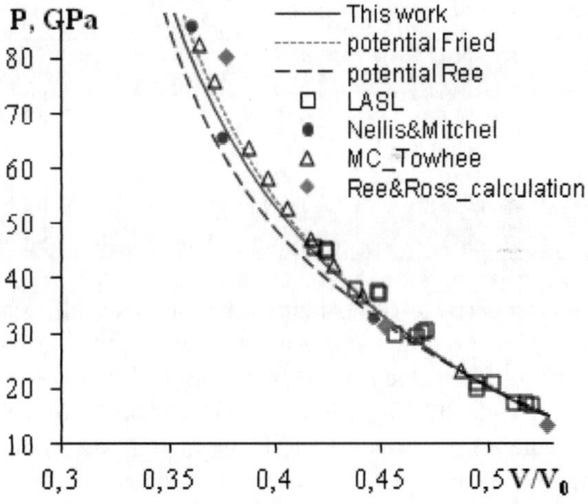

Figure 2. Shock Hugoniots for liquid oxygen in *P-V*. Symbols - experiments [17, 19] and MC simulation [18], lines – this work. Results of our calculations with potentials [13, 14] are showed also.

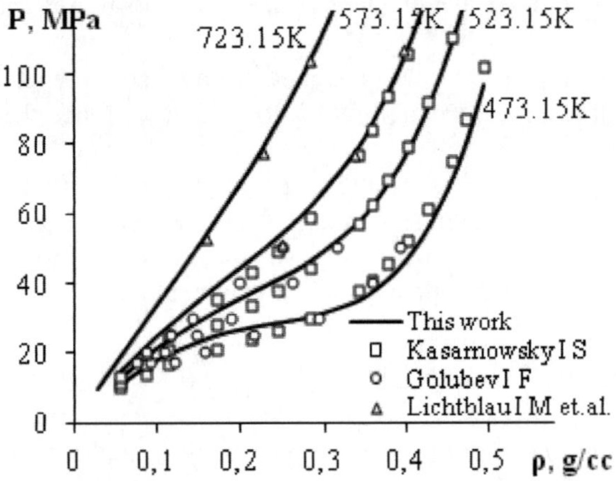

Figure 3. Isotherms for ammonia. Symbols - experiments [21-23], lines – this work.

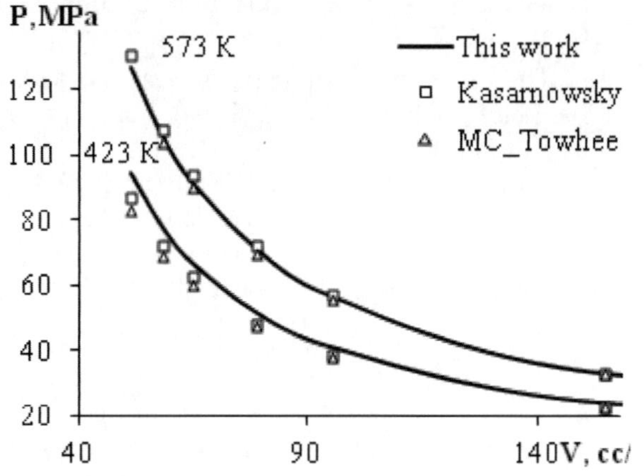

Figure 4. Isotherms for NH_3-H_2 mixture, mole fraction of NH_3:34.75%. Lines – this work. Symbols - experiment [24] and MC simulation. [7].

Table 2. Detonation parameters of different condensed high explosives.

HE	ρ, kg/m^3	H_0, kcal/mole	D_{exp}, m/s	T_{CJ}, K	P_{CJ}, GPa	D_{calc}, m/s	Δ*D*, %
$H_4N_2O_3$	1050	-87,28	4500	1667	5,56	4920	9,3
$H_5N_3O_3$	1630	-58,9	8690	2417	20	7913	8,9
$H_5N_5O_4$	1620	-6	8310	2481	28	8646	4,04
$H_{51.9}N_{29.1}O_{19.1}$	1420	-27,7	8640	3200	27	9163	6,05
HN_3	1127	64,4	7570	4631	16	7247	4,27
H_5N_5	1260	40	7600	1890	17,8	8090	6,5
NO	1294	17	5620	2652	8,9	5436	3,3

Comparisons of results of calculations with experimental data showed that the developed EOS for dense gas provide more exact consent with experiments, than the theoretical models of other authors.
It should be noted that for calibration of potential parameters experimental data on a detonation weren't used. Nevertheless, our model provided good consent with the measured detonation parameters in all considered cases. It is important to note that our model gives reliable forecasts of detonation parameters not only for HE considered above, but also for many other explosive systems. The last also HE in which detonation products besides gases significant amounts of the condensed substances are formed are among. The corresponding examples can be found in our works [1, 3, 8, 26].

Such success in a prediction of parameters of HE detonation having various temperatures, pressure and chemical compositions of detonation products was a consequence of physical validity of our model constructed on realistic intermolecular potentials and the exact theoretical EOS which is reliably reproducing data MC simulation in wide area of thermodynamic conditions of a fluid. The received results allow to conclude that our theoretical model reliably predicts thermodynamic properties of multicomponent gas mixes in the wide range of change of pressure, temperature and a chemical composition of mixtures, as provided successful forecasting of detonation characteristics of different HE with various density of charges.

4. Conclusion

One of the most significant results of work is the shown possibility of realistic thermodynamic modelling of complicate, multiphase and the multicomponent chemical reacting systems at high pressures and temperatures. It is shown that use of the developed technique of thermodynamic modelling and theoretically reasonable EOS of multicomponent fluid phases significantly increases reliability of thermodynamic modelling of difficult chemical systems at high pressures. Thus becomes possible to use results of thermodynamic calculations not only in the traditional field for this purpose connected with researches of shocks and detonation processes but also at the solution of a wide class of problems from other fields of science.

Acknowledgments

This work is supported by the Russian Science Foundation under grant 16-19-00188.

References

[1] Gubin S A, Korsunskii B L, Pepekin V I, Bogdanova Yu A 2009 *Comb. Expl. Sh. Waves* **45** 738.
[2] Shargatov V A, Bryakina U F, Gubin S A, Lyubimov A V 2012 *Rus. J. Phys. Chem. B* **6** 261.
[3] Maklashova I V, Dolgoborodov A Yu et al 2014 *J. Phys: Conf. Ser.* **500** 052010.
[4] Victorov S B, Gubin S A, Maklashova I V and Revyakin I I 2001 *Proc. Energetic. Materials, Ignition, Combustion and Detonation* 32nd Annual Conference of ICT (Karlsruhe, Germany) pp 69/1–69/15.
[5] Ree F H 1981 *Proc. 7th Int. Sympos. on Detonation* (Annapolis, Maryland) pp 646-661.
[6] Ree F H 1982 *J. Chem. Phys.* **76** 6287.
[7] Bogdanova Yu A, Gubin S A, Anikeev A A, Victorov S B, Maklashova I V 2015 *Rus. J. Phys. Chem. B.* **9** 392.
[8] Victorov S B, Gubin S A 2006 *Proc.13th Int. Detonation Symp.* (Norfolk: Los Alamos National laboratory) p 13.1118.
[9] Kang H S, Lee C S, Ree T, Ree F H 1985 *J. Chem. Phys.* **82** 414.
[10] Byers-Brown W, Horton T V 1988 *Mol. Phys* **63** 125.
[11] Reed T M, Gubbins K E 1973 *Applied Statistical Mechanics* (New York: McGraw-Hill) 132 p.
[12] Ree F H 1983 *J. Chem. Phys.* **78** 409.
[13] Fried L E, Howard W M *J. Chem. Phys.* 1998 **109** 7338.
[14] Van Thiel M, Ree F H *J. Chem. Phys.* 1996 **104** 5019 .
[15] Charlet F 1998 *J. Appl. Phys.* **84** 4227.
[16] Nellis W J et.al. 1991 *J. Chem. Phys.* **94** 2244.
[17] Nellis W J, Mitchell A C 1980 *J. Chem. Phys.* **73** 6137.
[18] Anikeev A A, Bogdanova Yu A, Gubin S A, 2015 *J. Phys: Conf. Ser.* **653** 012055.
[19] *LASL Shock Hugoniot Data* 1980 ed S. P. Marsh (University of California Press) p 653.
[20] Ross M, Ree F H 1980 *J. Chem. Phys.* **73** 6146.
[21] Kasarnowsky I S 1940 *Acta Physicochimica* **12** 513.
[22] Golubev I F, Kiyasheva V P, Perel'shtein I I, and Parushin E B 1978 *Thermophysical Properties of Ammonia* (Moscow : Izd-vo Standartov) 264 p[in Russian].
[23] Lichtblau I M, Bretton R H, Dodge B F 1964 *A.I.Ch. E. Journal* **10** 486.
[24] Kasarnovskiy Ya S, Simonov G B, Aristov G Ye 1968 *Technikal Translation* **142** AD 843492.
[25] Gubin S A, Odintsov V V and Pepekin V I 1986 *Thermodynamic calculations of condensed explosive detonations (*Preprint. Chernogolovka: IChPh USSR Academy of Sciences).
[26] Victorov S B, El-Rabii H, Gubin S A, Maklashova I V, Bogdanova Y A 2010 *J. En. Mat.* **28** 35.

Thermal analysis of the mixtures of paraffin with aluminum in wide temperature range

S A Gubin, I V Maklashova and I S Levitskaya

National Research Nuclear University MEPhI (Moscow Engineering Physics Institute)
Kashirskoye shosse 31, Moscow 115409, Russia

E-mail: ivmaklashova@mephi.ru

Abstract. The mixtures and composites of wax, paraffin and metals are widely used as energy efficient formulations and phase change materials for heat storage. Aluminum is frequently employed in the formulations of many composite explosives or propellants. Metal fuel additives are used in advanced explosive formulations to achieve higher combustion temperatures and longer pressure pulses. In this project, Al-paraffin wax composite materials were prepared and characterized. The thermal stability of the prepared powders was determined by differential scanning calorimeter, simultaneous thermogravimetry analysis-differential thermal analysis in the temperature range 30-1300°C at atmospheric pressure. The results of differential scanning calorimeter showed that the thermal performance and structure of the composite materials are stable up to 200°C.The paraffindecompositionwith an energy release is possible at temperatures over 200°C and the oxidation of aluminum may be at a temperature above its melting point.It is shown that the maximum total amount of heat generated by the thermal decomposition of the composition was at the mass fraction of aluminum of 16% - 18%.

1. Background and the purpose of work

The paraffin as phlegmatizer and the powder of aluminum for improving the energy are widely used in explosives mixtures and rocket propellants [1, 2].

The purpose of work was study the process of thermal decomposition of the mixture of paraffin and aluminum powder micron dispersion by a simultaneous thermal analysis; the estimation of the activation energy of decomposition of paraffin and the determining the optimal percentage of mixture components for using as energy efficient formulations.

Such mixtures are a promising material (phase change materials) in the field of energy saving for thermal batteries. To increase the heat capacity and thermal conductivity of the wax is added powder of aluminum [3, 4]. Thus, the heat of fusion of paraffin wax is used for accumulation of thermal energy in thermal batteries [4], and high values of heat of formation and phlegmatizing properties of wax to allow its use in high-energy formulations [5, 6].

Note that the mixtures of aluminum powder with paraffin are flammable. With the destruction or loss of integrity device possible ignition and combustion of the mixture.Therefore, the study of thermal decomposition of the compositions of paraffin wax and aluminum powder should be performed in a wide range of temperature changes. At low temperatures up to 200°C the information is needed to manage the storage of thermal energy at higher temperatures will be useful for applying such compositions in energy-intensive compositions for the understanding of their thermal decomposition and ignition of aluminum particles.

If thermal decomposition of paraffin has been studied rather thoroughly [7, 8], the patterns of thermal decomposition of mixtures of paraffin and aluminum powders is studied poorly. Introduction to different quantities of fine and ultrafine aluminum particles in the paraffin changes the density, thermal conductivity, specific heat and other thermophysical and thermodynamic characteristics of the composition. This causes a change in the regularities of thermal decomposition of mixtures of paraffin and aluminum particles.

2. The technique

The thermal stability of the composite materials was determined by differential scanning calorimeter, simultaneous thermogravimetry analysis-differential thermal analysis. The method of simultaneous thermal analysis, which brings together in one dimension thermogravimetric analysis (TGA / DTG), differential thermal analysis (DTA) and differential scanning calorimetry (DSC) gave information about the composition, thermal and oxidative stability of materials, phase transitions, and the kinetics of chemical reactions. The thermal decomposition and the possible interaction of decomposition products of wax and ultrafine aluminum upon heating these compounds was investigated in a wide range of temperature changes (from normal conditions to a temperature of 1300°C).

The heating rates varied from 2°C/min to 50°C/min to estimate the kinetics of processes. The experiments were carried out in air or argon atmosphere in a crucible with the opened or closed cover to estimate the oxidation in various oxidizing environments

3. Results

The analysis of the processes of melting and oxidation of the paraffin and the aluminum is carried out by different heating rate (2 K/min, 10 K/min 20K/min, 50K/min) in temperature range from 30°C to 1300°C in different atmospheres (air and Ar).

Figure 1 shows a typical thermogram of decomposition of a mixture of paraffin and aluminum. In the thermogram can be allocate four temperature ranges, where the Thermal decomposition, namely, there are peaks of heat flow: $T_1 = 40 \div 70$°C, T_2=200÷450°C, T_3=600÷700°C, T4 = 1000 ÷ 1100°C.

The first peak of the DSC curves associated with the melting of wax T_{melt} = 57 ° C. With increasing heating rate the peak a few shifts to higher temperatures.

Thermal decomposition of the wax begins after 200 ° C. In the temperature range $T = 200 \div 450$ ° C is an intensive mass loss of wax. The first range in said exothermic peak in the DSC curve is associated with a break C-C bonds in long hydrocarbon molecules and dehydrogenation. At the end of this temperature range is possible oxidation products of thermal decomposition of the wax in air [8]. At higher temperatures, there is an intensive $T \geq 800$K oxidation decomposition products wax in air.

Obtained by the method of Kissinger activation energy of decomposition of paraffin E= 28.2 kcal/mol in agreement with the experimental data [7- 9].

The kinetics of oxidation and combustion of the aluminum particulate were studied in [10 - 12]. Thermal decomposition aliminum particles decrease with increasing heating rate percentage burn-alumina, wherein while blowing air, this percentage is higher than argon purge. In the presence of air of aluminum powder melting begins at lower temperatures with increasing heating rate.

The figure 2 shows that the curve 4 (16.7% of the mass content of aluminum) is located above the rest in the investigated temperature range. Draws attention to the close location of the curves 4 (16.7%) and 6 (36%). But at high temperatures line 6 is located below the line 4. Therefore, the total amount of heat by thermal decomposition of a mixture of paraffin with a mass content of aluminium 16.7 % (line 4) is higher compared with the composition containing 36% aluminum (line 6).

Journal of Physics: Conference Series **751** (2016) 012019 doi:10.1088/1742-6596/751/1/012019

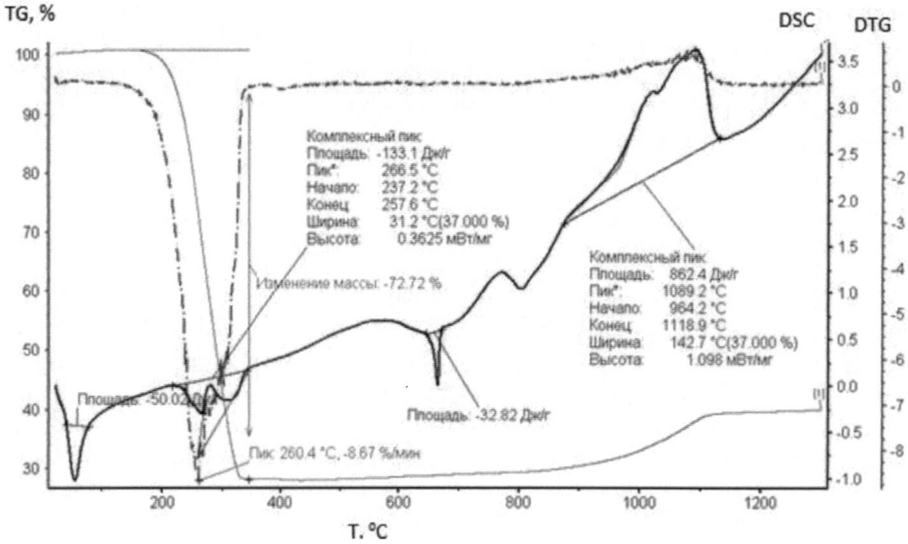

Figure 1. The typical thermogram of Al-paraffin mixture (18% Al) in air, heating rate 10K/min

The conclusion about the maximal condition of energy release of the mixture of paraffin with 16.7% of aluminum is confirmed by Figure 3, which shows the total amount of heat released during thermal decomposition of a mixture of paraffin depending on the mass number of aluminum in the composition.

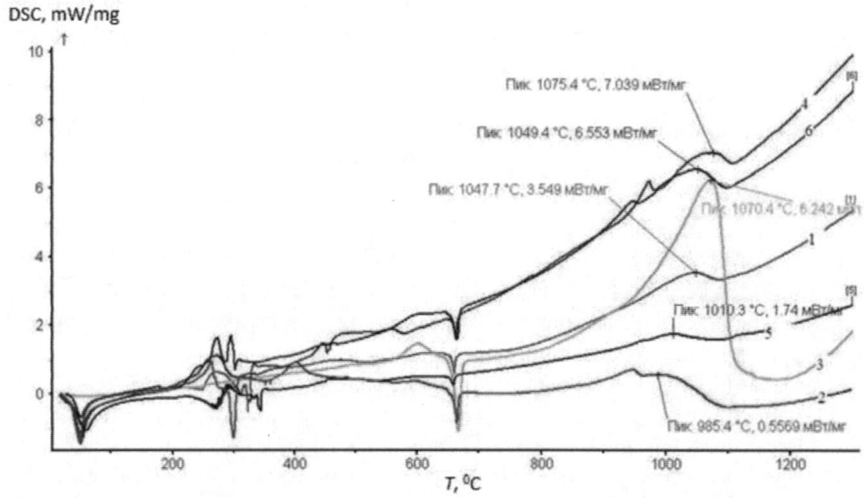

Figure 2. The thermograms of Al-paraffin mixture: 1– 10.2 % Al (closed crucible), 2– 10.9% Al (open crucible), 3– 100% Al, 4– 16.7% Al (closed crucible), 5– 5.2% Al (closed crucible), 6– 36% Al (closed crucible), hating rate 10K/min in air.

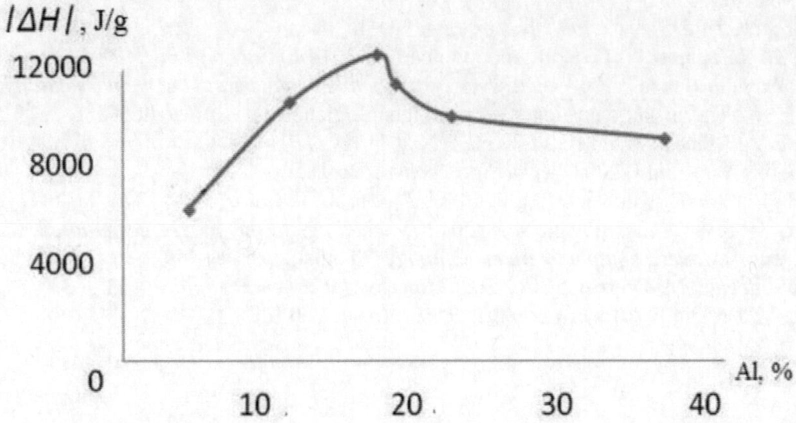

Figure 3. The total amount of generated heat depending on the weight amount of aluminum in the composition.

Mass percentage of aluminum 16.7% is optimal for use in energy-intensive compounds, as a consequence, improving their thermodynamic characteristics (thermal conductivity, heat capacity), due to a uniform warming of the mixture. If the mass of aluminum in the crucible is too large, the metal does not have time to warm up to the temperature of paraffin. When uneven heating of the aluminum particles in the surface layers of the sample are oxidized with the formation of refractory oxide on their surface. This reduces the area of contact of aluminum particles, degradation products of paraffin wax and air. The temperature of the wax becomes higher than the temperature of aluminum particles and there is no reaction of decomposition products of paraffin wax and aluminum particles. Therefore, mixtures with a mass content of aluminum 16%-18% have the highest dissipation by thermal decomposition in comparison with other compounds.

It can be noted that in the crucible without lid energy release is practically not observed. Paraffin vapors freely out of the crucible into the surrounding space. Thus, we can conclude that using the lid of the crucible helps to keep a pair of paraffin in a crucible, and contributes energy to the interaction of the vapour of paraffin with aluminum.

4. Conclusions

1. It is shown that with increasing heating rate melting of wax begins at somewhat higher temperatures.
2. According to the method of Kissinger activation energy of decomposition of paraffin E=28.2 kcal/mol.
3. It is shown that the total amount of heat generated by thermal decomposition of wax-aluminum mixtures has the maximum at aluminum content of 16 - 18% mass% due to the best energy performance degradation over a wide range of temperatures.
4. In closed spaces at high temperatures may be a considerable energy release during the decomposition of paraffin with aluminum. This aspect should be considered in designing termal storage.

Acknowledgments

The work is supported by the Russian Science Foundation under grant 16-19-00188

References

[1] Fedorov S G, Guseinov S L, Storozhenko P A 2010 *Nanotechnologies In Russia* **5** 565.
[2] Lipanov A M, Aliev A V 1995 *Proektirovanie raketnyh dvigateley tverdogo topliva* (Moskva: Mashinostroenie) p 400.

[3] Xiao X, Zhang P , Li M 2013 *Appl. Energy* **112** 1357.
[4] Babapoor A, Karimi G 2015 *Appl. Therm. Eng.* **90** 945.
[5] Mandilas C, Karagiannakis G, Konstandopoulos A G 2014 *Energy and Fuels* **28** 3430.
[6] N A Pokalyuhin et al 2008 *Smesevye energoemkie materialy: uchebno-metodicheskoe* (Kazan': Izdatel;stvo Kazanskogo gosudarstvennogo tehnologicheskogo universiteta 88p.
[7] Francina E V , Afanas'eva Yu I , Ivashkina E N 2011 *Izv. TPU* **318** 80.
[8] Liao K , Cong Y , Kang L 2008 *Petrol. Sci. Technol.* **26** 1852
[9] Gonen M , Balkose D , Ulku S , Ulku I 2008 *J. Therm. Anal. Calorim.* **94** 737.
[10] Short A G , Ionov I A , Karpovich M K, 2012 *Proceedings of the XVIII International scientific-practical conference "Modern equipment and technology"* (Tomsk, Russia) 359
[11] Zhang S, Schoenitz M, Dreizin E 2012 *Int. J. Energ. Mat. Chem. Propulsion* **11** 353.
[12] Schoenitz M, Dreizin E 2015 *Prog. Energy Combust. Sci.* **50** 81.

Journal of Physics: Conference Series **751** (2016) 012020 doi:10.1088/1742-6596/751/1/012020

Rigid polyurethane foam as an efficient material for shock wave attenuation

P V Komissarov [1,2], A A Borisov [1,2], G N Sokolov [2] and V V Lavrov [3]

[1] National Research Nuclear University MEPhI (Moscow Engineering Physics Institute), Kashirskoe highway 31, Moscow, 115409, Russia
[2] N.Semenov Institute of Chemical Physics, Russian Academy of Sciences, Moscow, 119991, Russia
[3] The Institute of Problems of Chemical Physics, Russian Academy of Sciences, Chernogolovka, Moscow region, 142432, Russia.

E-mail: dot-dot@inbox.ru

Abstract. A new method for reducing parameters of blast waves generated by explosions of HE charges on ground is presented. Most of the traditional techniques reduce the wave parameters at a certain distance from the charge, i.e. as a matter of fact the damping device interacts with a completely formed shock wave. The proposed approach is to use rigid polyurethane foam coating immediately the explosive charge. A distributed structure of such a foam block that provides most efficient shock wave attenuation is suggested. Results of experimental shock wave investigations recorded in tests in which HE charges have been exploded with damping devices and without it are compared.

1. Introduction

At present, the majority of available blast wave localizers use predominantly mechanical energy and momentum absorption and dissipation techniques based on multiple reflections of shock waves from an assembly of obstacles and on entrainment of condensed phase of a shield by the gas flow behind shock front. In particular, water-based or polymer foams of various density are used for the shock wave attenuation as described in [1, 2, 3, 9, 10].

There are two basic ways to use foam for explosion localization - coating the charge with a large amount of low density water-mechanical or polymer foam or shielding objects to be exposed to a shock wave with a layer of polymer foam. For example, in [2], blast effect of explosion of a 285 g TNT charge was significantly reduced with a cubic meter of water foam. The second approach assumes that the object exposed to blast wave must be known in advance. Therefore, this approach is not applicable to emergency explosion localization tasks (e.g., when surrogate explosive is to be destroyed). It is well known that the shock wave attenuation mechanism in polyurethane foams is similar to that in water-mechanical foams.

The main physical principle underlying functioning of all devices that damp blast waves is transfer of both energy and momentum from the gas flow to condensed phase of foams [3, 4, 5]. This process includes mechanical acceleration, heating, evaporation, and thermal decomposition of the condensed phase, which result in gas flow deceleration and cooling, i.e., in blast wave attenuation. In some situations, acceleration of the condensed phase can cause an opposite effect, namely, kinetic energy of the accelerated condensed phase is converted into pressure when it stops at a reflecting solid surface, increasing thereby the pressure behind the reflected wave to a level exceeding that inherent in the reflection wave in gas (with no foam) [5]. This mostly relates to foam shields adjacent to objects protected. The higher the shock wave amplitude the more prominent is the effect of accelerated condensed foam phase on the reflected shock wave amplitude. Weak shock waves behave as nearly acoustic waves when reflected from solid surfaces. Thus, factors that reduce the condensed phase acceleration would diminish the reflected wave amplitude. It would seem that shields adjacent to an object to be protected are inefficient. Inasmuch as the energy losses within the foam are quite significant the reflected wave pressure can nevertheless be lower than in gas with no foam shield under certain conditions. For example, destruction of thin partitions between cells in water foams, which increases gas permeability of foam, reduces the condensed phase acceleration but intensifies the energy exchange. This is true for polymer foams as well. On the other hand, entrainment of condensed phase by the gas flow attenuates the incident shock wave, therefore a layer of foam coating the charge would significantly reduce the amplitude of shock wave leaving the layer not only because of energy losses within the layer but due to difference in the acoustic impedances at the foam-air boundary. Explosion localizer efficiency depends on many factors, such as rigidity of foam, its density and structure, ability to resist intense (high speed and dense) gas flows, and thermal stability of the condensed phase. Unfortunately so far there is not too much information available in literature on how to select a proper material for foam, where it is to be disposed with respect to the charge and objects to be protected, and what foam density is needed to attain the most efficient suppression of the blast effect in each particular situation.

Porous structure of foams combines two important qualities, namely, gas permeability and high aerodynamic drag. Explosion products flow through it, allowing thereby them to expand, but at a low velocity, i.e., the amplitude of a shock wave entering the ambient air drops. This implies that to interact with the detonation products in the near zone around the charge the required density and rigidity of the foam material must be as close to those of the detonation products as possible. Low density water foams are least efficient in attenuation of shock waves in this zone. For the efficient energy exchange between the detonation products and foam material a sufficient quantity of the total mass of the rigid foam obstacle is also needed.

To solve the problem of the blast wave attenuation one should refer to polymer foams that have an elastic structure of sufficient rigidity. Authors of [6] studied attenuation of shock waves with triangular shaped pressure profiles by a cylindrical polyurethane foam block placed in a shock tube. It has been found that the shock wave amplitude behind the block diminishes when the block length is close to the incident wave pulse length. It is concluded also in [6] that as the foam mass and the amplitude of the wave increase so does the efficiency of the wave attenuation. Application of polyurethane foam to reduce the intensity of shock waves has been increasingly discussed in the literature in recent years [9, 10].

As shown in [9] the highest efficiency of attenuation is observed in a shock tube when the foam as a whole moves with the wave. The degree of wave amplitude reduction varied from 85 to 95% when foam blocks moved freely whereas it dropped to 18 - 60% for fixed foam blocks. That means that a greater attenuation effect can be achieved by setting a foam block as a barrier between the charge and the object to be protected as close to the charge as possible rather than by protecting the object with a foam layer coating its surface. Of course this is true provided only a part of the block hits the object while its major part is dissipated in the flight to it. This conclusion is supported by the fact that in some cases the shock wave pressure rises when porous foam layer coats the object [11]. That the shock wave attenuation is the higher the greater the Mach number of incident wave is another observation reported in [5] which confirms findings of authors of [6]. This leads one to conclude that the closer the foam block is positioned to the charge the more efficient its damping action on the shock wave amplitude. The mechanism of energy absorption in the course of polyurethane compression is examined comprehensively in [10]. We allocate the findings most important for further consideration. The foam 0.409 kg/m^3 in density is destroyed completely being compressed at an initial velocity exceeding 141 m/s, its interaction with a shock wave is mainly controlled by foam material viscosity and heat conductivity.

As the compression velocity decreases to 43 m/s, the dissipative processes associated with the viscoplastic flow within the foam structure begin to dominate in absorption of the compression wave energy. This implies that an increase in the heat capacity of the material (including heat expenditure for evaporation) in the close vicinity of the charge, and, at the same time, in the density of the foam structure at greater distances would favor the absorption of the explosion energy. Therefore the best way to absorb explosion energy by rigid polyurethane foam is to coat a charge with layers of different grades of foam material with various mechanisms of energy absorption depending on the distance from the charge.

Polyurethane based foams are wide-spread; their properties (density and size of cells) vary in a wide range depending on polymerization regime. Polyurethane foam formation requires water as an agent reacting with the isocyanate group (one of the components a) and producing carbon dioxide that fills the foam cells. Normally, reactants of such foams in the initial state consist of a mixture of polyol and isocyanate [8 p. 81]. When the mixture is activated by water, an exothermic polymerization reaction attended by gas evolution starts. To accelerate this reaction it is sufficient to increase the temperature of the reactants, for example, spraying the foam producing reactants together with hot water. The rate of elastic foam formation has been verified experimentally.

The reagents were sprayed together with boiling water. As a result, a 50-liter hemispherical foam block was obtained. The amounts of reagents and water consumed were 5 liters and 8 liters respectively. Water in this case was in excess compared to its amount needed to polymerize the reactants. The excessive water was introduced intentionally into the foam block in order to fill the foam pores closest to the charge zone in order to rapidly reduce temperature of the detonation products. The resulting material resembles a heavily wetted sponge. The average density of the material is about 280 kg/m^3. In the central part of the foam block, the material density was higher - up to 500 kg/m^3.

Such density distribution in the foam block has been provided deliberately and would be discussed below. These experiments show that a material consisting of polyurethane foam and water can be prepared in sufficient quantities in a short period of time. The block formation time can be reduced introducing a polymerization catalyst in the mixture of reagents [8].

Below we discuss plausible approaches to optimizing localization of the blast effect. In the area around the charge, keeping in mind the fact that the flow of products in this zone is laminar, it is necessary to place a foam with coarse open pores and density as high as possible (up to the use of metal foams). Such a foam material would enhance instability of the expanding flow of detonation products that generate the shock. The jets induced by the instability perturbations intensify drastically the energy and momentum exchange between the localizer material and gas flow, the temperature of the products and shocked gas is decreased additionally due to thermal decomposition of the foam material.

The first zone of interaction, let's call it an "instability zone", should be of an order of magnitude of the charge size or slightly thicker. The “instability zone” should be followed by water-saturated foam of lower density. Jets of the hot products turbulized in the “instability zone” would penetrate into that foam layer and mix intensively with water reducing thereby the gas flow temperature. The kinetic energy of moving products would induce a viscoplastic flow in the foam structure in the course of its deformation, which would further reduce the gas flow temperature. In addition, a multiple reflections of the shock wave within the foam enhances the dissipation of the energy and momentum of the gas and its mixing with water present in pores. All the aforementioned interactions would reduce the driver ability of detonation products.

Thus, the major damping processes occur in this second zone. At the end of interaction of the products with foam material in this zone (we call it "attenuation zone"), the foam block fragmented into separate small pieces would produce secondary dangerous high speed fragments. To mitigate action of this factor, as well as to further reduce the shock wave energy one more zone of interaction is needed, where the damping material should be a low-density fine-cell foam. Secondary fragments formed in the "attenuation zone" would get stuck in the outer layer of the light foam loosing their kinetic energy because of high aerodynamic drag.

2. Experiment

To estimate the efficiency of shock wave attenuation by the foam block, a comparative test was carried out. First, a 500-g cylindrical (L/D=2) charge of granulated TNT was exploded on a flat sandy area. In the second experiment, an identical charge was covered with a polyurethane foam block prepared from the AKVIDUR[tm] [7] foaming agent. The block was nearly semispherical and had a distributed structure - the inner 5-10 cm layer of coarse foam, a 25-cm middle foam layer with a cell size of about 6-10 mm containing some amount of water, and an 8-10-cm outer light foam layer (see. Fig. 1).

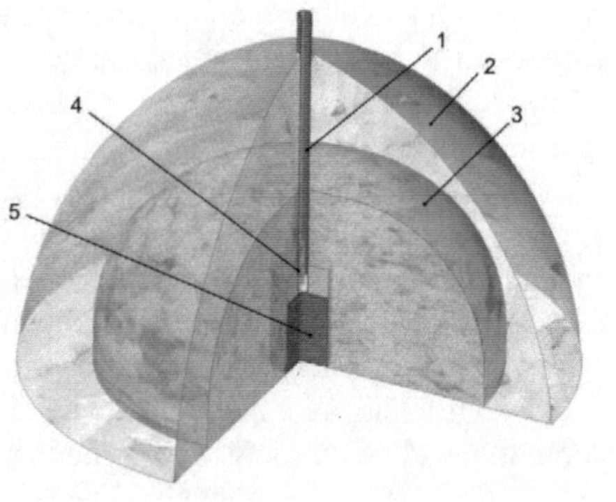

Figure 1. Schematic of the foam block: 1 – tube for detonator wires; 2 – outer light foam layer; 3 – middle water-saturated layer; 4 – detonator cap, 5 – HE charge.

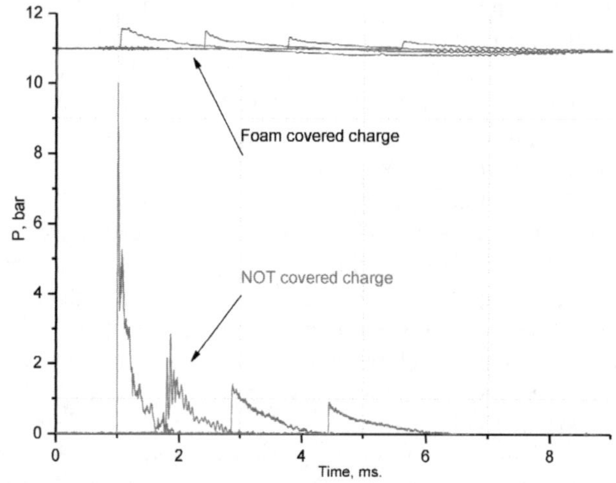

Figure 2. Comparison of pressure profiles recorded in a test with and without foam block.

Geometrical dimensions of the block were selected keeping in mind that the greatest efficiency of the blast wave attenuation would be in the case when the foam layer is comparable with the shock wave pulse length (see above). A charge 100 mm in ID consisted of 430 g of granulated TNT and 50 g of ammonite 6JV, was installed into a recess precut in the foam block. A 12-g ammonite charge 20 mm in ID with a 8-g RDX cap was used as a booster to initiate the TNT charge. The charge unit together with the foam block was installed on a steel plate (see. Fig. 1).

As a result, the pressure traces at different distances from the charge were obtained. The resulting pressure-time histories of the processes studied are displayed in Fig. 2. The records demonstrate a significant reduction of the shock wave amplitude and duration caused by the foam block. TNT equivalency of the explosion in terms of pressure amplitude was calculated based on pressure records, using the Sadovsky formula modified for granulated TNT [12], and on the times of shock wave arrival at the sensors, according to the method described in [15]. The coefficient of ground explosion energy absorption was set at 0.7. The results are shown in Table 1. The two methods of TNT equivalency assessment yield similar results. Slightly greater TNT equivalency value at shorter distances from the charge in the test with no foam is due to the cylindrical, rather than semispherical shape of the charge. The behavior of recorded signals changes qualitatively in the test with foam. The TNT equivalency values assessed based on the pressure records is much lower in the vicinity of the foam block whereas its values calculated using measured arrival times behave oppositely. It is clear that shock wave is damped by the foam quite efficiently, which follows from the low TNT equivalency value at the position of the first pressure gauge. The shock wave pulse duration at this distance is slightly increased. However the detonation products partially "locked" within the block accelerate the foam layers and increase the shock wave amplitude. At later stages, the foam block fragments move obviously at a velocity which is close to that of the gas, therefore the TNT equivalency remains constant. TNT equivalency as determined based on arrival times is controlled by the shape of the shock front, which is certainly oblique because the charge is cylindrical, break through of the gas at the foam-metal plate interface can not be excluded either. These factors are able to increase the velocity of the leading shock wave edge affecting only slightly the overpressure value. At larger distances, all the transient processes (including oblique shock wave and break through of the detonation products along the metal plate) fade out and the TNT equivalency values as determined by the two methods are identical and remain constant. It is clear that assessment of the blast effect in a zone close to the charge employing the Sadovsky formula and pressure amplitudes yields more correct results.

Table 1. Results of measurements of the shock wave parameters.

Charge type	*R*, m	*T*, ms	*P*, bar	TNTp, kg	TNTt, kg
NO FOAM	1.053	0.65	9.8	0.58	0.46
	1.585	1.42	2.7	0.42	0.5
	2.096	2.5	1.4	0.4	0.44
	2.766	4.08	0.79	0.4	0.42
FOAM	1.053	1.7	0.6	0.015	0.085
	1.585	3.04	0.56	0.045	0.075
	2.096	4.42	0.36	0.05	0.05
	2.766	6.28	0.23	0.05	0.05

3. Discussion

To compare efficiency of explosion localizer based on rigid polyurethane foam with other types of explosion localization devices and methods we plotted the overpressure versus reduced distance ($R'=R/m^{1/3}$, R – distance from the charge in meters, m – charge energy expressed in kg of TNT) .for various cases of the blast wave attenuation (see Fig. 3). Unfortunately the graph provides incomplete information about efficiency of localizers because it demonstrates only its ability to reduce the shock wave overpressure not related to the amount of damping material It is noteworthy that all the methods of blast wave attenuation yield the curves that lie below the Sadovsky curve for TNT explosion with no damping devices within the entire range of reduced distances. First of all, two types of damping devices are to be distinguished. Damping devices of the first type, like water curtains, are inevitably arranged at some distance from a charge where the wave amplitude is fairly low. They reduce the wave amplitude by a factor of two and lower. The localizers surrounding charges are much more efficient, particularly at high wave amplitudes. Light water foams coating charges reduce the high wave amplitude quite efficiently; the overpressure drops steeply as the distance increases. However this is certainly true within the foam block, that is, to attenuate the shock wave to a harmless level one needs foam blocks of very large volumes. The shock wave that escapes the block would decay slower than the blast wave spreading in air with no foam, because the water mass accelerated by the shock wave in the block would support it as a piston. Plastic bags filled with water, sand, and CaCl that cover a charge reduce significantly the shock wave overpressure in the range of few bars at reduced radii not exceeding 0.5. Although no data on the behavior of blast waves beyond these distances are available the curves suggest that the "blankets" are much more efficient than water curtains. The common feature of the above discussed localizers is that they absorb the major fraction of the kinetic energy and momentum of gas mechanically and the contribution of heat transfer from the gas to the condensed phase is insignificant when the filling agent is powder. This contribution is appreciable when the "blanket" is filled with water because water is brought up by the shock wave into fine droplets that evaporate, therefore this localizer is the most efficient among the "blanket"-type ones.

Unfortunately there is no information about how the expanding dense mass of the "blankets" can affect the surrounding objects due to its kinetic energy and how its further motion can influence the blast wave parameters. The data for the stratified foam lie within the same group, although the damper mass-to-charge mass in this case is lower (see below). The data however show some promising trend, namely, the damper efficiency tends to be better compared to that for the "blanket" devices both at shorter (the curve is less steep) and longer (the curve is steeper) distances from the charge. A possible explanation of this trend is higher energy losses at high shock wave amplitudes and lighter fragments of the foam material less supporting the wave at longer distances.

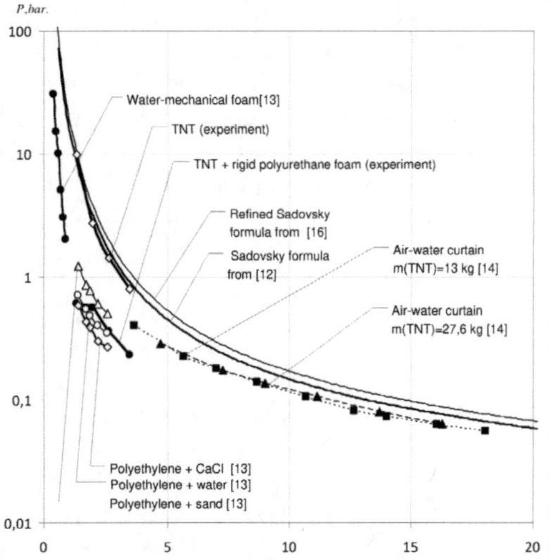

Figure 3. Comparison of the present pressure vs reduced distance data with data available in literature on tests in which various techniques of shock wave attenuation have been employed.

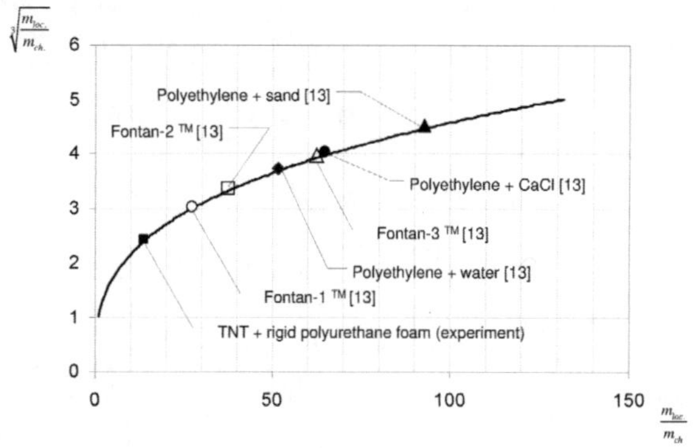

Figure 4. Comparison of attenuation coefficients (data partially taken from [13]).

The range of overpressures to which the shock wave is to be attenuated in practice is fairly wide. Depending on the situation their level is dictated by necessity of protecting surrounding equipment or buildings. But most frequently it should not exceed lethality or injury threshold for people or animals. In any case the blast wave overpressure should be below few tens of bars or even below 1 bar. It is well known that overpressure is not the only wave parameter that defines the damage and lethality thresholds the other important characteristic is wave impulse, therefore as far as the safety problems are concerned, these two parameters are to be considered together. Results of our experiment show that the foam block reduces significantly the wave impulse (determined as an integral of pressure with respect to time) as well, however by a factor of 2 or 1.5 less than it decreases the overpressure. As expected, acceleration of the condensed foam phase by the gas results in formation of a kind of a gas-permeable piston that partially supports the blast wave and makes it longer. Therefore the wave impulse drops with distance from the charge slower than does overpressure. The foam increased the duration of the positive pressure pulse phase by a factor of about nearby the chare and about 1.5 at longer distances.

The damper mass-to-charge mass ratio which is the most important characteristic that defines the damper efficiency, can conveniently be presented as suggested in [13] by the explosion damper parameter (the cubic root of this ratio). As Fig. 4 shows, the stratified polyurethane foam device proposed in this work is superior to all the localizers information about which we have been able to find in literature. They include also the data on shock wave attenuation with the aid of standard polymer foam blocks piled up around a charge (Fontan)[13].

4. Conclusion

As a summary we can state that an efficient method for localization of ground explosions in vicinity of a charge is proposed. The method is based on application of stratified polyurethane foam blocks with an optimized structure A distinctive feature of the method is fast in situe fabrication of the block and mobility of the equipment, possibility of remote control of fabrication. The optimal mechanism of interaction between the damping material expanding detonation products is discussed. Based on this mechanism, a possible construction of the foam blocks with differentiated porosity, composition and strength is suggested. The block should consist of several layers of foam with different characteristics: the inner one must be mechanically strong foam with coarse open pores and of high density in order to induce highly perturbed flow of the detonation products.

The middle layer, of lower density and saturated with water serves as the main energy and momentum absorbing (damping) foam block unit. The function of the outer layer, made of light fine-cell polymer foam, is to decelerate fragments of the inner layers and reduce their number. In the comparative test performed, the TNT equivalency of the explosion was reduced by the block by a factor of 5 to 10 depending on distance from the charge. The proposed explosion localization method is shown to be superior to the known damping techniques in virtually all parameters.

References

[1] Raspet R, Griffits S, Powers J, Krier H, Panczak T, Butler P, Jahani F 1988 Attenuation of blast waves using foam and other materials. *USA-CERL technical manuscript N-89/01*. US Army Corps of Engineers.
[2] Raspet R, Butler P B, Jahani F 1987 *Appl. Acoustics* **22** 243.
[3] Borisov A A , Gelfand B E et al 1978 *Acta Astronautica* **5** 1027.
[4] Gelfand B E, Gubanov A V, Timofeev E I 1981 *Fiz. Goreniya i Vzryva* **17** 129 (in Russian).
[5] Gelfand B E, Gubanov A V, Kogarko S M , Popov O E 1975 *Prikl. Matem. Tekhn. Fiz* **6** 74.
[6] Gelfand B E, Gubanov A V, Timofeev E I 1993 *Mekh. Zhidk. Gas.* **4** 79 (in Russian)
[7] URL: http://www.akvidur.ru
[8] *Handbook of polymeric foams and foam technology* 2004 Ed. Klempner D, Sendijarevic V (Munich: Hanser publishers)
[9] Kitagawa K, Takayama K, Yasuhara M 2006 *Shock Waves* **15** 437
[10] Woodfin R L 2000 Using rigid polyurethane foams (RPF) for explosive blast energy absorption in applications such as anti-terriorist defenses *Sandia report* SAND2000-0958
[11] Zhu F, Chou C, Yang K 2011 *Composites B* **42** 1202
[12] Sadovsky M A 1952 *Fizika Vzryva* (Moscow: Acad. Sci. Publisher) **1**
[13] Gelfand B E, Silnikov M V 2006 *Barometric effect of explosions* (St. Petersburg: Asterion) p 551 (in Russian)
[14] Buzukov A A 2000 *Fiz. Goren. Vzryv* **36** 120 (in Russian).
[15] Borisov A A, Komissarov P V, Sokolov G N, Kaplyukov G V 2010 *Gorenie I Vzryv* **3** 161 (in Russian).
[16] *Explosion Physics* 2002 Ed. L.P. Orlenko **3** vol 1 (Moscow: Fizmatlit) Publ. 133

Journal of Physics: Conference Series **751** (2016) 012021 doi:10.1088/1742-6596/751/1/012021

An approximate method for calculating composition of the non-equilibrium explosion products of hydrocarbons and oxygen

V A Shargatov, S A Gubin and D Yu Okunev

National Research Nuclear University MEPhI (Moscow Engineering Physics Institute), Kashirskoe Shosse 31, Moscow 115409, Russia

E-mail: shargatov@mail.ru

Abstract. We develop a method for calculating the changes in composition of the explosion products in the case where the complete chemical equilibrium is absent but the bimolecular reactions are in quasi-equilibrium with the exception bimolecular reactions with one of the components of the mixture. We investigate the possibility of using the method of "quasi-equilibrium" for mixtures of hydrocarbons and oxygen. The method is based on the assumption of the existence of the partial chemical equilibrium in the explosion products. Without significant loss of accuracy to the solution of stiff differential equations detailed kinetic mechanism can be replaced by one or two differential equation and a system of algebraic equations. This method is always consistent with the detailed mechanism and can be used separately or in conjunction with the solution of a stiff system for chemically non-equilibrium mixtures replacing it when bimolecular reactions are near to equilibrium.

1. Introduction

The chemical kinetics models are important for analysis and design in many chemistry and industrial processes. The chemical reactions that occur in the explosion products change their composition and, consequently, thermodynamic parameters. During cooling, expansion and mixing with air the internal energy, temperature and density are changed. The changed thermodynamic parameters correspond to a different chemical equilibrium composition of the mixture. The time evolution of the composition is calculated using a stiff system of chemical kinetics equations. This is the most general approach. At a large number of individual species in the explosion products and a large number of chemical reactions involved, calculation of the mixture composition requires a long computation time. Typically, the time it takes for calculating the mixture composition is much longer than that spent on solving the gasdynamics equations.

The applicability of the chemical equilibrium mixture model to the products of explosion of fuel–oxygen and fuel–air mixtures was discussed in detail in [1–6]. The applicability of the chemical equilibrium model to hydrogen–oxygen mixtures in the case of heat supply at constant density was assessed in [1]. In [2, 5], a sufficient condition for the applicability of the chemical equilibrium mixture model to gasdynamic calculations was obtained. The authors of [6] compared the results of gasdynamic calculations performed using the system of equations of chemical kinetics for calculating the composition of the explosion products with the results of gasdynamic calculations within the framework of the chemical equilibrium mixture model and the constant composition mixture model.

It was demonstrated [1, 2, 4, 5] that the process of establishment of chemical equilibrium in the products of explosion of gas mixtures proceeds in several stages. First, partial chemical equilibrium is established, without changes in the total number of molecules in the system. That is, even in cases where there is no chemical equilibrium, it suffices to know the molar mass of the mixture and two required thermodynamic parameters (for example, specific internal and density) to determine the composition of the mixture.

The existence of quasi-equilibrium in exchange reactions was discussed in [6]. This same assumption underlies an approximate model for calculating the molar mass of gas mixtures at high temperatures [7]. The author of [7, 8] proposed an approximate kinetic equation for calculating the molar mass of the mixture and then the thermodynamic parameters.

In [9–11] we made an effort to speed up directly chemical integrator for some special case. In [9–11] we developed a method for calculating the changing composition of the explosion products in the case where there is no complete chemical equilibrium, but the bimolecular reactions are in equilibrium. A detailed kinetic mechanism is used, but without the need to solve the stiff system of equations. The core of our approach is to replace the complete system of chemical kinetics equations by an equivalent system composed of a differential and a few algebraic equations. The proposed method is based on a physically reasonable assumption that equilibrium in the bimolecular reactions is established much faster than the complete chemical equilibrium, being applicable when this assumption holds. According to [9–11] the method includes only one differential equation for calculating the change in the number of molecules per unit mass of the mixture due to the reactions of recombination and dissociation. This equation is used instead of dozens of equations of chemical kinetics. In the present work, we propose to supplement this equation with an algebraic equations for calculating the equilibrium composition at given values of the density, internal energy, molar mass and the linear combinations of logarithms of the concentration. The equations are derived based on the characteristic function (entropy or Helmholtz free energy) extremum method.

The applicability of the model was demonstrated by solving the test problems in [9 –11]. It was shown that the proposed model can be used to calculate the characteristics of the explosive transformation process after the induction period in hydrogen-oxygen mixtures.

The aim of the work described here was to develop a method for calculating the changes in composition of the explosion products in the case where the complete chemical equilibrium is absent but the bimolecular reactions are in quasi-equilibrium with the exception bimolecular reactions with one of the components of the mixture. In addition we have investigated the possibility of using the method of "quasi-equilibrium" for mixtures of hydrocarbons and oxygen.

2. Equation for calculating the partial chemical equilibrium composition of the mixture

According to [9] we obtain equations to calculate the mixture composition assuming that the entropy of the mixture reaches its maximum value at a given temperature, density, number of molecules per unit volume and the linear combinations of logarithms of the concentration. The entropy of a mixture of ideal gases with known dependence of the heat capacity on the temperature is given by

$$S=\sum_{i=1}^{k}\left\{S_i-RT\ln\left[RT\,M_i/(v\,P_0)\right]\right\}M_i$$

The total amount of each element j, internal specific energy and the number of molecules in the system are known

$$\sum_{i=1}^{k}\alpha_{ij}M_i-\beta_j=0,\ \ (j=1,..,m),$$

$$U_0-\sum_{i=1}^{k}U_i\,M_i\ =\ 0,$$

$$\sum_{i=1}^{k} M_i - M_0 = 0, \tag{1}$$

where k is number of individual compounds in the mixture; M_i is the number of moles of the i-th substance per 1 kg of mixture, R is the universal gas constant, T is the temperature of the mixture, v is the specific internal volume of the mixture, S_i is the standard entropy, U_i is the internal energy of one mole of mixture component i at a given temperature T and $P_0 = 10^5$ Pa, $M_0 = c/\rho$ and c is the total number of moles of all mixture components per unit volume, m is the number of different chemical elements in the mixture, β_j is the number of kilogram-atoms of the j -th element in 1 kg of mixture, α_{ij} is the number of atoms of the j -th element in a molecule of the i -th substance.

If A+B $\leftrightarrow$ C+D is the nonequilibrium bimolecular reaction we can use additional condition on the concentrations

$$\ln(M_A) + \ln(M_B) - \ln(M_C) - \ln(M_D) - \ln(W_K) = 0, \tag{2}$$

where W_k is the additional independent variable and is calculated by solving differential equation. K is the index of the reaction.

If all reactions with the specie A are nonequilibrium the differential equation for the concentrations of A is used. The additional condition on concentrations is

$$M_A = W_A. \tag{3}$$

Following the Lagrange procedure, we write the function

$$L = \sum_{i=1}^{k}\{S_i - R\ln[R\,T\,M_i/(v\,P_0)]\}\,M_i + \sum_{j=1}^{m}(\sum_{i=1}^{k}\alpha_{ij}\,M_i - \beta_j)\lambda_j$$
$$+ (U_0 - \sum_{i=1}^{k}U_i\,M_i)\lambda_u + \sum_{i=1}^{k}(M_i - M_0)\,\lambda_\mu + G\,\lambda_R\,,$$

where λ_j , λ_μ , λ_u and λ_R are the Lagrange multipliers, which, along with Mi, are the independent variables of the Lagrangian. If the additional condition (2) is used

$$G = \ln(M_A) + \ln(M_B) - \ln(M_C) - \ln(M_D) - \ln(W_K),$$

and G is found from

$$G = M_A - W_A$$

for additional condition (3).

At equilibrium, the derivatives of the Lagrangian with respect to the independent variables must be equal to zero. This condition allows us to obtain $k + m + 3$ algebraic equations (1-3) for $k + m + 3$ unknowns

$$S_i - R\ln[R\,T\,M_i/(v\,P_0)] + \sum_{j=1}^{m}\alpha_{ij}\lambda_j - H_i/T + \lambda_\mu + (\partial G/\partial M_i)\,\lambda_R = 0,\ (i=1,\dots k) \tag{4}$$

$$\sum_{i=1}^{k}\alpha_{ij}M_i - \beta_j = 0,\ (j=1,..,m) \tag{5}$$

$$U_0 - \sum_{i=1}^{k}U_i\,M_i = 0, \tag{6}$$

$$\sum_{i=1}^{k} M_i - M_0 = 0, \tag{7}$$

$$G(M_1,...,M_k,W)=0 \tag{8}$$

This system of algebraic equations has no analytical solution, so it is solved approximately by the Newton method.

When the chemical kinetics equations are applied to calculating the composition of the mixture, the following a system of equations is used

$$\frac{\partial c_i}{\partial t}=\omega_i,\ i=1,...,k. \tag{9}$$

$$\sum_{i=1}^{N} v'_{ij}\chi_i \rightleftarrows \sum_{i=1}^{N} v''_{ij}\chi_i,\ j=1,...,J,$$

$$\omega_i=\sum_{j=1}^{J} k_j(v''_{ij}-v'_{ij}),$$

$$\frac{\partial c}{\partial t}=\sum_{i=1}^{k}\omega_i, \tag{10}$$

$$\frac{\partial \ln(W_K)}{\partial t}=\frac{\omega_A}{c_A}+\frac{\omega_B}{c_B}-\frac{\omega_C}{c_C}-\frac{\omega_D}{c_D}, \tag{11}$$

$$\frac{\partial c_A}{\partial t}=\omega_A \tag{12}$$

where ω_i is the rate of production of species i per unit volume; c_i are the molar concentrations of the mixture components; c is the rate of production of the total number of moles of all species per unit volume, k_j is the rate of reaction j calculated according to the law of mass action, χ_i is the chemical formula of species i participating in the reaction j ; v''_{ij} and v'_{ij} are the stoichiometric coefficients of the product or reactant i in reaction j, J is the number of reactions in the kinetic scheme. W_k is the parameter of nonequilibrium of selected reaction, $W_A=c_A/\rho$.

3. Results of calculations

The proposed approximate method for solving stiff systems of kinetic equations was applied to a test problem. Consider the chemical equilibrium explosion products of stoichiometric acetylene–oxygen. The composition of the mixture given in Table 1 correspond to the so called instantaneous explosion of the initial mixture. The composition is calculated for an initial mixture at 0.1 MPa and 298 K.

Table 1. Equilibrium products composition (mole fraction) at T=3985 K and ρ=1.24 kg/m^3.

O_2	H_2	OH	H_2O	CO	H	O	CO_2	HO_2
0.120	0.0364	0.109	0.0963	0.343	0.0712	0.120	0.102	0.000216

We reduce the temperature of the mixture to 2800 K at constant volume and constant composition. The mixture relaxes to the new state of chemical equilibrium at constant density and constant specific internal energy. It is assumed that there are no expansion and heat loss. The calculations were performed by numerical solving the system of differential equations of chemical kinetics, corresponding to the selected detailed mechanism (blue lines in figures 1 – figures 4).

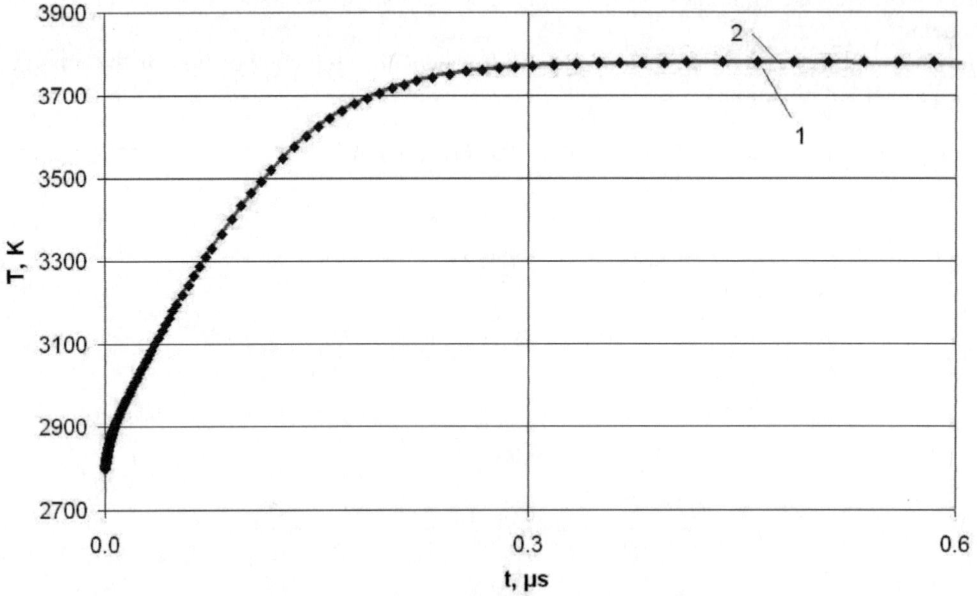

Figure 1. Temperature vs time: 1—exact method including 10 differential equations, 2—proposed method with one differential equation (6) for *c*.

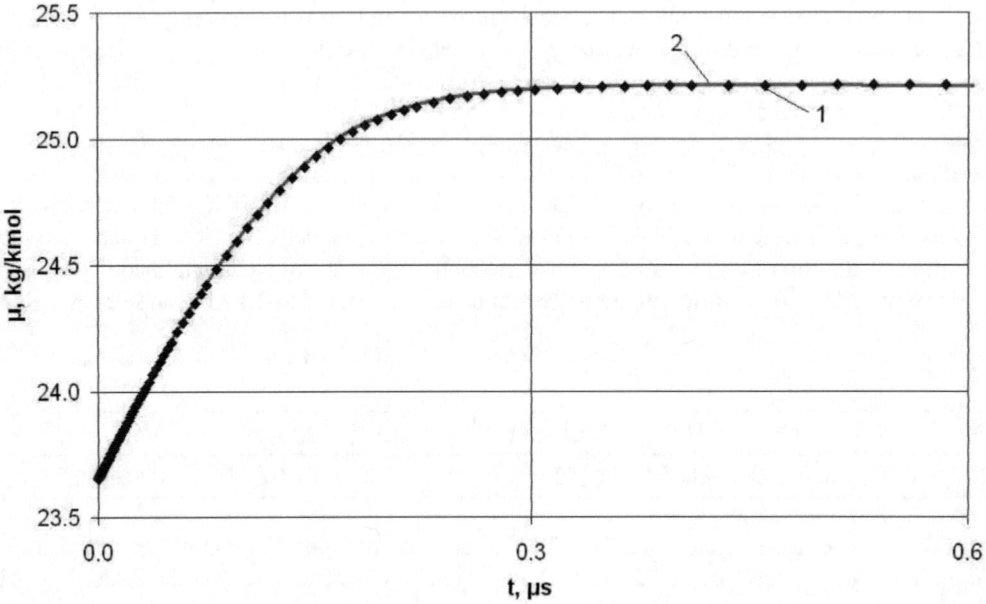

Figure 2. Molar mass of the explosion products vs time: 1—exact method including 10 differential equations, 2—proposed method with one differential equation (6) for *c*.

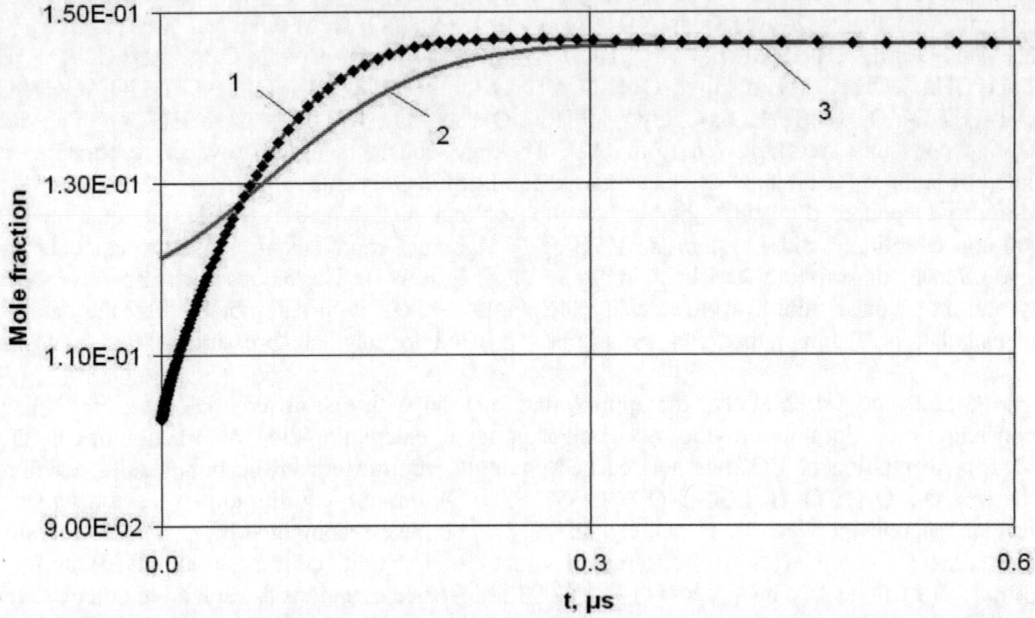

Figure 3. Molar mass CO_2 vs time: 1—exact method including 10 differential equations, 2—proposed method with one (6) differential equation for *c,* 3 -proposed method with two differential equations (6) for *c* and (8) W_{CO2}.

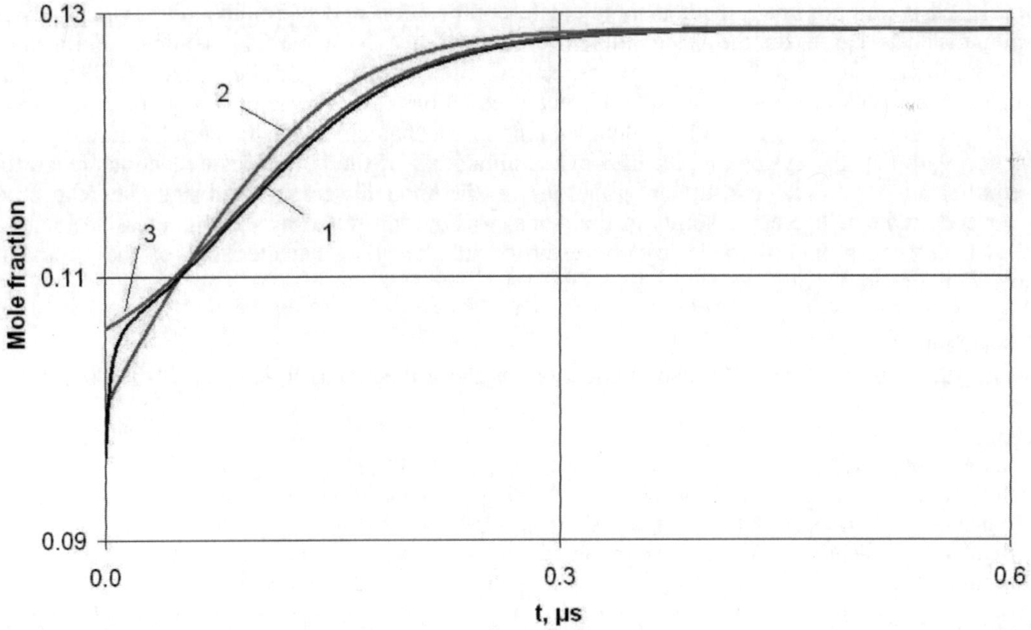

Figure 4. Molar mass of H_2O products vs time: 1—exact method including 10 differential equations, 2—proposed method with one (6) differential equation for *c,* 3 -proposed method with two differential equations (6) for *c* and (8) W_{CO2}.

It was necessary to solve a stiff system of ten differential equations (the system (8) of 9 equations for the concentrations and one equation for the temperature). The calculation was performed until the state of the mixture was close to the equilibrium. The species that are taken into account in the explosion products are CO, CO_2, H, H_2O, HO_2, H_2, O, OH and O_2. We use the detailed kinetic mechanism including 18 reactions: O+CO+M=CO2+M, O+H+M=OH+M, O+O+M=O_2+M, H+OH+M=H_2O+M, H+H+M=H_2+M, H+O_2+M=HO_2+M, O_2+CO=O+CO_2, H+O2=O+OH, O+H2=H+OH, OH+CO=H+CO_2, O+H_2O=OH+OH, H+H_2O=OH+H_2, OH+HO_2=O_2+H_2O, OH+OH=H+HO_2, O_2+H_2=H+HO_2, O+H_2O=H+HO_2, OH+O_2=O+HO_2, HO_2+CO=OH+CO_2. The data on the rate constants are borrowed from [12]. The rate constants of the reverse reactions were calculated from those for direct reactions using the equilibrium constants

In addition, we conducted calculations by the proposed methods. A single differential equation (10) for total concentration c and a system (4) – (7) of 14 algebraic equations are solved to calculate the time evolution of the mixture (red lines in figures 1 – figures 4). The results for temperature-time history and molar mass-time histories are in excellent agreement with data obtained by the detailed kinetic calculations. Temperature and molar mass is used to calculate pressure of the explosion products.

Figure 3 and Figure 4 show that the approximate method with one differential equation (10) for total concentration c does not provide very good accuracy in calculating the mole fraction of CO_2 and H_2O. The concentration of CO_2 has not reached complete equilibrium in the bimolecular reactions O_2+CO=O+CO_2, OH+CO=H+CO_2, HO_2+CO=OH+CO_2. A partial equilibrium is established for bimolecular reactions between H, H_2, O, O_2, OH, H_2O. We have recomputed the test problem using the approximate method with two differential equations for total concentration c (10) and for concentration of CO_2 (12). The system (4) – (8), of 15 algebraic equations are solved to calculate the quasi-equilibrium composition and temperature. The calculation of CO_2 and H_2O mole fractions - time histories based on the approximate method with two differential equation improves the agreement with exact solution (curves 3 in figure 3 and in figure 4).

4. Conclusions

We developed the approximate method to calculate composition and thermodynamic parameters of explosion products based on the assumption of the existence of a partial chemical equilibrium. Without significant loss in accuracy, the respective stiff system of detailed kinetics differential equations can be replaced by one or two differential equations and a system of algebraic equations. This method is always consistent with the detailed kinetic mechanism and can be used separately or in conjunction with the stiff system, replacing it when almost all of the bimolecular reactions reach the quasi-equilibrium state. The constituent equations of the model were derived and the respective computer code written. The applicability of the model was demonstrated by solving a test problem. It was shown that the proposed model can be used to calculate the characteristics of the explosion products of hydrocarbons and oxygen.

Acknowledgments

This work was supported by the Russian Foundation for Basic Research, project No. 14-03-00420.

References

[1] Nikolaev Y A and Fomin P A 1982 *Fiz. Goreniya Vzryva* **18** 66-72
[2] Borisov A A, Gubin S A and Shargatov V A 1991 *Prog. Astronaut. Aeronaut.* **134** 138–53
[3] Wintenberger E, Cooper M, Pintgen F and Shepherd J E 2004 *J. Propuls. Power* **20** 957-59
[4] Bryakina U F, Gubin S A, Tereza A M and Shargatov V A 2010 *Russ. J. Phys. Chem. B* **4** 969–76
[5] Bryakina U F, Gubina T V and Shargatov V A 2011 *Russ. J. Phys. Chem. B* **5** 482–90
[6] Bryakina U F, Gubin S A, Shargatov V A and Lyubimov A V 2012 *Russ. J. Phys. Chem. B* **6** 261–70
[7] Nikolaev Y A 2001 *Combust. Explos.* **37** 4–12
[8] Nikolaev Y A 1978 *Combust. Explos.* **14** 468–71
[9] Gubin S A and Shargatov V A 2013 *Russ. J. Phys. Chem. B* **7** 154–60
[10] Shargatov V A, Gubin S A and Okunev D Y 2015 *J. Phys.: Conf. Ser.* **653** 012054 1-8
[11] Shargatov V A, Gubin S A and Okunev D Y 2015 *Phys. Proc.* **72** 342–6
[12] Dautov N G and Starik A M 1997 *Kinet. Catal.* **38** 185–208

Modelling of non-equilibrium flow in the branched pipeline systems

S I Sumskoi[1], A M Sverchkov[2], M V Lisanov[2] and A F Egorov[3]

[1]National Research Nuclear University MEPhI (Moscow engineering physics institute), Kashirskoye shosse 31, Moscow 115409, Russia
[2]CJSC Scientific technical center of industrial safety problems research, Block 14, Perevedenovsky Pereulok 13, Moscow 105082, Russia
[3]D. Mendeleyev University of Chemical Technology of Russia

E-mail: sumskoi@mail.ru

Abstract. This article presents a mathematical model and a numerical method for solving the task of water hammer in the branched pipeline system. The task is considered in the one-dimensional non-stationary formulation taking into account the realities such as the change in the diameter of the pipeline and its branches. By comparison with the existing analytic solution it has been shown that the proposed method possesses good accuracy. With the help of the developed model and numerical method the task has been solved concerning the transmission of the compression waves complex in the branching pipeline system when several shut down valves operate. It should be noted that the offered model and method may be easily introduced to a number of other tasks, for example, to describe the flow of blood in the vessels.

1. Introduction

When the valve actuates in the pipeline, as well as when the mode of operation of the transfer pump stops or changes the slowing down or complete stop of the medium movement in the pipeline takes place.In its turn, the breaking of the medium before the closing of valves or in front of the pump leads to the increase in pressure, the formation of the area with higher pressure in the pipeline and the subsequent spread of the area with higher pressure along the pipeline route. This propagation of waves with increasing pressure in pipes is called water hammer [1].

The water hammer is a significant threat to the integrity of the pipelines, as the growth of pressure can lead to rupture of the pipe or the formation of dangerous defects [2].

Thus, the water hammer has served as one of the causes of the accident with rupture of the pipeline at the tank farm "Sheskharis" (Novorossiysk) and pollution of the portaquatorium [3].

The danger of water hammer in pipeline systems is aggravated by the fact that the one-dimensional flows are usually realised in the pipes. Since there is a one-dimensional geometry, the pressure waves, which are propagating through the pipeline, attenuate rather slowly, and this leads to the creation of areas of potential danger for tens of kilometres.

The research of the water hammer has been carried out for decades, starting up with the work of Zhukovsky N.E.[1] and ending with modern monographs and scientific theses [2, 4-7]. It is common to consider water hammer in single line pipeline system. However, the real pipeline systems often have a more complex topology, which includes branching and looping. Branching structure of the pipeline may significantly impact on the development of water hammer.

The setting of the problem about the water hammer in a branching system is extremely essential from the practical point of view. Indeed, the effects of water hammer are the most dangerous in cases of rapid response of the valves. Such rapid (3-5 seconds) actuation of valves and complete shutoff of the flow often occur during shipment at the sea terminals. It is important to quickly cut off the flow, since emissions can get into the aquatic environment, which is fraught with huge ecological damage. The shipment of the large volumes (and at high velocity) often is carried out through several pipes in order to reduce the total time of shipment. As a result, when loading the tankers the occurrence of water hammer is possible in the branched system.

The purpose of this paper is to describe the models and numerical methods designed for modelling the water hammer in a branched pipeline system, as well as to illustrate the opportunities of the formulated approach.

2. Mathematical model

In the isothermal approximation the motion of the fluid in the tube of variable cross section is described by the system of following equations for one-dimensional non-stationary flow:

- continuity equation

$$\frac{\partial(A\cdot\rho)}{\partial t}+\frac{\partial(\rho\cdot A\cdot u)}{\partial x}=0, \tag{1}$$

- equation of momentum conservation

$$\frac{\partial(A\cdot\rho\cdot u)}{\partial t}+\frac{\partial(A\cdot\rho\cdot u^2)}{\partial x}=-A\cdot\frac{\partial p}{\partial x}-A\cdot\frac{\lambda(\mathrm{Re})}{2\cdot D}\cdot\rho\cdot u\cdot|u|-A\cdot g\cdot\rho\cdot\beta, \tag{2}$$

- relation between the pressure and density (equation of state)

$$p-p_0=c^2(\rho-\rho_0) \tag{3}$$

here ρ, p, u - density, pressure and velocity of the fluid, averaged according to the cross section; t - time; x - the distance from the start of the pipeline; $\lambda(Re)$ - the coefficient of friction as a function of the Reynolds number $Re = Du/v$; A - cross sectional area of the pipeline, D- the diameter of the pipeline, g - acceleration of gravity power; β - sine of the angle of inclination of the route, which is determined by the elevations points of the pipeline $h(x)$, v - kinematic viscosity; ρ_0 - density of the fluid at pressure p_0 and transportation temperature (usually p_0 = 10325 Pa), c - the speed of sound in the fluid.

The considered model takes into account the following factors:

- non-stationary processes;
- variable cross-section of the pipeline;
- convective motion of the medium (the second temss are in the left parts (1) and (2);
- the appearance and circulation of the waves at a stop/start up of the pumps, closing valves (first term in the right part (2);
- the presence of friction on the pipe wall (the second term is in the right part (2);
- effect on the flow of gravity when the pipeline's route goes along the complex terrain (the third term is in the right part (2).

The equations (1) - (3) are closed by initial and boundary conditions. As the boundary conditions the pressure is set on the inlet and outlet of the pipeline, this pressure corresponds to the pressure of the pumps or vessels at the ends of the pipeline. As an initial data the parameters are set for the stationary pumping, which can be obtained analytically from the solution of the system (1) - (3).

When the flow is shutoff by the valve the zero speed is set as a boundary condition.

The system (1) - (2) describes the motion in one linear sector. In the case of a branched pipeline system when the individual linear sectors are combined in a specific sequence, the system of equations (1) - (2) is recorded for each linear sector. The boundary conditions at the connection of two linear sectors are set in such a way that the flows of mass and impulses could be saved from one sector to another.

In this paper to determine the $\lambda(Re)$ the dependence of Colebrook - White is applied [8], linking friction coefficient λ with Reynolds number *Re* and the characteristics of the pipeline (diameter *D* and roughness *k*)

$$\frac{1}{\sqrt{\lambda(\mathrm{Re})}} = -2\lg\left[\frac{2{,}51}{\mathrm{Re}\sqrt{\lambda(\mathrm{Re})}} + \frac{k}{3{,}71\cdot D}\right] \tag{4}$$

here *k* – the size of the pipeline roughness.

Thus, the system of equations (1) - (4) allows a fairly full account of the real characteristics of theequipment functioning in the oil trunk pipeline: pipes, pumps, valves, etc .

3. Numerical Techniques

As a rule, when modelling the flows in pipelines the method of characteristics is applied [1], which is based on the assumption of constancy of the disturbance velocity of propagation in the pipeline. However, where this is no longer a constant speed, the use of the method of characteristics is problematic. In addition, under the common standard implementation the method of characteristics has extremely low accuracy - first-order accuracy. There are some works [8], in which the finite-difference methods are used to calculate the flows in pipelines. In these papers the derivatives in differential equations are replaced by difference analogues. The arithmetic expressions, obtained in such a way, can be more or less easily solved. These methods are more versatile compared with the method of characteristics. However, the difficulty with their practical implementation is that not all properties of the original differential equations are kept in the used arithmetic expression. This is why one needs to comply with complex procedures for providing even a rough match of numerical solutions with the exact ones.

The Godunov type method [9] has a higher versatility in comparison with the method of characteristics. It allows transferring the properties of exact solutions into numerical solutions by a natural way without major complications. Until recently, the major drawback of this method was the relatively long time needed for making calculations. But modern algorithmic and hardware/software solutions have made this method more comprehensible. As for now it is possible to confirm that the calculation time for tasks using the Godunov type method is not a limiting factor, as a rule.

The Godunov type method is offered in [10, 11] for solving the system of equations (1) - (4). The Godunov type method belongs to a class of integral difference methods, in which the main element is the calculation of fluxes (mass, impulse and energy) across the borders of certain elementary volume (difference cells). Summing up all flows along the borders for a certain period of time the values of the corresponding amounts after this interval of time are determined. Actually these fluxes are usually calculated on the basis of the solution of the Riemann problem for the decay of a discontinuity at the boundary of difference cells.

In practice Godunov type method is implemented as follows. The linear sector of the pipeline is divided into cells of size Δx, in which the state at any moment of time $n\Delta t$ (pressure p_i^n, density ρ_i^n, fluid velocity u_i^n, velocity of sound c_i^n,and velocity of perturbations transmission c_{wi}^n) was assumed to be constant in the entire volume of the cell.

One essential point is to be specified in this situation: the point, which is not always clearly perceived. Generally speaking, the flow in the pipeline, even considered in the one-dimensional formulation, is substantially determined by a combination of multi-dimensional effects. Such factors as the operation of the valves, change of the diameter pipelines, branching, and etc. make the dimensiality obvious.At that gaps may even occur in the one-dimensional flow, which are the consequence of the real multi-dimensional effects. However, even simple propagation of the compression wave in the pipeline is accompanied by certain two-dimensional effects. As far back N.E.

Zhukovsky [1] showed that due to the change of the diameter of the pipe because of the pressure change in the pipe, the velocity of disturbance propagation in the system "liquid-thin-walled tube" (c_w) changes, and does not coincide with the velocity of disturbance propagation in an infinite volume of fluid (the velocity of sound c). According to Zhukovsky formula [1, 5] c_w the velocity of disturbance propagation along the pipeline is determined by the following equation:

$$c_w = \left(\frac{\rho}{K} + \frac{\rho D}{E\delta} \right)^{-0.5} \tag{5}$$

here δ - pipe wall thickness, E - Young's module of the pipe material, K - the elastic module of the liquid medium, transported through the pipeline.

Therefore, it is the speed of the waves, which is calculated by the formula (5), is applied in the numerical method used when solving the Riemann problem. It is clear that the use of (5) corresponds to the solution of the Riemann problem in the acoustic approximation, which is quite acceptable for the flows of the liquid medium within the pressure of several tens of atmospheres.

As a matter of fact the numerical method for solving problems consists of two stages, built on the principle of splitting in physical processes. That is, the equations, obtained from the system (1) - (2), are solved consequently, and only some physical processes are described. Primarily the flow is modelled without taking into account the friction and the gravity, and at the stage II the action of friction and gravity is modelled.

Stage I. The systems of equations (1) - (3) are solved, in which there are no components that are responsible for the action of viscous forces and the forces of gravity. At this stage the flows of mass and impulses are directed in each i-cell, and after that there is a new state in the cell. The flows are calculated with the help of solving the problem of the decay of discontinuity. As the liquid, in specified circumstances, can be considered as little compressible, the solution of the problem of decay of a discontinuity at the boundary of two cells is allowed in a sound approximation. Writing down the equations for waves on the right and left borders, we'll get the following system for finding the parameters of cells on the border of the cells:

$$\begin{cases} p_r - p_{i+1}^n = \rho_{i+1}^n \cdot c_{wi+1}^n \cdot \left(u_r - u_{i+1}^n\right) \\ p_r - p_i^n = -\rho_i^n \cdot c_{wi}^n \cdot \left(u_r - u_i^n\right) \end{cases} \tag{6}$$

$$\begin{cases} p_l - p_i^n = \rho_i^n \cdot c_{wi}^n \cdot \left(u_l - u_i^n\right) \\ p_l - p_{i-1}^n = -\rho_{i-1}^n \cdot c_{wi-1}^n \cdot \left(u_l - u_{i-1}^n\right) \end{cases} \tag{7}$$

where the index "r" means belonging to the right parameters and the index "l" - to the left border of the i-cell.

The systems of equations (6) and (7) uniquely determine the flow parameters on the left and right boundaries of the i-cell of the n- time step:

$$\begin{cases} u_r = \dfrac{p_i^n - p_{i+1}^n + \rho_{i+1}^n \cdot c_{wi+1}^n \cdot u_{i+1}^n + \rho_i^n \cdot c_{wi}^n \cdot u_i^n}{\rho_{i+1}^n \cdot c_{wi+1}^n + \rho_i^n \cdot c_{wi}^n} \\ p_r = p_{i+1}^n + \rho_{i+1}^n \cdot c_{wi+1}^n \cdot \dfrac{p_i^n - p_{i+1}^n + \rho_i^n \cdot c_{wi}^n \cdot \left(u_{i+1}^n - u_i^n\right)}{\rho_{i+1}^n \cdot c_{wi+1}^n + \rho_i^n \cdot c_{wi}^n} \end{cases} \tag{8}$$

$$\begin{cases} u_l = \dfrac{p_{i-1} - p_i^n + \rho_i^n \cdot c_{wi}^n \cdot u_i^n + \rho_{i-1} \cdot c_{wi-1} \cdot u_{i-1}}{\rho_i^n \cdot c_{wi}^n + \rho_{i-1} \cdot c_{wi-1}} \\ p_l = p_i^n + \rho_i^n \cdot c_{wi}^n \cdot \dfrac{p_{i-1}^n - p_i^n + \rho_{i-1}^n \cdot c_{wi-1}^n \cdot \left(u_i^n - u_{i-1}^n\right)}{\rho_i^n \cdot c_{wi}^n + \rho_{i-1}^n \cdot c_{wi-1}^n} \end{cases} \tag{9}$$

Using the obtained according to (8) - (9) the pressure values p_l and p_r to (3) are values of densities ρ_l and ρ_r are calculated at the boundaries of the cell. Knowing p_l, ρ_l, u_l and p_r, ρ_r, u_r, one can obtain the total flows of mass Φ_{mi} and impulse Φ_{pi} in each cell:

$$\Phi_{mi} = u_l \cdot \rho_l - u_r \cdot \rho_r, \; \Phi_{pi} = u_l^2 \cdot \rho_l + p_l - u_r^2 \cdot \rho_r - p_r. \tag{10}$$

The final values of the parameters in i-cell at $n+1$ time step are calculated with the following formula:

$$\rho_i^{n+1} = \frac{m_i^{n+1}}{\Delta x} = \frac{m_i^n + \Phi_{mi} \cdot \Delta t}{\Delta x}, \; u_i^{n+1} = \frac{R_i^{n+1}}{m_i^{n+1}} = \frac{R_i^n + \Phi_{pi} \cdot \Delta t}{m_i^{n+1}}, \tag{11}$$

where m_i^{n+1} and m_i^n - masses, R_i^{n+1} and R_i^n - impulses in i-cell at $n+1$ and n time steps. The pressure in i-cell at $n+1$ time step is determined according to the correlation (3).

Stage II. The forces of friction and gravity are taken into account. The velocity in the i-cell is recalculated according to the formula

$$u_i^{n+1} = u_i^{n+1} - \lambda(\mathrm{Re}_i^{n+1}) \cdot \frac{\left|u_i^{n+1}\right| u_i^{n+1}}{2D} \cdot \Delta t - g\beta \cdot \Delta t \tag{12}$$

It should be noted that this article considers the isothermal flow of poorly compressible fluid, but the approach of the Godunov type method can be generalized to the cases of multiphase flows with heat exchange, as well as of more complicated equations of the medium state. The described option of the numerical calculation method is characterized by the first order of accuracy; however, the accuracy can be improved easily. Finally, the offered method is easily extended to non-regular grid.

4. Test calculations and comparison with exact solutions

The proposed method has been implemented as a computer program for a personal computer.

When using any numerical method, a natural question arises - what is the accuracy of the obtained decisions? One of the ways to assess the quality of the solution of any given numerical method is the comparison with the solutions of the problems that have analytical solution. To check the accuracy of the obtained numerical solutions and the applicability of this approach several problems have been solved that have the analytical solution:

- about stationary motion in a horizontal pipe with friction against the walls;
- about stationary motion in the pipe without wall friction but with altitude difference on the track;
- about filling the tube with liquid when one end is closed.

The numerical solutions received for these test problems coincide with a high degree of accuracy with the analytical solutions.

For branching systems there are few problems that have analytical solutions. One of such problem is described in [6]. In [6] the exact solution is presented for the flow near the branching when the wave of water hammer comes out from the pipe in the diverge of two pipes; in one of these pipes the medium moves, and in the other – does not. The friction against the pipe wall is not taken into account in this decision. The scheme of this problem according to [6] is shown in figure 1.

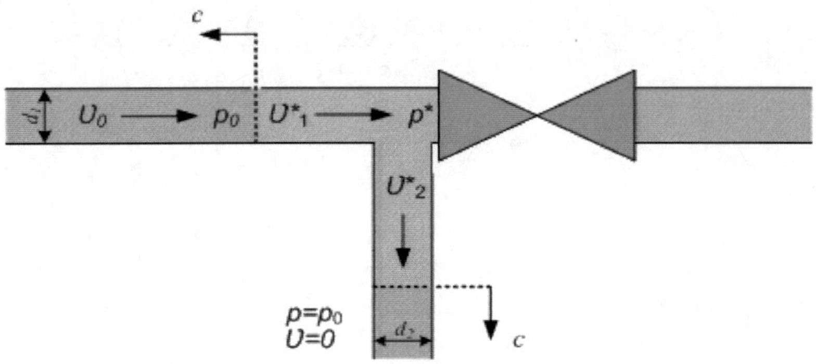

Figure 1. The scheme of water hammer in the branching system where in one of the pipes (bottom bend) the medium does not move initially.

The flow in the pipe with the velocity u_0 of the medium with the density ρ_0 is considered. The diameter of this pipe is d_1 (at Fig. 1 this pipe is represented horizontally). The pipe has a divergence with the diameter d_2 (Fig. 1 shows this pipe downward). The medium in the divergence is calm, its density is ρ_0. At some point in time right behind the divergence the valve operates, which immediately cuts off the flow. As a result, the flow on the valve stops; and two shock waves propagate along the main pipe and the divergence.

According to [6] two waves are formed propagating in the two pipes. Pressure p^* in the main pipe and the branching pipeline, which occurs under such setting of the problem, is described by the following equation for the pressure increase:

$$\Delta p = p^* - p_0 = \rho_0 u_0 c \frac{d_1^2}{d_1^2 + d_2^2} \tag{13}$$

The results of solving this problem, obtained numerically using the described above method, are shown in Table 1. Table 1 shows the calculated Δp_{calc}, and defined by the formula (13) Δp_{exact} of the pressure increase. The calculations were performed for the medium with a density ρ_0 = 840 kg/m^3 and the velocity of propagation of perturbations c = 1300 m/s.

Table 1. Exact and calculated value of the pressure increase.

№	u_0, m/s	d_1, m	d_2, m	Δp_{calc}, kPa	Δp_{exact}, kPa	Error, %
1	0.245	1	0.7071	181.21	178.36	1.59
2	0.5	1	1	271.91	273.00	0.40
3	1	1	1.4142	359.7	364.00	1.19

As it is seen from the Table 1 the error of calculation makes approximately 1%, and this may be referred to as pretty low inaccuracy.

5. Calculations of water hammer: one valve actuation

Let us consider the following model configuration of the branching pipeline (see figure 2). From the beginning of the pipeline (point A) to the branching point (point O) the pipe with the length of 20 km has a diameter of 1 m, at the point O the pipeline is diverged into two identical pipes with the length of 20 km and the diameter of $1/(2)^{0.5}$ m each, i.e., the cross-sectional area of the two divergences equals to the cross-sectional area of the main pipe. All pipes have the same roughness - 0.3 mm, and are laid on a flat terrain. At the beginning of the pipe has the pressure of 10 atmospheres. The pipe-end branches have the pressure of 3 atm., and the slide valves are installed, which completely stop the flow within 10 seconds.

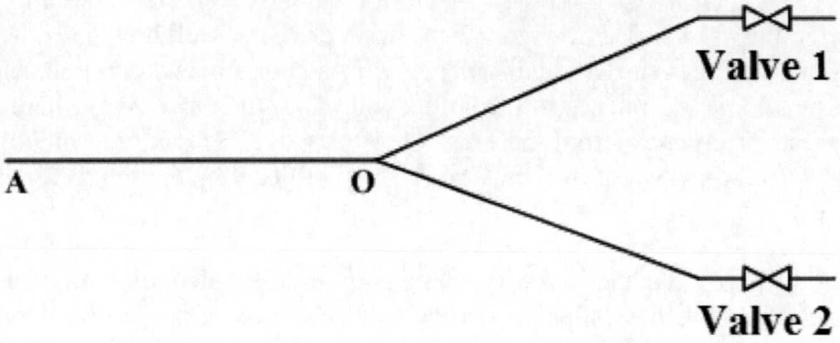

Figure 2. The scheme of branching pipeline system.

The following scenario of the valves shutdown is considered. Initially the valve number 2 operates, and then, after a while, the valve №1 operates. When making calculations the time count down starts from the beginning of the valve №2closing.

It was assumed when making calculation that the speed of disturbancepropagation coincides with the speed of sound and is 1300 m/s.

The figure 3 shows the pressure profiles at different time points in the case where only the valve№2 operates (the response time of the valve №1 is referred to the infinity). At this figure the main pipe corresponds to the section from the beginning (point A) to the valve №1, the divergence in this figure corresponds to the section from the point O to the valve №2.

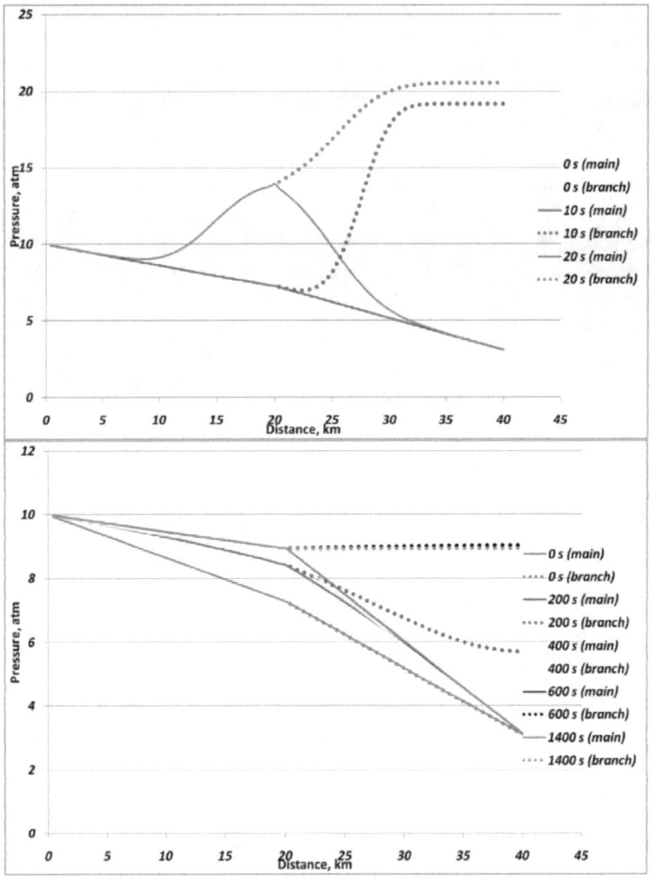

(a)

(b)

Figure 3. Pressure profiles in the pipeline at time points of 10, 20 seconds (a) and 200, 400, 600, 1400 seconds (b) after operation of the valve №2 at the end of the divergence.

From the figure 3 a) it is clearly seen as the wave formed in the divergence (see. Fig. 3 a), 10) goes into the main pipe (see Fig. 3 a), 20). The Figure 3 b) shows perfectly well how after the circulation of the waves in the system the pressure gradually increases in all parts of the pipeline. Most significant increase of the pressure comes on the section from the point O to the valve №2, where the flow stops completely. In the other divergence (from the point O to the valve №1) and the main pipe (from the beginning to the point O) the growth of the pressure is less significant by the relative value.

6. Conclusion

The paper proposes to apply the Godunov type method for the calculation of water hammer in branched pipeline systems. The possibility to disseminate this method among the branched pipeline systems has been successfully demonstrated.

The example that has an analytic solution of the problem of water hammer in the diverging of the pipeline system shows high accuracy of the numerical method used.

The example of water hammer in branched pipeline system was presented. This example demonstrates the good quality of solution using the proposed numerical method.

Acknowledgments

This work is supported by the Russian Science Foundation under grant 16-19-00188

References

[1] Zhukovsky N E 1949 *About water hammer in water pipes* (Moscow-Leningrad: State Publishing of Thecnical and Theoretical Literature)

[2] Guseynzade M A 1999 *Characteristic features of wave flow in the tubes. Water hammer* (Moscow: Oil & Gas)

[3] Gushchin V V 1998 *Bezopasnost Truda v Promislennosti* **1** 8

[4] 2004 *Pipeline transportation of oil.* (Moscow: Nedra-business center)

[5] Lurie M V 2003 *Mathematical Modeling of Gas, Oil and Oil Product Transport* (Moscow: Gubkin University Press)

[6] Arbuzov N S 2014 *Providing of technological safety of the hydraulic system of marine oil terminals in the process of loading vessels at mooring facilities (theory and practice)* (Moscow: Thesis for the degree of Doctor of Technical Sciences)

[7] Mirkin A Z and Usinsh V V 1991 *Pipeline systems* (Moscow: Chemistry)

[8] Seleznev V E, Aleshin V V and Pryalov S N 2009 *Basics of numerical modeling of trunk pipelines* (Moscow: MAX Press)

[9] Godunov S K 1959 *Matematicheskiy sbornik* **47** 271

[10] Gubin S A, Gubina T V, Sumskoi S I and Lisanov M V 2013 *Bezopasnost Truda v Promislennosti* **10** 66

[11] Sumskoi S I and Sverchkov A M 2015 *Phys. Proc.* **72** 347

Journal of Physics: Conference Series **751** (2016) 012023 doi:10.1088/1742-6596/751/1/012023

Simulation of systems for shock wave/compression waves damping in technological plants

S I Sumskoi[1], A M Sverchkov[2], M V Lisanov[2] and A F Egorov[3]
[1]National Research Nuclear University MEPhI (Moscow engineering physics institute), Kashirskoye shosse 31, Moscow 115409, Russia
[2]CJSC Scientific technical center of industrial safety problems research, Block 14, Perevedenovsky Pereulok 13, Moscow 105082, Russia
[3]Mendeleev University of Chemical Technology of Russia, Miusskaya Square 9, Moscow 125047, Russia

E-mail: sumskoi@mail.ru

Abstract. At work of pipeline systems, flow velocity decrease can take place in the pipeline as a result of the pumps stop, the valves shutdown. As a result, compression waves appear in the pipeline systems. These waves can propagate in the pipeline system, leading to its destruction. This phenomenon is called water hammer (water hammer flow). The most dangerous situations occur when the flow is stopped quickly. Such urgent flow cutoff often takes place in an emergency situation when liquid hydrocarbons are being loaded into sea tankers. To prevent environment pollution it is necessary to stop the hydrocarbon loading urgently. The flow in this case is cut off within few seconds. To prevent an increase in pressure in a pipeline system during water hammer flow, special protective systems (pressure relief systems) are installed. The approaches to systems of protection against water hammer (pressure relief systems) modeling are described in this paper. A model of certain pressure relief system is considered. It is shown that in case of an increase in the intensity of hydrocarbons loading at a sea tanker, presence of the pressure relief system allows to organize safe mode of loading.

1. Introduction

During a pipeline system operation the pumps can stop and the valves can shut down. As a result, the flow in the pipe slows down and compression waves appear. This phenomenon is called water hammer [1, 2].

The water hammer can have various negative impacts [3]:

- first, high pressure, which can destroy the wall of the pipeline;
- secondly, large pressure impulse (pressure action for a certain time) affects various structural elements, that leads to the flange damage and a leak appearance;
- thirdly, the rate of pressure rise; the difference in the loading rates of different parts of the pipeline can lead to excessive loads that results in pipeline displacement, the destruction of its mountings;
- fourth, the pressure oscillations resulting from the water hammer negatively affect the pipeline if cyclical pressure increase/decrease takes place.

The most dangerous situations occur when the flow is stopped quickly. This rapid flow cutoff often takes place in an emergency at sea terminals when liquid hydrocarbons are loaded into sea tankers. The special loading lines are used for this loading. When an accident takes place, it is necessary to stop the loading quickly. An urgent cutoff prevents hydrocarbon releases into the

Journal of Physics: Conference Series **751** (2016) 012023 doi:10.1088/1742-6596/751/1/012023

environment. For this purpose, various kinds of valves are used to cut off the flow and to stop further loading. At sea terminals there are several types of valves to ensure safety (during shipment, for example) [3]:

- safety valves (ball valves or butterfly type valves) are located on the route of a loading pipeline; if there is any threat of spills, they quickly cut the flow off;
- safety valves are located in the drift stender, they quickly cut off the flow directly to a tanker (stander - special pipeline connection to the hinge ensures the supply of oil/oil products from the pier to the tanker); these valves can be triggered automatically when the threat of separation tanker from the dock at high wind load, the wave rolling;
- ship valves are located on the tanker; they are triggered when the staff of the tanker immediately arises an urgent need to stop the loading (such as overflow tanks).

All of these valves are fast. Time of complete shut-off is between 3-5 seconds to 8-30 seconds. This quick cutoff is necessary that, on the one hand, does not create more pollution in marine waters: huge penalties are charged for water pollution. On the other hand, the rapid cutoff of release to the environment is necessary to prevent the appearance of large amounts of fire and explosion hazard substance near such a valuable object as a marine tanker: the larger the volume of release, the higher the probability of a tanker accident, and the sequences of tanker accident are already really catastrophic.

Thus, the closing of the safety valves is extremely fast (for 3-30 s), much faster than in the trunk pipelines: there complete valve closing takes up to a few hundred seconds. Obviously, the faster the flow is stopped, the more intense is pressure growth and situation became more dangerous.

Pressure rise in the pipeline can be estimated by using Zhukovsky formula [1], which is obtained by solving the problem of stopping the flow with velocity u on a rigid wall When the instantaneous flow stopping takes place, pressure, according to the formula of Zhukovsky, increases by an amount Δp:

$$\Delta p = c\rho\Delta u \tag{1}$$

here c - the velocity of propagation of the disturbance in the liquid medium, ρ -flow density, u – flow velocity.

From (1) it follows that for characteristic values of c = 1300 m/s and ρ = 900 kg/m^3 u=1 m/s flow stopping leads to pressure rise on 1170000 Pa, i.e. almost 12 atm.

It should be noted that flow velocity in pipelines is about 1-2 m/s, so for water hammer hypothetical pressure increase should not exceed 12-24 atmospheres. In practice, because of the slow response valve that growth is significantly lower. This increase in pressure can be compensate by increasing the strength of the pipe. If strength characteristics of the pipe can not ensure the integrity of the systems require special protection against water hammer.

Velocity of oil and oil products while loading in marine terminals can reach 4-5 m/s. If the valve shuts down quickly, the pressure may increase to tens of atmospheres (50-60 atmospheres or more). In this case, almost always a system of protection against water hammer (SPAWH) is necessary.

The general principle of SPAWH is to take away part of oil or oil products from the pipeline, when a water hammer is developed. This enables the pressure reduction, accordingly.

There are various types of safety devices:

- bursting discs;
- air caps;
- spring valve;
- gas spring valve.

As mentioned above, the general principle of SPAWH operation is to take away all the excess liquid phase from the pipeline, commonly this liquid phase is placed into a special vessel.

Safety valves may have a significant effect on the flow and, above all, reduce the pressure, preventing unacceptable loads. It is therefore extremely important to have an accurate and reliable model to describe the operation of a pipeline system containing safety valves.

In this paper, the approach, that allows to consider the work of the pipelines in the presence of safety valves on them, is proposed. This approach is based on the model of isothermal flow of weakly compressible liquid.

2. Mathematical model

In the isothermal approximation the motion of the fluid in the tube of variable cross section is described by the system of following equations for one-dimensional non-stationary flow:
Continuity (mass conservation)

$$\frac{\partial(A\cdot\rho)}{\partial t}+\frac{\partial(\rho\cdot A\cdot \mathrm{u})}{\partial x}=-M_0, \tag{2}$$

Momentum conservation

$$\frac{\partial(A\cdot\rho\cdot \mathrm{u})}{\partial t}+\frac{\partial(A\cdot\rho\cdot \mathrm{u}^2)}{\partial x}=-I_0-A\cdot\frac{\partial p}{\partial x}-A\cdot\frac{\lambda(\mathrm{Re})}{2\cdot D}\cdot\rho\cdot \mathrm{u}\cdot|\mathrm{u}|-A\cdot g\cdot\rho\cdot\beta, \tag{3}$$

Link between the pressure and density (equation of state)

$$p-p_0=c^2(\rho-\rho_0) \tag{4}$$

here ρ, p, u – density, pressure and velocity of the fluid, averaged according to the cross section; t – time; x – the distance from the start of the pipeline; $\lambda(Re)$ – the coefficient of friction as a function of the Reynolds number $Re=Du/v$; A – cross sectional area of the pipeline, D – the diameter of the pipeline, g – acceleration of gravity power; β – sine of the angle of inclination of the route, which is determined by the elevations points of the pipeline $h(x)$, v – kinematic viscosity; ρ_0 – density of the fluid at pressure p_0 and transportation temperature (usually p_0 = 101325 Pa), c – the speed of sound in the fluid, I_0 – intensity of the loss of momentum with the release of the transported product at the site of relief valve, M_0 – intensity of the loss of mass with the release of the transported product at the site of relief valve.

In this paper to determine the $\lambda(Re)$ the dependence of Colebrook - White is used [4], relation friction coefficient λ with Reynolds number Re and the characteristics of the pipeline (diameter D and roughness k)

$$\frac{1}{\sqrt{\lambda(\mathrm{Re})}}=-2\lg\left[\frac{2{,}51}{\mathrm{Re}\sqrt{\lambda(\mathrm{Re})}}+\frac{k}{3{,}71\cdot D}\right], \tag{5}$$

here k – the size of the pipeline roughness.

The function of mass release M_0 is a function of several parameters and especially:

- the pressure at which relief valve starts working pr, (the pressure of relief valve opening); this pressure is set by adjusting the relief valve;
- the time delay in the opening of the relief valve τr; time from start of opening of the valve to the start of flow through relief valve; the existence of this time lag is due to the inertia of process of valve opening, in fact, this is the time it takes to rupture a disk or to compress spring, etc .; in practice, it ranges from several tens of milliseconds to a few seconds;
- the dependence of the flow rate through the relief valve as degree of relief valve opening; usually in practice flow rate dependent on the ratio of overpressure to the pressure that relief valve starts working.

The most simple and rather accurate approximation for discharge rate calculation is the use of the Bernoulli formula. The outflow velocity U_0 through relief valve was calculated using such Bernoulli formula:

$$U_0 = \sqrt{2\frac{p - P_0}{\rho}}, \tag{6}$$

here p – pressure in pipeline at relief valve location, ρ - liquid density, P_0 – outer pressure. The equation (6) is used when p more p_r it true during the time period longer than τ_r. The mass rate through S_0 square relief valve can be found as:

$$F = \alpha S_0 U_0 \rho, \tag{7}$$

here α— coefficient equal 0,6.

The pulse rate through S_0 square relief valve can be found as the product of the mass ejected from the pipeline and the liquid velocity within the pipeline.

The equations (2) - (4) are supplemented by initial and boundary conditions. As the boundary conditions the pressure is set on the inlet and outlet of the pipeline, this pressure corresponds to the pressure of the pumps or vessels at the ends of the pipeline. As an initial data the parameters are set for the stationary pumping, which can be obtained analytically from the solution of the system (2) - (4). When the flow is cut off by the valve the zero speed is set as a boundary condition.
Godunov type method [5, 6] was used to solve equations (2-7).

3. The calculation results and discussion

3.1. The flow calculation without safety system
Let us consider the following example of a pipeline system. There is a 2 km long pipeline to pump liquid product. The vessel with 2 atmospheres pressure is placed at the inlet of the pipeline The vessel with 1.5 atm pressure is placed at the end of the pipeline. The first vessel is raised above the second to the height of 200 m. The tube diameter is 500 mm. The density of the product is 860 kg/m^3 and the viscosity of the transported product is 10^{-4} m^2/s. The valve is installed at the end of the pipeline (before the vessel). The time of the complete flow shutdown is 0,1 seconds. The perturbation propagation velocity is taken equal to 1200 m/s, and the velocity of sound in the liquid is taken equal to 1250 m/s for this system.

This pipeline is a realistic model, for example, for a process like a sea tanker loading from a shore tank.

The flow in this pipeline occurs due to both pressure gradient and gravity. The pressure profile for the steady-state flow is shown in figure 1. The flow shut down at the end of the pipeline is the event triggering the appearance of non-equilibrium. A high pressure area is formed before the valve, then this area extends over the entire pipeline. The dynamics of such pressure increase is shown in figure 2 (lines 1-8). As a result, the system goes into a new state, corresponding to hydrostatic equilibrium.
As one can see in figure 2, relatively high pressure can be achieved during non-equilibrium transformations in the system and its transition to a state of rest (up to 84 atm). The distribution of the maximum pressure achieved is shown in figure 2 (line "max pressure"). Special relief valve is used to prevent the high pressure rise. The results of calculations of the same problem, but in the presence\ absence of pressure relief system are presented below in sections 3.2-3.4.

The following variations in the safety system are considered below:
- different pressure of safety valves operations;
- various placements of safety valves;
- different diameter of relief valves.

3.2. The flow simulation in the presence of safety system. (different pressure of the safety valves operations)
As noted above, one of the basic settings for the safety valve operation is the opening pressure of the valve, that is the pressure at which the valve opens and the discharge through it begins. The influence of this pressure is considered in this section.

The results of solving the problem stated above (see sec. 3.1) are presented in figures 3-4. These calculations take into account the presence of the relief valve, that is located 500 meters to the end of

Journal of Physics: Conference Series **751** (2016) 012023 doi:10.1088/1742-6596/751/1/012023

the pipeline. The diameter of the relief valve is 100 mm. Figure 3 shows pressure profiles at different time points (line 1-8) and the maximum pressure profile achieved (line "max pressure") if the valve is activated, when pressure reaches 5 atmospheres. Figure 4 shows the same profiles if the valve is activated, when pressure reaches 10 atmospheres.. As one can see in figures 3 and 4, the working pressure has minor effect on the pressure reached in the pipeline. This is due to the rapid increase in pressure. As a result the relief release starts at about the same time, so there is practically no difference in the data presented in figures 3 and 4.

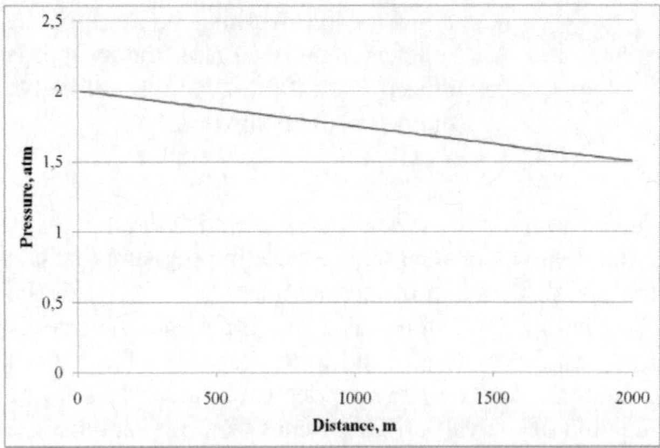

Figure 1. Pressure vs. distance at stationary state.

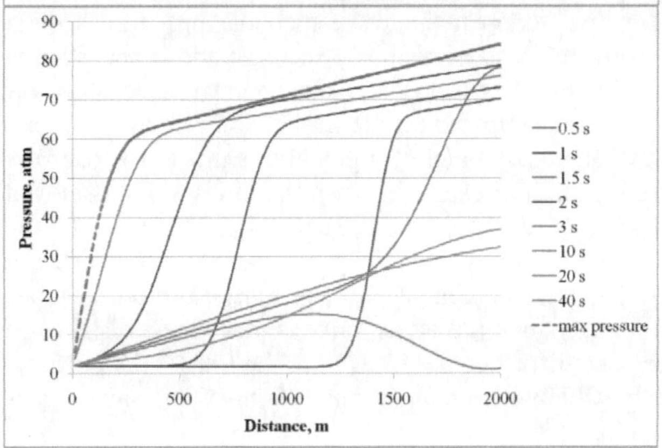

Figure 2. Pressure vs. distance at different time points after valve closing starts: 1 – 0.5 s; 2 – 1 s; 3 – 1.5 s; 4 – 2 s; 5 – 3 s; 6 – 10s; 7 – 20 s; 8 – 40 s (in the absence of pressure relief system).

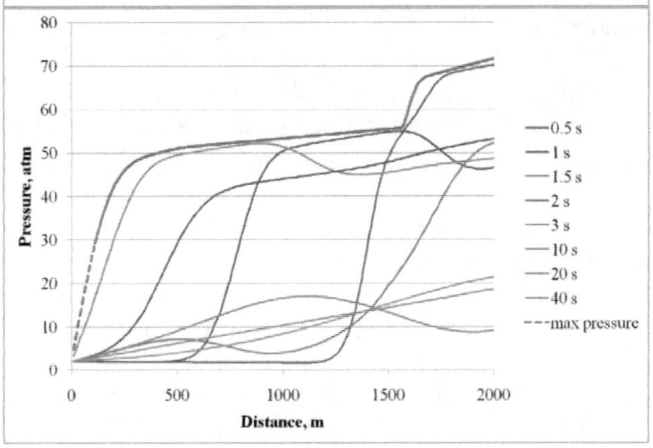

Figure 3. Pressure vs. distance at different time points after valve closing starts: 1 – 0.5 s; 2 – 1 s; 3 – 1.5 s; 4 – 2 s; 5 – 3 s; 6 – 10s; 7 – 20 s; 8 – 40 s (the working pressure of relief system is 5 atm, the relief system is placed at 1500 m, the relief valve diameter is 100 mm).

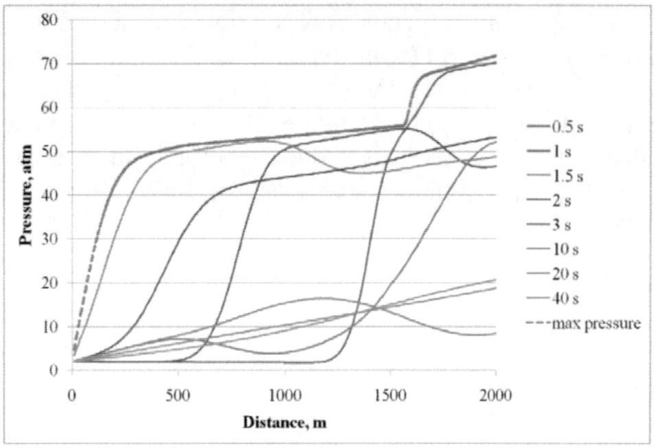

Figure 4. Pressure vs. distance at different time points after valve closing starts: 1 – 0.5 s; 2 – 1 s; 3 – 1.5 s; 4 – 2 s; 5 – 3 s; 6 – 10s; 7 – 20 s; 8 – 40 s (the working pressure of the relief system is 10 atm, the system is placed at 1500 m, the relief valve diameter is 100 mm).

3.3. The flow simulation in the presence of safety system. (various placements of safety valves;)
It is clear in figures 2-4 that the presence of the relief valve considerably reduces the pressure that can be reached in the pipeline. It should be noted that the distribution of the maximum pressure achieved in the pipe can be clearly divided into two areas: before the safety valve and after one. The safety valve helps to reduce the pressure in the most powerful way (compared to the case of safety valves absence) in the upstream region (0-1500 m in figure 3, 4). The pressure decreases much less in the downstream area (1500-2000 meters from the location of the valve). This comes from the fact that the downstream section pressure may grow longer time to open the valve than the upstream section pressure, i.e. the valve is closed when the downstream pressure begins to increase and it opens (or is opened already), when the pressure in the upstream point begins to rise. Thus, the mass discharge will contribute to reducing the pressure growth rate in the upstream section. In this connection it is clear, that the displacement of the safety valve towards the closed end of a pipeline (shift downstream) is able to increase the area of intensive pressure reduction. For example, when the safety valve is moved to the mark of 1700 m, this area naturally expands up to 1700 meters. This fact is confirmed by the results of the calculations shown in figures 5, 6. These figures show the pressure profiles for different moments of time after the valve closing, as well as the maximum pressure reached in the process. Figure 5 shows the results of calculations for the case when the safety valve is placed at a distance of 1,700 meters, and figure 6 shows the one, placed at a distance of 1900 m from the start of the pipeline. The calculations in figures 4-6 deal with a safety valve with a diameter of 100 mm. This valve opens when the pressure in the pipeline is equal of 10 atm.

As one can see in figures 4, 5 and 6, the displacement of the safety valves downstream extents the zone of strong pressure decrease (it lies upstream of the valve). It is interesting to note that the value of maximum pressure reached at the end valve is only slightly sensitive to the displacement of the safety valve As one can see in figures 4, 5 and 6, the pressure at 2000m (at the end valve) is about 70 atm. The pressure in the vicinity of the relief valve in upstream direction is just slightly changed: in all the three cases, it is about 55 atm.

It should also be noted that when the safety valve is moved, the flow pattern is still low-sensitive to the opening pressure of the safety valve. This is confirmed by comparing figures 7 and 6, where corresponding pressure lines are virtually indistinguishable (while one safety valve opens at 10 atm (figure 6), the other one - at 5 atm (figure 7)) for the same safety valve position (1900 m from the start of the pipeline) and the same diameter of the safety valve (100 mm). One can argue that in the case of closer position of safety valve to the closed pipeline end (figure 6, 7), the pressure difference is less noticeable than in the case of larger valves removal from the end of the pipeline (figures 3, 4).

3.4. The flow simulation in the presence of safety system. (various release area of safety valves;)
As one can see from the calculations, the results of which are presented in sections 3.1-3.3, no starting pressure, nor the location of the relief valve does not allow considerably reduce the pressure for a given diameter of discharge hole of the relief valve (100 mm under consideration in this article

configuration). It is clear that this result is mainly due to high-speed of valve cut-off operation - 0.1 s. At longer cut-off importance discussed in Sections 3.2-3.3 factors is likely to grow.

However, it seems appropriate, as before, assuming the fast opening relief valve, explore another factor - the influence of cross-sectional area of the relief valve. Obviously, the increase in the area of the relief valve hole through which fluid is discharged, will lead to higher discharge rate, at least immediately after the opening of the valve until the pressure in the pipeline has not much changed.

Figures 8 and 9 show the pressure profiles for case of discharge through the relief valve 200 mm, that starts to operates at 10 bar (previously we consider a case for 100 mm). Figure 8 shows the profiles of pressure for relief valve, located at 1,900 meters from pipeline inlet, and figure 9 at 1500 meters.

Figures 8 and 9 show us that increasing in the area of the relief valve hole substantially reduces the pressure in the pipeline section from its beginning to the place of installation of the relief valve. According to the obtained results, the magnitude of reduction of pressure is about 20 atm, as for the case of the relief valve at 1500 m (Figures 4,9) and for the case of the relief valve location at 1900 m (Figures 6,8). Thus, increasing the cross-sectional area of the relief valve is an extremely efficient way to protect the upstream pipeline section.

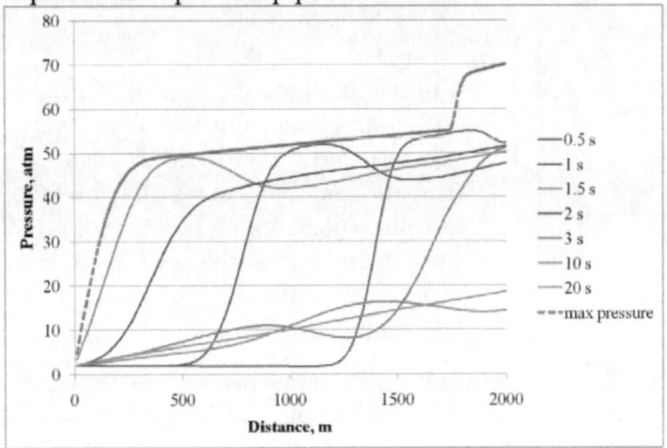

Figure 5. Pressure vs. distance at different time points after valve closing starts: 1 – 0.5 s; 2 – 1 s; 3 – 1.5 s; 4 – 2 s; 5 – 3 s; 6 – 10s; 7 – 20 s; 8 – 40 s (the working pressure of the relief system is 10 atm, the relief system is placed at 1700 m,the relief valve diameter is 100 mm).

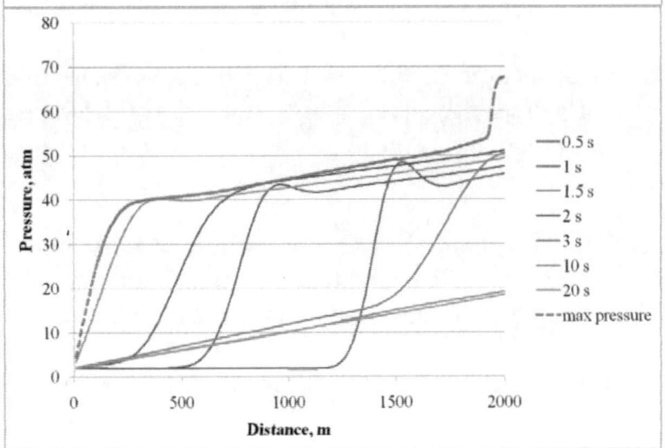

Figure 6. Pressure vs. distance at different time points after valve closing starts: 1 – 0.5 s; 2 – 1 s; 3 – 1.5 s; 4 – 2 s; 5 – 3 s; 6 – 10s; 7 – 20 s; 8 – 40 s (the working pressure of the relief system is 10 atm, the relief system is placed at 1900 m, the relief valve diameter is 100 mm).

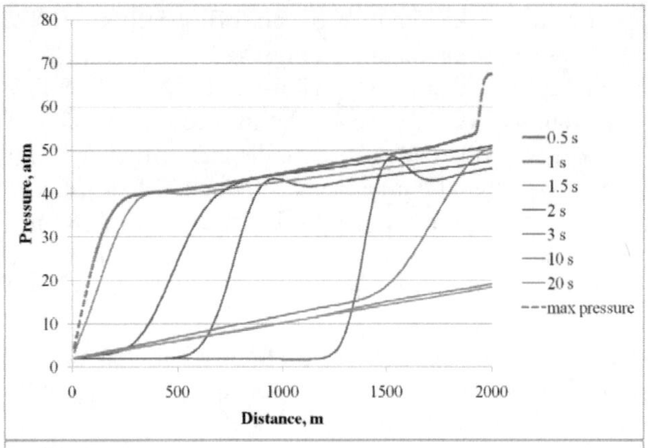

Figure 7. Pressure vs. distance at different time points after valve closing starts: 1 – 0.5 s; 2 – 1 s; 3 – 1.5 s; 4 – 2 s; 5 – 3 s; 6 – 10s; 7 – 20 s; 8 – 40 s (the working pressure of relief system is 5 atm, the relief system is placed at 1900 m, the relief valve diameter is 100 mm).

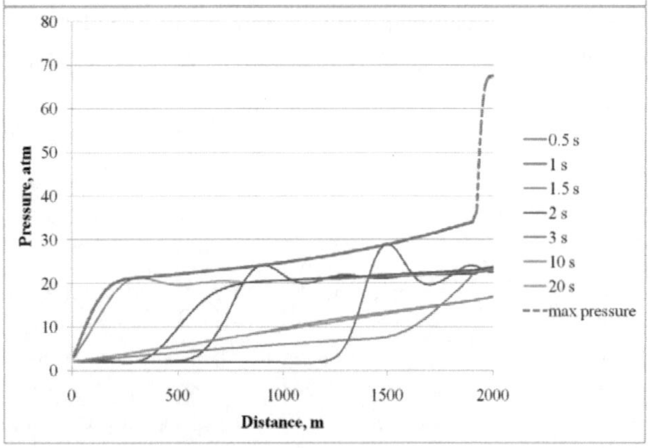

Figure 8. Pressure vs. distance at different time points after valve closing starts: 1 – 0.5 s; 2 – 1 s; 3 – 1.5 s; 4 – 2 s; 5 – 3 s; 6 – 10s; 7 – 20 s; 8 – 40 s (the working pressure of relief system is 10 atm, the relief system is placed at 1900 m, the relief valve diameter is 200 mm).

However, as one can see from Figures 6 and 8, 4 and 9, an increase in discharge capacity of the relief valve has practically no effect on the value of the pressure attained in the area downstream from the safety valve. In our opinion this is due to the rapid, for 0.1 s, the time of closing the valve at the end of the pipeline.

It should also be noted that the investigated above (section 3.2-3.3) the effect of the pressure that relief valve starts to work takes place for the case of increasing valve capacity (the area of discharge hole) also.

4. Conclusion

The paper examines the approaches to the modeling of systems of protection against water hammer (SPAWH). A model of SPAWH is discribed. The results of calculations of piping systems with SPAWH are presented. It is shown that SPAWH allows under certain conditions to reduce the pressure in the case of water hammer.

The article presents the results of calculation of the effects of water hammer in the single pipeline in the presence of relief valve.

In the case of fast-acting valves and the appearance of water hammer in the pipeline the influence of three factors on the pressures achievable in the pipeline is considered: a pressure relief valve, its location and flow when it is reset.

It was shown that the presence of the relief valve in the pipeline divides this pipeline into two parts with different character of pressure rise:

- upstream, with a significant reduction in the pressure;
- downstream, with a weak reduction in pressure.

At displacement of relief valve downstream the first zone length is increased, and the second region is reduced. The magnitude of the decrease in pressure in each of these two sections weakly depends on the opening pressure of the safety valve and on relief valve location (for the case of rapid response valve at the end of the pipeline). It is interesting to note that the value of maximum pressure reached at the end of the valve in the pipeline is slightly sensitive to the displacement of the relief valve.

The biggest influence on the value of the pressure reduction in the water hammer has a relief area of a relief valve. The more its area, the greater the pressure drop. However, a noticeable reduction in pressure occurs only in the upstream section.

Thus, in case of the valve at the end of the pipeline urgent cut-off, it is almost impossible to reduce water hammer pressure and to protect the section between this end valve and the relief valve against water hammer: This is because the response time of the SPAWH is much more, than the pressure rise time. On this basis, it must be concluded that the safety of the end section can be ensured only by proper choice of the time of end valve shutdown, primarily by the time increase.

5. Acknowledgments
This work is supported by the Russian Science Foundation under grant 16-19-00188

References
[1] Zhukovsky N E 1949 *About water hammer in water pipes* (Moscow-Leningrad: State Publishing of Thecnical and Theoretical Literature)
[2] Guseynzade M 1999 *Characteristic features of wave flow in the tubes. Water hammer* (Moscow: Oil and Gas)
[3] Arbuzov N S 2014 *Provision of technological safety of the hydraulic system of marine oil terminals in the process of loading vessels at mooring facilities (theory and practice)* (Moscow: Thesis for the degree of Doctor of Technical Sciences)
[4] Mirkin A and Usinsh V 1991 *Pipeline systems* (Moscow: Chemistry)
[5] Gubin S A, Gubina T V, Sumskoi S I et al 2013 *Bezopasnost Truda v Promyshlennosti* **82** 66–71
[6] Sumskoi S I and Sverchkov A M 2015 *Phys. Proc.* **72** 347–50

Journal of Physics: Conference Series **751** (2016) 012024 doi:10.1088/1742-6596/751/1/012024

Simulation of compression waves/shock waves propagation in the branched pipeline systems with multi-valve operations

S I Sumskoi[1], A M Sverchkov[2], M V Lisanov[2] and A F Egorov[3]

[1]National Research Nuclear University MEPhI (Moscow Engineering Physics Institute), Kashirskoye shosse 31, Moscow 115409, Russia
[2]CJSC Scientific technical center of industrial safety problems research, Block 14, Perevedenovsky Pereulok 13, Moscow 105082, Russia
[3]D. Mendeleyev University of Chemical Technology of Russia

E-mail: sumskoi@mail.ru

Abstract. With the help of the developed model and numerical method the task has been solved concerning the transmission of the compressive wave complex in the branching pipeline system when several shut down valves operate. It is shown that for branches of the pipeline the highest peak of pressure can be achieved in them under the asynchronous operation of the shutdown valves, but not under the simultaneous operation of all valves, as it has been assumed in previous studies.

1. Introduction

The shock waves/compression waves (SW/CW) can have a negative impact and cause destructions. Typically, the SW/CW attenuate during their propagation and their damage effects decrease. A significant factor of the waves' attenuation is the SW/CW divergence in multidimensional space, when the flow is divergent. However, in practice, there are flows that do not have significant multidimensional effects, leading to the attenuation of SW/CW. In this case, the SW/CW can propagate over long distances, with maintaining its damage potential. Since such flows take place in real technical devices and structures the methods are needed for modelling the pressure wave's propagation processes.

The most famous example of the weak attenuating pressure waves' propagation is the flow in pipeline systems known as water hammer. The problem about the SW propagation under the explosion in the tunnel system, for example, in mine workings may serve as another example. Both of these tasks in a fairly accurate approximation can be set and resolved in the one-dimensional formulation, naturally, taking into account the multidimensional effects, if any.

This article presents results for solving the task of water hammer in the pipeline system. The task is considered in the one-dimensional non-stationary formulation taking into account the realities such as the change in the diameter of the pipeline and its branches.

2. Calculations of water hammer: operation of two valves

Let us consider the following model configuration of the branching pipeline system (see Fig. 1). From the beginning of the pipeline (point A) to the branching point (point O) the pipe with the length of 20 km has a diameter of 1 m. At the point O the pipeline is diverged into two identical pipes with the length of 20 km and the diameter of $1/(2)^{0.5}$ m each, i.e. the cross-sectional area of the two branched pipes equals to the cross-sectional area of the main pipe. All pipes have the same roughness - 0.3 mm, and are laid on a flat terrain. The pressure at the beginning of the pipeline is 10 atmospheres. The pressures in the end of each branched pipeline equal 3 atmospheres. There are two valves in the end of each pipeline. These valves completely stop the flow within 0,1 seconds.

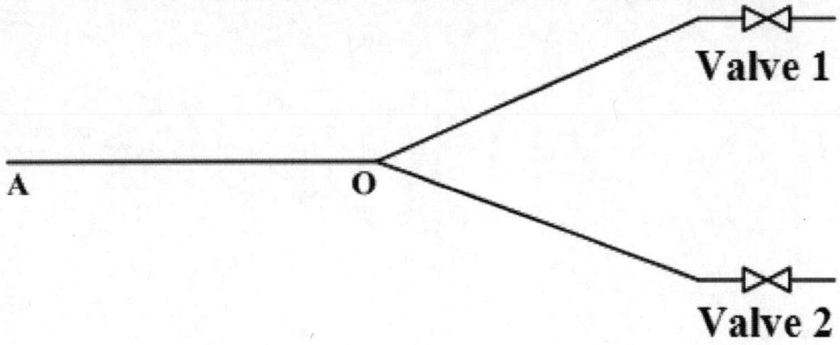

Figure 1. The scheme of branching pipeline system.

The following scenario of the valves actuation is considered. Initially the valve №2 triggers, and then, after a while, the valve №1 operates. When making calculations the time count starts from the beginning of the valve №2 closing. It was assumed when making calculation that the speed of disturbance propagation coincides with the speed of sound and is 1300 m/s.

Obviously, the operation of the two valves can result in a substantial increase in pressure in the pipeline system, as the CW from the valve № 1 will be propagated through the liquid, where the pressure increase has occurred as a result of the valve №2 closing.

To solve this problem we used the physical model and numerical method offered in [1, 2]

The shutdown of valves may happen simultaneously or with a certain degree of asynchrony. In this regard the question arises: which of the options of the valves shutdown will be the most dangerous relating to the maximum peaks of the pressure? To answer this question the calculations have been made with different variants of operation of valves. The general scheme of the flow stopping looked as follows. At first the valve shuts down in the main section, and then, with some time delay, the valve shuts down in the branch. The calculation results in the form of distribution of the maximum achievable values of the pressure in the main pipeline and in the branch pipeline are shown in the Fig. 2 (a-e). The Fig. 2 a) shows the profiles for the cases of the second valve shut down with delay in 1, 2.5 and 5 seconds; Fig. 2 b) - 10, 15, 20, 25 and 30 seconds; Fig. 2 c) - 35, 40, 45 and 50 seconds; Fig. 2 d) - 60, 70, 90 and 110 seconds; Fig. 2 e) – 240 seconds. Each of these figures shows the profiles of the maximal pressure reached in the pipeline system when only one valve (red lines) operates and when simultaneous operation of two valves (yellow line) takes place.

From Figure 2 a) it is clear that for a small time delay the pressure profiles in branches are close to the pressure, reached in the case of simultaneous operation of valves. At that for 1 second delay the deviations are observed only in the zone of the branching. As for the deviations themselves, they are small: in the main pipe the pressure decreases, and it increases in the branch. As the time delay grows the difference in pressures is becoming more visible, and the pressure growth in the branch is more significant than the drop of the pressure in the main pipe: so for 5 second delay (brown lines) the pressure drop in the main pipe is about 1.5 atmosphere (brown solid line in comparison with the yellow line); whereas as the pressure growth in the branch is greater than 5 atmospheres (dashed brown line in comparison with the yellow line). When the time-synchronization grows the decrease in the maximal peaks of the pressure achieved at the beginning of the pipeline is also observed.

With the growth of the time delay the mentioned trends are kept also for the time interval of 10-30 s in the shut down delay of valves. As can be seen from Fig. 2 b), the maximum pressure, achieved in the main pipeline, continue to decrease, tending to the pressure level characteristic for the case of operation of only one valve (solid black line comes close with a solid red line). At the same time the

Journal of Physics: Conference Series **751** (2016) 012024 doi:10.1088/1742-6596/751/1/012024

pressure in the branch increases significantly: under the 30 second delay the maximum pressure (black dotted line) exceeds the pressure in the case of simultaneous valve shut down (yellow line) by about 10 atm.

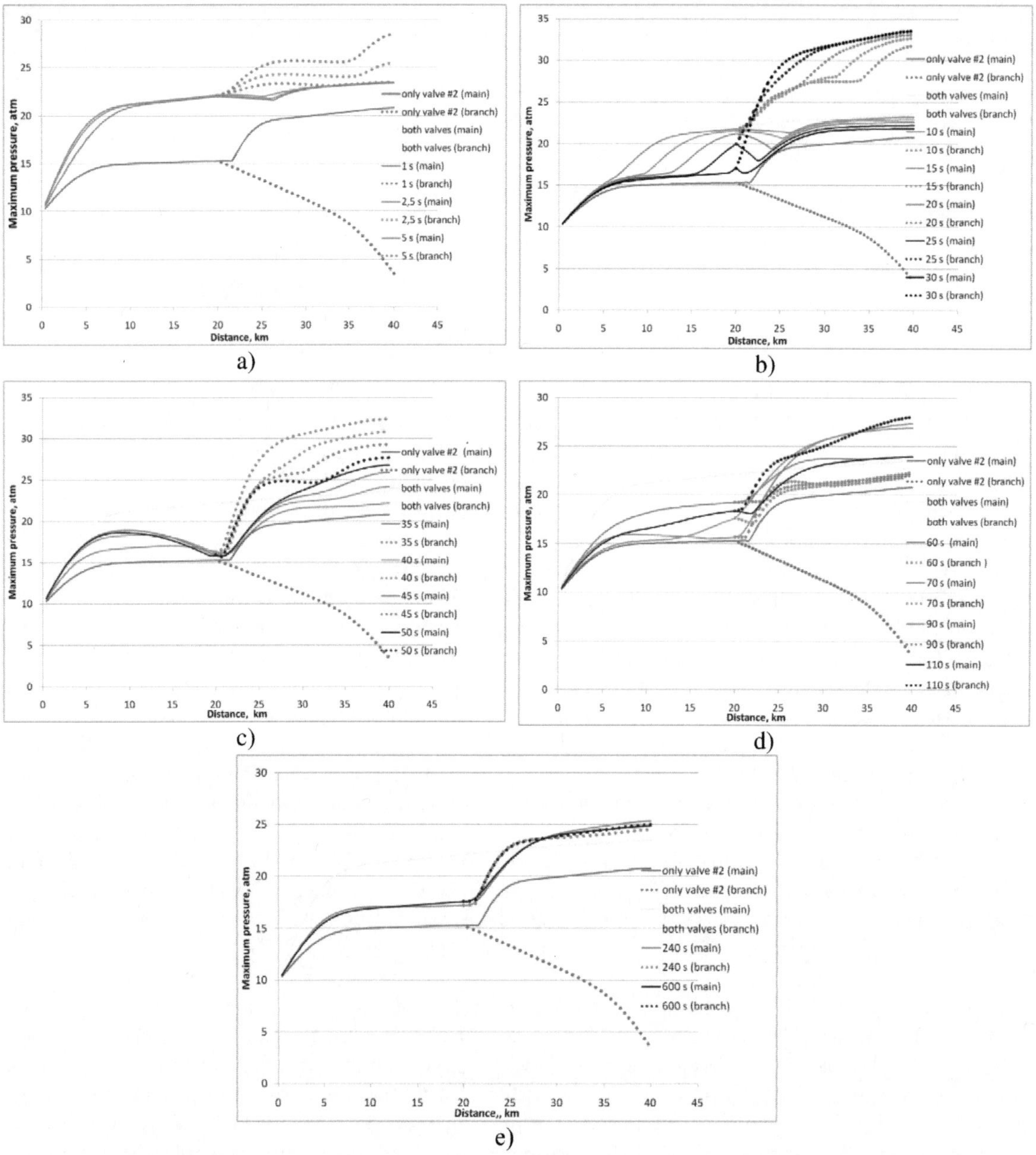

Figure 2. The maximal peaks of the pressure achieved inside the main pipeline and in the branch under operation of two valves with a time difference of 1, 2.5 and 5 s (a); 10, 15, 20, 25 and 30 s (b); 35, 40, 45 and 50 s (c); 60, 70, 90 and 110 s (d); 240s (e). All figures also show the maximum pressure when only one valve operates and when simultaneous operation of two valves takes place.

At 30 second delay on the response time of the second valve, as it is seen in Fig. 2 b) and c), the maximal peaks of the pressure, achieved inside the main pipe, are at the minimum level, and in the branch – at the maximum level. This may be explained by the fact that for 30 with a little over seconds the compression wave, generated at the moment of closing the first valve (valve #2), succeeds, on the one hand, to reach the start of the pipeline and to begin unloading the tube of the main section up to the branching point, and, on the other hand, to achieve the second valve (valve #1, in the branch) and by this pressing the liquid in this section of the branch up to the highest values. As a result of this, the second compression wave, generated when the second valve (valve #1) operates, will be propagated along the maximum compressed liquid in the branch and among the most unloaded liquid in the main section.

As can be seen from Fig. 2 b) and c), the further increase in the time-synchronization causes the growth of the maximum pressure attained in the main section, and, in the opposite, they decrease in the branch. This is reasoned by the fact that now the first compression wave passes through the branch, and the pressure decreasing begins in it, respectively, in the branch the second compression wave propagates among the less compressed liquid. Owning to this factor starting with approximately 50-55 seconds the maximal peaks of the pressure in the main section after the branching point start exceeding the maximal pressure in the branch (see line "50 s" in Fig. 2 c) and green line "60 s" Figure 2 d).

In case of delay in the activation of the second valve for more than 50 seconds the maximal peaks of the pressure, achieved in the main section downstream to the branch point, firstly fall down (to 60 s), then increase somewhat (to 90 s) and then fall again (to 110 s) (see Fig. 2 d)-e). In the end, starting with a certain value of time delay, the maximum achievable pressure comes to a certain level, which is not affected by a further increase in the delay time until the second valve actuates. Similar oscillations and stabilization on the some stationary level of maximum achievable pressure are observed also for the sections after the branching points (see Fig. 2 d) to e).

The maximal achievable peaks of pressure corresponding to the situation when these peaks of the pressure do not change due to the time delays are shown in Fig. 2 e), for example, with the delay of 240 seconds. At the same time, as shown in Fig. 2 e), for large delays in the activation of the second valve the maximal achievable pressure up to the branching point exceed a little bit the corresponding values for the case of only one valve shut down, and the maximum pressure reached in the areas after the branching point, slightly exceed the pressure in the case of simultaneous operation of valves.

If we analyze the received results shown in Fig. 2, we can draw the following conclusion: for different sections of the pipe the greatest danger (i.e., the greatest growth of the pressure) is implemented in a variety of scenarios. This implies that when calculating the branching of pipelines one cannot be limited by the single scenario of closing valves.

Most often, for example in [3], the scenario of simultaneous shut down of all valves is considered as the most dangerous scenario. The safety of marine shipping terminals is analyzed in this formulation in [3]. According to the results presented in this paper, such scenario is the most dangerous only for the section before the branching point. This is reasoned by the fact that in this case, when the waves are simultaneously propagating, the maximum compression is reached in this waves; and the desynchronization in valve shut down allows to start the unloading of the separate sections and, accordingly, the value of maximal achievable pressure decrease. For the sections after the branching points the situation with a simultaneous closing of the two valves is, as it may seem paradoxical, the best option, as in this case for all sections the maximum pressure is the least one; in all other cases, and it can be clearly seen from Fig. 2, in the sections after the branching point there is always the area where the greater peaks of the pressure are achieved (and sometimes very much significantly) than the maximum pressure for the simultaneous valve shut down.

It is clear that in the case of varying diameters and lengths of pipelines even more diverse range of situations may develop. In this connection, the important conclusion should be made – the analysis of loads on the branching pipeline in the event of water hammer phenomenon should be carried out across the entire spectrum of possible scenarios of closing valves.

Journal of Physics: Conference Series **751** (2016) 012024 doi:10.1088/1742-6596/751/1/012024

3. Conclusion

The paper proposes to apply the Godunov Type Method for the calculation of water hammer in branched pipeline systems. The possibility to disseminate this method among the branched pipeline systems has been successfully demonstrated.

The problem of the water hammer under different variants of the valves operation for the model system with one branch and valves at the ends of the pipeline system has been solved. The numerical modeling results show that in the given case the maximum load on the main pipeline (from its start to the point of branching) occurs at the simultaneous closing of the valves.

And in the pipelines after branching point the maximum loads on the pipeline under the water hammer happen not at the simultaneous operation of the shutdown valves at the ends of the pipeline, but at some delay in the shutdown of valves at the ends.

It should be noted that even a small desynchronization in the operation time of valves can lead to a significant growth of the pressure in the branches compared to the case of simultaneous operation of valves. Moreover, the case of simultaneous operation of valves in the considered system provides the smallest increase in pressure in the branches comparing to any asynchronous option of closing the valves.

With that in mind, one can make an important methodological conclusion: when making the analysis of branching systems to determine the maximal peaks of pressure on the branching sections it is needed to consider the spectrum of events associated with the different options of closing valves.

The proposed method of calculating of the liquid medium motion can be used (with appropriate modification) and for other tasks, similar topology. The dissemination of this methodology to the problem of shock waves in tunnels or the flow in the blood vessels seems to be of current concern.

Acknowledgments

This work is supported by the Russian Science Foundation under grant 16-19-00188

References

[1] Gubin S A, Gubina T V, Sumskoi S I and Lisanov M V 2013 *Bezopasnost truda v promislennosti* **10** 66

[2] Sumskoi S I and Sverchkov A M 2015 *Phys Proc* **72** 347

[3] Arbuzov N S 2014 *Providing of technological safety of the hydraulic system of marine oil terminals in the process of loading vessels at mooring facilities (theory and practice)* (Moscow: Thesis for the degree of Doctor of Technical Sciences)

Journal of Physics: Conference Series **751** (2016) 012025 doi:10.1088/1742-6596/751/1/012025

Developing the model of non-stationary processes of motion and discharge of single- and two-phase medium at emergency releases from pipelines

S I Sumskoi[1], A S Sofin[2] and M V Lisanov[2]

[1]National Research Nuclear University MEPhI (Moscow Engineering Physics Institute), Kashirskoye shosse 31, Moscow 115409, Russia
[2]CJSC Scientific technical center of industrial safety problems research, Block 14, Perevedenovsky Pereulok 13, Moscow 105082, Russia

E-mail: sumskoi@mail.ru

Abstract. The system of equations describing gas dynamic processes with extremal gradients of velocity, temperature, concentration and pressure in pipeline systems is presented. The proposed model most completely takes into account the actual processes in pipeline systems transporting various media. The model is formulated in one-dimensional approach. However it takes into account the multidimensional effects affecting the flow, including the development of non-equilibrium processes. These multidimensional effects are due to the branching pipeline, diameter change, valve shut down. This model can be used for modeling an emergency situation, that includes single- and two-phase mediums release.

1. Introduction

Trunk pipelines are used to transport liquid or gaseous products (typically hydrocarbon raw materials) over distances of hundreds or even thousands of kilometres. The great length of piping systems and their large diameter (up to 1200 mm) requires presence of up to several hundred or even thousand tons of product in a pipeline system. It is obvious, that such a huge mass of products can produce a considerable damage if this mass is released in case of accident.

On the other hand, there are a lot of events producing considerable changes in flow through creating large gradients. Examples of such actions are:

- changes in pump operation;
- shut down of valves;
- releases through pressure relief systems;
- rupture of the pipeline;
- heating of the transported product;
- injection of a product with different composition or temperature into the pipeline system;
- cavitation etc.

As a result, gradients of different nature (velocity, temperature, concentration and phase) may appear in the pipeline. These gradients can initiate various processes, provoking a new mode of operation in the pipeline system.

Thus, the pipeline system can be treated as a system, turning in its functioning from one state to another through series of gas dynamic processes. These processes may cause stress that could destroy the pipeline, releasing hundreds or even thousands tons of its content into the environment.

So modelling pipeline systems as real (non-idealized) ones is an urgent task and appropriate models for this description are needed. However, in most cases the pipelines are described in significantly simplified interpretation. Typically the simplified model is used for modelling a pipeline. This system of equations describes an isothermal single component flow in the relief, taking into account the losses due to friction against the walls [1,2]. This approach narrows the range of consideration of real-life situations, many pipeline systems and modes of their operation can not be modelled in the framework of models of the type [1,2].

In this paper the system of equations, describing the behavior of a pipeline system with velocity, temperature, concentration, pressure and phase gradients, is presented. This system of equations allows to simulate a wide range of practical problems.

2. Mathematical model

2.1. Main assumptions

To describe the flow in the pipeline we propose to use the following scheme for one-dimensional stationary equations expressing conservation laws in one-velocity approximation for all transportable components. It is assumed that all the parameters are averaged over the cross section. It is believed that there is no flow rotation.

It is assumed that the pipe system has arbitrary branched configuration (with looping, by-pass from the output to the output of the pump, etc.). The mixture of N different substances with a mass fraction Y_m ($\sum_{m=1}^{N} Y_m = 1$) is transported in the pipeline system. It is assumed that the components of the mixture do not react and do not fall down onto the inner surface of the pipeline.

We consider flow under head gradient, when the tube cross section is blocked by the transported medium completely, i.e. gravity-fed areas (shallow water flow) are not considered.

2.2. Conservation equations

System of equations using assumptions stated above, is as follows.

The law of mass conservation (continuity equation)

$$\frac{\partial(\rho \cdot A)}{\partial t} + \frac{\partial(\rho \cdot A \cdot w)}{\partial x} = M_k - M_0 \tag{1}$$

The law of individual components conservation:

$$\frac{\partial(\rho \cdot A \cdot Y_m)}{\partial t} + \frac{\partial(\rho \cdot A \cdot Y_m \cdot w)}{\partial x} = Y_{mk} M_k - Y_m M_0 + \frac{\partial}{\partial x}\left(\rho \cdot A \cdot (D_m + D_{turb}) \cdot \frac{\partial Y_m}{\partial x}\right) \tag{2}$$

Conservation of momentum:

$$\frac{\partial(\rho \cdot A \cdot w)}{\partial t} + \frac{\partial(\rho \cdot A \cdot w^2)}{\partial x} = I_k - I_0 - A \cdot \frac{\partial p}{\partial x} - A \cdot g \cdot \rho \cdot \frac{\partial z}{\partial x} - \frac{\pi}{2} \cdot \left(\frac{\lambda}{2 \cdot R} + \xi_l\right) \cdot \rho \cdot w \cdot |w| \cdot R^2 \tag{3}$$

The law of energy conservation:

$$\frac{\partial}{\partial t}\left(\rho\cdot A\cdot\left(\varepsilon+\frac{w^2}{2}\right)\right)+\frac{\partial}{\partial x}\left(\rho\cdot A\cdot w\cdot\left(\varepsilon+\frac{w^2}{2}\right)\right)=$$

$$E_k-E_0+\frac{\partial}{\partial x}\left(\rho\cdot A\cdot\sum_{m=1}^{N}\varepsilon_m\cdot\left(D_m+D_{turb}\right)\cdot\frac{\partial Y_m}{\partial x}\right)+$$

$$-\frac{\partial(A\cdot p\cdot w)}{\partial x}-A\cdot w\cdot g\cdot\rho\cdot\frac{\partial z}{\partial x}-\frac{\pi}{2}\cdot\left(\frac{\lambda}{2\cdot R}+\xi_l\right)\cdot\rho\cdot w^3\cdot R^2 \quad (4)$$

$$+Q\cdot A+\frac{\partial}{\partial x}\left(\left(k+k_{turb}\right)\cdot A\cdot\frac{\partial T}{\partial x}\right)-\Theta\left(T,T_{sur}\right)$$

here t – time, g – the acceleration of gravity, k – the coefficient of thermal conductivity, k_{turb} – the coefficient of turbulent heat transfer, x – distance from the pipeline beginning, p – pressure averaged over the cross section of the pipe, z –leveling mark track, w – velocity averaged over the cross section of the pipe, ε – specific internal energy, ε_m – specific internal energy of the m-th component, ξ_l – local friction losses, λ – the friction coefficient, which depends on the flow regime in the pipe (Reynolds number $\mathrm{Re}=\frac{wD}{\nu}$), ν – kinematic viscosity ($\nu = \mu/\rho$), μ – dynamic viscosity of transported product (in general, depending on the temperature of the transported medium), π – a number equal to 3.14 ..., ρ – density averaged over the cross section of the pipe, A – cross sectional area of the pipeline, D – inner diameter of the pipeline, D_m – the diffusion coefficient of the m-th component of the transported product; D_{turb} – the coefficient of turbulent diffusion, E_0 – the intensity of the energy loss with the release of the transported product at the site of the accident, E_k – flow rate of additional energy in/out of k-th pipeline, I_0 – intensity of the loss of momentum with the release of the transported product at the site of the accident, I_k – flow rate of additional momentum in/out of k-th pipeline, M_0 – intensity of the loss of mass with the release of the transported product at the site of the accident, M_k – flow rate of additional mass in/out of k-th pipeline, Q – intensity of heat exchange with the environment (heating of the transported product), R – the inner radius of the pipe, T – temperature, T_{sur} – soil temperature, Y_m – mass fraction of m-th component, Y_{mk} – mass fraction of m-th component in flow of mass in/out of k-th pipeline, $\Theta\left(T,T_{sur}\right)$ – the function of heat exchange with the ground.

If a product is transported at a constant temperature (an isothermal process) the equation (4) can be neglected. However, in some cases in the pipeline system thermal nonequilibrium can occur. Then it is necessary to solve the equation (4) to:

- take into account the transportation of initially pre-heated products;
- take into account additional injection of products with different temperature in the pipeline;
- take into account the variation of the soil temperature;
- take into account the change in temperature when boiling of unstable liquids or gas expansion in rarefaction waves takes place.

It should also be noted that in the general case, the value ε_m in equation (4) consists of two parts: elastic energy and heat energy (in the presence of a liquid phase). But in the case of transporting of fluids through pipelines elastic energy is small and can be neglected in ε_m. Accordingly, equation (4) can actually be considered as a heat balance equation.

2.3. Initial and boundary conditions

The equations (1)-(4) are supplemented by initial and boundary conditions. As the boundary conditions the pressure is set on the inlet and outlet of the pipeline, this pressure corresponds to the pressure of the pumps or vessel at the ends of the pipeline. As an initial data the parameters are set for the stationary pumping, which can be obtained analytically from the solution of the system (1)-(4).

In a more general case, boundary conditions are set by specifying more complex dependency parameters corresponding to real devices located at the ends of the linear section. This approach considers boundaries of the following types:

- presence of diverse valves in various stages of closing at the ends of the pipeline;
- pumps/compressors (or group of pumps/compressors) taking or injecting the transported product;
- safety relief valve;
- interconnections with other pipelines of different diameters;
- pipeline branching;
- presence of release on the site of the accident;
- presence of a vessel at a pipe inlet or outlet.

For example, for a closed valve zero velocity is a boundary condition.

The system (1)-(4) describes the flow in one linear sector. In the case of a branched pipeline system when the individual linear sectors are combined in a specific sequence, the system of equations (1)-(4) is recorded for each linear sector. The boundary conditions at the connection of two linear sectors are set in such a way that the flows of mass and impulses could be correctly transferred from one sector to another (according to the conservation equations).

The considered model takes into account the following factors:

- non-stationary processes;
- variable cross-section of the pipeline;
- convective motion of the medium;
- appearance and circulation of the waves at a stop/start up of the pumps, closing valves;
- presence of friction on the pipe wall;
- gravity force effect on the flow, when the pipeline's route goes along complex terrain.
- taking away/injection of mass, momentum and energy out/into the pipeline system;
- loss of mass, momentum and energy in the system due to release of the product at rupture point;
- molecular and turbulent diffusion of individual components of the mixture along the axis of the pipeline;
- heating of the transported product;
- molecular and turbulent transport of heat;
- heating of the product transported by the pipeline to the ground (via the pipe wall).

Some of physical processes are not taken into account in the equations (1)-(4), because these processes have a negligible effect. For example, in equation (3) we don't consider turbulent viscosity in the direction of fluid flow. This is due to the fact that the velocity gradient along the axis of the flow in the absence of a shock wave is very insignificant, so the action of turbulent viscosity is also low.

The system of equations (1)-(4) enables to describe the motion of the liquid, gaseous or two-phase medium, in the latter case for different phases equilibrium velocity approximation is used.

2.4. Closer equations

The system of equations (1)-(4) is not closed, it includes the value for which you need to specify methods of calculation: pressure, friction coefficients , the laws of heating and heat exchange with the soil, source (and boundary) terms - M_0, M_k, I_0, I_k, E_0, E_k..

The closure of system (1)-(4) is carried out by different means for different mediums in various phase states.

For example, to calculate the pressure it is necessary to know the equation of state - the relationship between pressure, density and specific internal energy: In the most general case, this relationship can be obtained by thermodynamic calculation (solution of UV task).

This approach is versatile and can be used for both liquid and gaseous, and two-phase media. In the latter case, the solution of UV-problem is especially attractive, since it will determine not only the pressure but also the ratio of gas and liquid phases, as well as their composition.

In the case of transportation of liquid, neglecting the elastic component of the energy for the calculation of the pressure and temperature, one can use the following equation:

- the relation between pressure, density and temperature (equation of state):

$$P_c - P_0 = c^2(\rho - \rho_0) - c^2\rho_0\varsigma(T_0 - T), \tag{5}$$

- equation for termal part of energy:

$$\varepsilon_m = C_v T\,, \tag{6}$$

here ρ_0 – density at standard conditions p_0=101325 Pa, T_0=293 K, ς –coefficient of volume expansion, C_v – average specific heat/.

In a similar way (i.e. taking into account the phase states) other physical quantities in the system (1)-(4) are calculated. For liquids closer assumptions about these variables, can be found in [3].

3. Calculation of water hammer flow without and with mass release

A typical example of a flow in the pipeline system is water hammer - appearance and propagation of a high pressure wave in a pipeline. In this case the pipeline system can evolve through several states.

Let us consider the following example of a pipeline system. There is a 2 km long pipeline to pump liquid product. The vessel with 2 atmospheres pressure is placed at the inlet of the pipeline The vessel with 1.5 atm pressure is placed at the end of the pipeline. The first vessel is raised above the second to the height of 200 m. The tube diameter is 500 mm. The density of the product is 860 kg/m^3 and the viscosity of the transported product is 10^{-4} m^2/s. The valve is installed at the end of the pipeline (before the vessel). The time of the complete flow shutdown is 3 seconds. The perturbation propagation velocity is taken equal to 1200 m/s, and the velocity of sound in the liquid is taken equal to 1250 m/s for this system.

This pipeline is a realistic model, for example, for a process like a sea tanker loading from a shore tank.

The flow in this pipeline occurs due to both pressure gradient and gravity. The pressure profile for the steady-state flow is shown in figure 1. The flow shut down at the end of the pipeline is the event triggering the appearance of non-equilibrium. A high pressure area is formed before the valve, then this area extends over the entire pipeline. The dynamics of such pressure increase is shown in figure 2 (lines 1-8). As a result, the system goes into a new state, corresponding to hydrostatic equilibrium. Thus, the proposed model allowed us to simulate a nonequilibrium process initiated by the closing of the valve.

As one can see in figure 2, relatively high pressure can be achieved during non-equilibrium transformations in the system and its transition to a state of rest (up to 84 atm). The distribution of the maximum pressure achieved is shown in figure 2 (line "max pressure"). Special relief valve is used to prevent the high pressure rise. This is another element of the pipeline system, which may contribute to the development of non-equilibrium process. The results of solving the problem stated above are presented in figures 3-4. These calculations take into account the presence of the relief valve, that is located 500 meters to the end of the pipeline. The diameter of the relief valve is 100 mm. Figure 3 shows pressure profiles at different time points (line 1-8) and the maximum pressure profile achieved (line "max pressure") if the valve is activated, when pressure reaches 5 atmospheres. Figure 4 shows the same profiles if the valve is activated, when pressure reaches 10 atmospheres.. As one can see in figures 3 and 4, the working pressure has minor effect on the pressure reached in the pipeline. This is due to the rapid increase in pressure. As a result the relief release starts at about the same time, so there is practically no difference in the data presented in figures 3 and 4.

It is clear in figures 2-4 that the presence of the relief valve considerably reduces the pressure that can be reached in the pipeline. Thus, a relief valve can be regarded as a controller of non-equilibrium by creating other non-equilibriums.

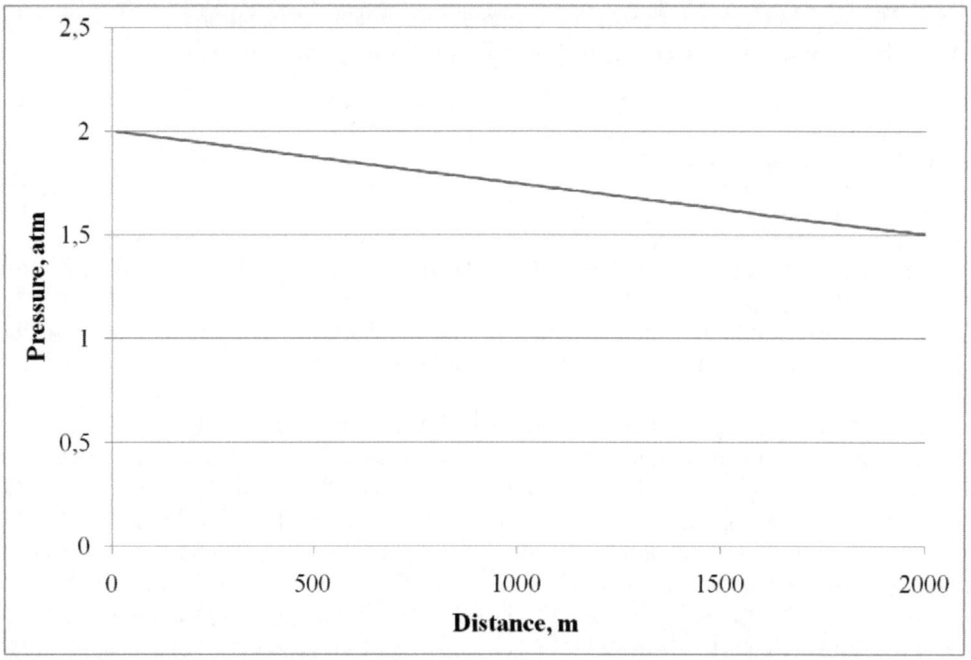

Figure 1. Pressure vs. distance at stationary state.

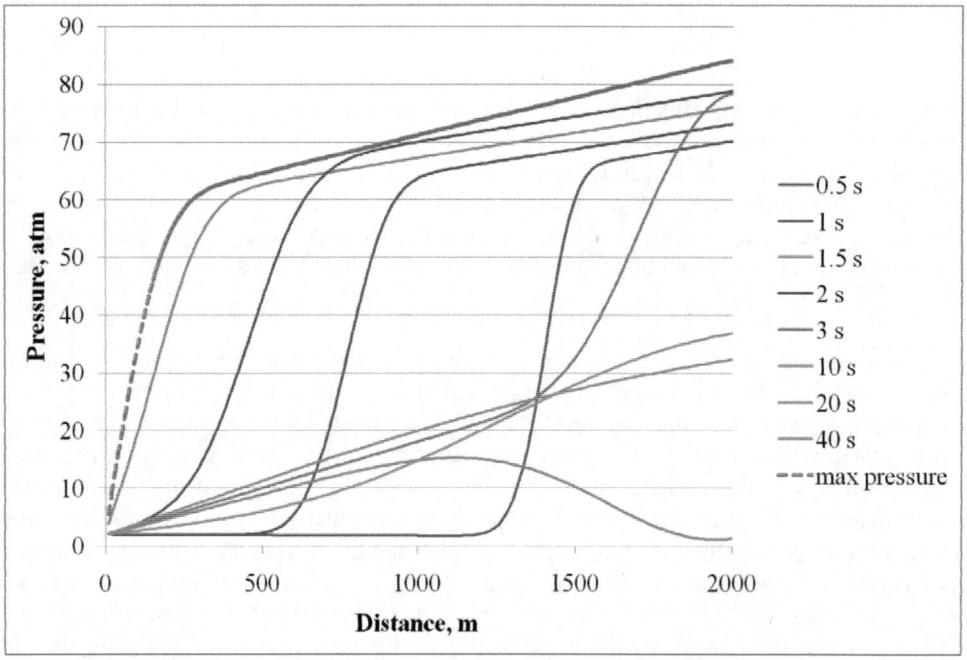

Figure 2. Pressure vs. distance at different time points after valve closing starts: 1 – 0.5 s; 2 – 1 s; 3 – 1.5 s; 4 – 2 s; 5 – 3 s; 6 – 10s; 7 – 20 s; 8 – 40 s (in the absence of pressure relief system).

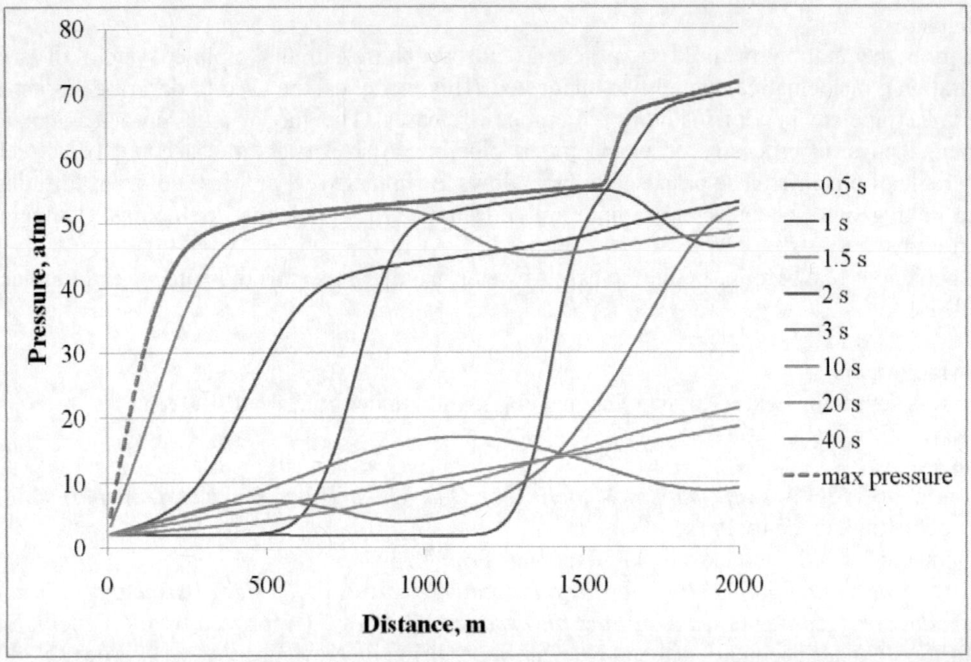

Figure 3. Pressure vs. distance at different time points after valve closing starts: 1 – 0.5 s; 2 – 1 s; 3 – 1.5 s; 4 – 2 s; 5 – 3 s; 6 – 10s; 7 – 20 s; 8 – 40 s (the working pressure of relief system is 5 atm).

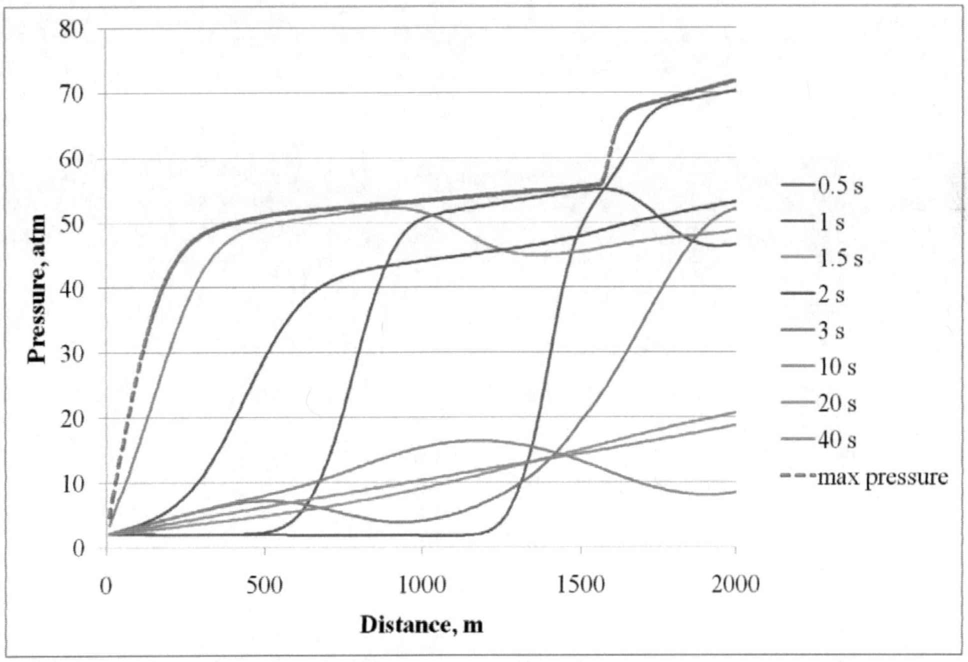

Figure 4. Pressure vs. distance at different time points after valve closing starts: 1 – 0.5 s; 2 – 1 s; 3 – 1.5 s; 4 – 2 s; 5 – 3 s; 6 – 10s; 7 – 20 s; 8 – 40 s (the working pressure of relief system is 10 atm).

4. Conclusion

To describe the motion of fluids (gas, liquid or two-phase) in the pipline system of arbitrary configuration a mathematical model was suggested. This model can be used to describe any operation regime of a pipeline system including emergency releases. The model takes into account all the fundamental physical processes occurring in the pipeline. Also it takes into account basic technical characteristic of pipelines. The proposed model allows to simulate real processes of transportation by a pipeline taking into account appearance of different type gradients (velocities, temperatures, concentrations).

The behaviour of a pipeline system in case of water hammer flow was modelled taking into account mass release.

Acknowledgments

This work is supported by the Russian Science Foundation under grant 16-19-00188

References

[1] Lurie M V 2003 *Mathematical Modeling of Gas, Oil and Oil Product Transport* (Moscow: Gubkin University Press)
[2] Sumskoi S I and Sverchkov A M 2015 *Phys Proc* **72** 347
[3] 2014 *Safety Guide "Guidelines for the quantitative analysis of the risk of accidents at hazardous production facilities of trunk oil and oil product pipelines "* (Approved by the Federal Service for Ecological, Technological and Nuclear Supervision on November 7, 2014, order № 500)

Journal of Physics: Conference Series **751** (2016) 012026 doi:10.1088/1742-6596/751/1/012026

Coating formation at laser irradiation of a dusty gas medium

A F Glova, A Yu Lysikov, S S Nelyubin, P I Peretyatko, Yu F Ryzhkov and V B Turundaevskii

State Research Center of Russian Federation Troitsk Institute for Innovation and Fusion Research (SRC RF TRINITI), Moscow, Troitsk, 142190 Russia

E-mail: afglova@triniti.ru

Abstract. A free vertical jet of microparticles in a gas is illuminated by a horizontal focused laser beam. The products of the interaction of particles with radiation are deposited on a heated substrate and form a coating. The substrate is a cathode or an anode at the application of an auxiliary discharge. The parameters of the jet have been calculated and the threshold intensity of laser radiation needed for the evaporation of particles has been estimated. The radiation spectra of the dusty gas medium have been measured and the temperature of particles has been determined. Samples of AlN, Ti, and Ti/AlN coating on steel St.3 have been obtained.

1. Introduction

The service life of parts of machines and mechanisms can be increased by deposition of functional coating with required operating properties on their surfaces. A most important property of coating is their hardness. High hardness is characteristic of binary compounds of light elements with short covalent bonds. Among them are, e.g., AlN, BN, TiN, SiC and some other compounds.

At present time, there are many methods of deposition of coatings. Of particular interest is the laser method. This is due to its relative simplicity and the possibility of evaporating of almost any material for forming coatings. There are reports on preparing of AlN coating by the laser method at irradiation of a solid targets [1,2]. In a number of works coatings are obtained using a combination of laser ablation of target and gas discharge. For example, a discharge with a frequency of 12.45 GHz was used in [3]. This made it possible to decrease the synthesis temperature of AlN on a silicon substrate surface to lower than 80°C at laser ablation of pure Al in a nitrogen atmosphere at a pressure of $2 \cdot 10^{-4}$ mmHg by radiation of second harmonic of a Nd:YAG laser. It should be noted that the role of auxiliary gas discharge in those and other experiments is not completely understood and further studies are necessary. Actually, existence of the discharge changes the ion composition of the plasma and the velocity of ion movement near the surface; microfragments outgoing from the target surface that were not evaporated during the ablation take a positive charge due to the thermal emission or a negative charge due to the charge with plasma electrons and, therefore, they participate in the drift movement and in plasmachemical reactions in the volume and on the surface, etc.

We propose a method of deposition of coating by laser ablation of particles of a dusty gas medium that also makes it possible to evaporate almost any material, like the method of laser ablation of solid targets. In this case, the power and the power density of laser radiation can be noticeably decreased, since the energy losses of laser radiation by particles are lower than the losses in the case of irradiating a solid target. A flow of charged particles directed to the substrate can be formed in an electric field or

Journal of Physics: Conference Series **751** (2016) 012026 doi:10.1088/1742-6596/751/1/012026

in a gas discharge which in combination with the flow of heated particles that take velocities of reactive motion as a result of partial evaporation, can increase the deposition rate as compared to the rate of deposition from a gaseous phase during ablation of a solid target.

The aim of this work is to study the possibility of obtaining coating on metallic substrates using the method of ablation of particles of a dusty gas medium in a form of free vertical jet of microparticles in a gas when the jet is irradiated by a horizontal laser beam.

2. Experimental setup

The schematic of the experimental setup is shown in figure 1. A focused beam *1* enters into interaction chamber *3*. Weigher *4* is filled with a powder of depositing material or a powder mixture and forms a dusty gas medium as a free vertical jet of microparticles *5* in a gas at a controlled pressure. Products of interaction of the particles with radiation in interaction region *6* are deposited on substrate *7* and form a coating. Auxiliary electrode *8* is used to study the influence of a dc electric field or a dc glow discharge on the deposition process. The substrate and the electrode have central holes for transmitting the beam. A part of the laser radiation passes through the interaction region *6* and with the help of mirrors *9* and *10* is used to heat the substrate. We used steel St.3 substrates 20 mm in diameter and 1.5 mm in thickness. The preliminary preparation of their working surfaces was carried out by chemical etching or by quartz sand jet.

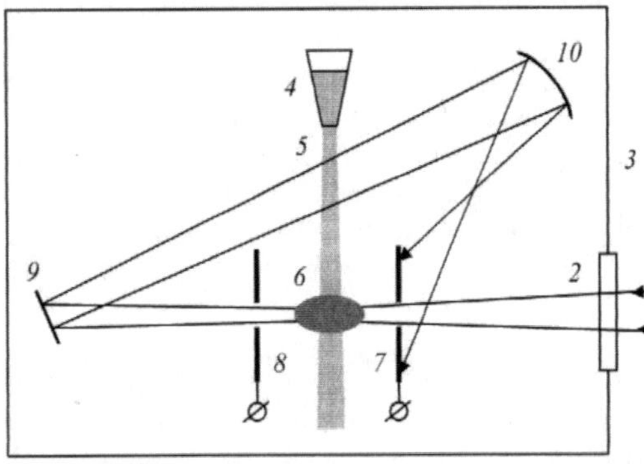

Figure 1. Schematic of the experimental setup. (*1*) laser beam, (*2*) input window, (*3*) interaction chamber, (*4*) weigher, (*5*) microparticle jet, (*6*) interaction region, (*7*) substrate, (*8*) auxiliary electrode, (*9,10*) rotating mirrors.

The experiments were performed using a fiber Yb laser with radiation wavelength $\lambda = 1.07$ µm and the maximum radiation power of 2 kW. The laser can operate in cw, pulse or pulse-periodic mode with varying pulse duration and pulse repetition rate. The laser radiation is multimode. As a multimode laser radiation is focused, the effective beam radius $w(z)$ at the distance z on both sides from the focal plane ($z = 0$) is described by the expression

$$w(z) = 2\mathrm{BPP}\frac{F}{D}\left[1+\left(\frac{z}{4\mathrm{BPP}}\left(\frac{D}{F}\right)^2\right)^2\right]^{1/2}, \tag{1}$$

where F is the lens focal length, D is the output laser beam diameter, $\mathrm{BPP} = D\gamma/2$ is the beam quality, γ is the radiation divergence. Radiation intensity I_0 averaged over a spot with radius $w(z)$ is

$$I_0 = \frac{\Delta\delta W}{\pi w(z)^2}, \tag{2}$$

where W is the radiation power, $\delta = 0.86$ is the power fraction in a spot, and $\Delta = 0.792$ is the power loss for reflection from the optical elements. For caustic length we have

$$z_\mathrm{k} = 8\mathrm{BPP}\left(\frac{F}{D}\right)^2. \tag{3}$$

As the lens focus length increases the caustic length increases markedly and this makes it possible to decrease requirements to the accuracy of the lens placing with respect to the jet. In accordance with our data (BPP = 2 mm·mrad, $D = 10$ mm) $z_k = 45$ mm for $F = 53$ cm and $z_k >> d_c$ where $d_c = 5$ mm is the jet diameter.

3. Parameters of the dusty gas medium

3.1. The initial velocity of motion and the density of particles in jet

The main parameters of a microparticle jet in the absence of laser irradiation are the velocity of particle motion and the density or the number of particles in the interaction region. To estimate these parameters, we will consider a particle as a sphere with radius r. Velocity v of the vertical motion of a particle without collisions with other particles can be found from the solution of equation $dv/dt = g - av$ where g is the acceleration due to gravity, $a = 6\pi\eta r/m$, η is the gas dynamical viscosity, and m is the particle mass. Figure 2 shows the results of calculations of the dependence of v on r for three values of h for Al particles moving in nitrogen at normal conditions at the initial condition $v = 0$ (here h is the distance from the weigher outlet to the laser beam axis). Comparatively coarse particles ($r >$ 30 μm) demonstrate noticeable difference in dependences $v(h)$ at a fixed radius, since the velocities of such particles are still far from the constant velocity limit at the chosen h. The results of calculation for other pressures will be slightly different from the results shown in figure 2, because of a weak pressure dependence of the dynamic viscosity of the gas [4].

The density of particles with almost the same size in a given cross section of the jet with area S_c can be estimated from their flow rate G using relationship $n_p = G/(mvS_c)$. For the interaction region, we have $S_c = \pi d_c^2/4$. Because the interaction region volume is also known ($V \approx \pi\, w(z)^2 d_c$), the number of particles in the region will be

$$N_p = n_p V. \tag{4}$$

Figure 3 shows the results of calculations of N_p as a function of r at $h = 3.5$ cm, $d_c = 5$ mm, $w(z) =$ 0.23 mm and $G = 20$ mg/s typical in the experiments.

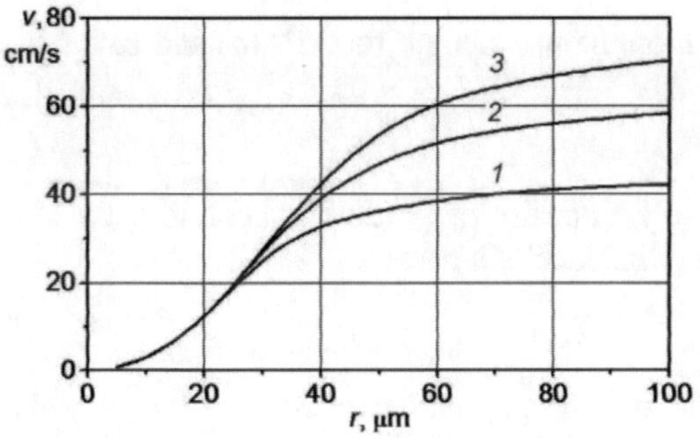

Figure 2. Dependences of v on r for Al particles moving in N_2 at $h = 1$ (*1*), 2 (*2*) and 3 cm (*3*).

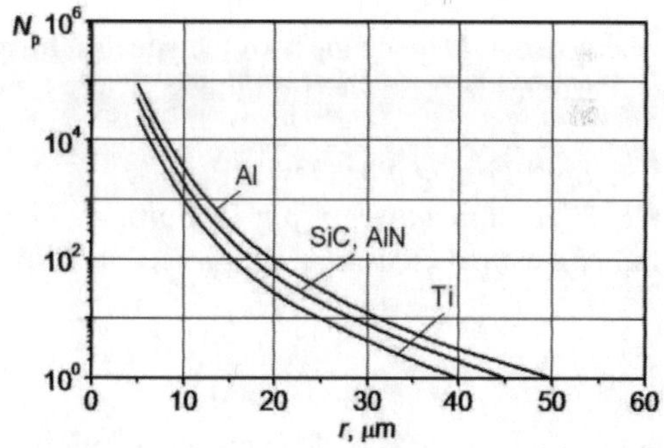

Figure 3. Dependences of N_p on r at $h = 3.5$ cm, $d_c = 5$ mm, $w(z) = 0.23$ mm and $G = 20$ mg/s.

3.2. Temperature of particles

The irradiation of a jet by laser radiation results in the formation of bright illumination in the interaction region with sizes that noticeably larger than the geometric size of the interaction region. The illumination center is slightly shifted along the beam with respect to the jet axis. The spectrum measurements were performed in the wavelength range of 400-800 nm. The radiation was focused by a lens with focal length $F = 53$ cm. The initial laser radiation intensity in a spot with radius $w(z) = 0.23$ mm was $I_0 \approx 7{\cdot}10^5$ W/cm^2, $h = 3.5$ cm; the nitrogen pressure in the interaction chamber was $p_{N2} = 1.5$ or 100 mmHg.

A feature of the measured spectra was the increase in the spectral brightness of illumination with the wavelength. The spectra of compounds such as AlN and B_4C are quite smooth at the irradiation conditions noted (figure 4), the spectra for Ti and TiC particles have clear bands and individual lines. The smooth spectra are more preferable to determine the temperature of particles of the dusty gas medium, since they can be interpreted using the standard formula for the Planck spectrum (e.g., [5]). The temperature was calculated in the assumption that the emissivity factors are independent of the wavelength. The temperatures found at p_{N2} = 1.5 mmHg for AlN and B_4C particles were 2470 K ± 15% and 2770 K ± 15%, respectively. The fact that, at the same irradiation conditions, the temperature of B_4C particles is higher seems to be due to the difference in the radiation absorption coefficient: the B_4C powder is darker than the AlN powder. Note that, at p_{N2} = 100 mmHg, the spectra for AlN and B_4C particles are similar to the spectra shown in figure 4, and the temperature is within the limits of the measurement error.

With allowance made for the measurement accuracy, the maximum temperature for AlN particles is 2840 K and close to the evaporation temperature of AlN (2790 K). Because of this, it is interesting to estimate the intensity of the laser radiation required to evaporate these and other particles in the dependence on their radii and to compare it with fixed value $I_0 \approx 7{\cdot}10^5$ W/cm^2 chosen during the measurements.

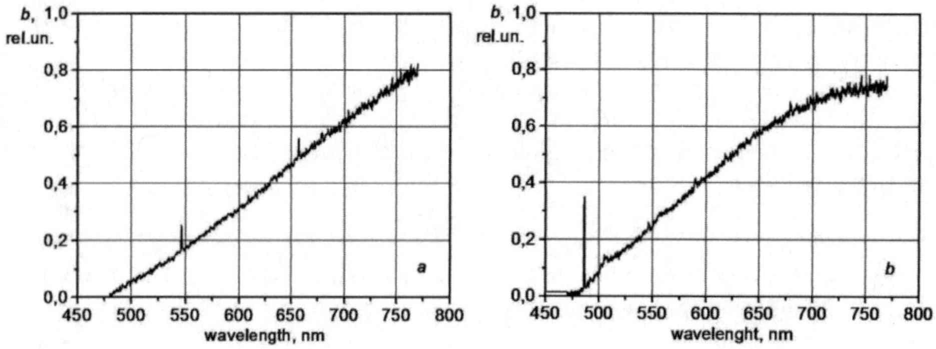

Figure 4. Dependences of the spectral brightness *b* on the wavelength for AlN (*a*) and B_4C particles (*b*) at p_{N2} = 1.5 mmHg.

3.3. Evaporation of particles

3.3.1. Threshold intensity for evaporation. The threshold intensity of particle evaporation by laser beam was estimated using a simplified equation of energy balance of a particle

$$ASI_v\delta t = m(Q_m + Q_v + cT_v) + \delta t(\sigma AST_v^4 + a_T ST_v), \quad (5)$$

where *A* is the absorption ability of a material, *S* is the surface area of a particle, I_v is the threshold intensity of evaporation, T_v is the evaporation temperature, Q_m is the specific melting energy, Q_v is the specific evaporation energy, σ is the Stephan-Boltzmann constant, a_T is the heat loss coefficient and $\delta t = 2w(z)/v$ is the flying time of a particle through the interaction region. The heat loss coefficient was calculated using the expression for laminar flow of spherical bodies taken from [6]. The model used does not take into account the change in particle mass during its evaporation and the change in δt, and actually gives the upper estimation of the intensity.

The calculations were performed for Ti, AlN and B_4C particles at $w(z)$ = 0.23 mm, h = 3.5 cm, p_{N2} = 1.5 mmHg and A = 0.3 that is the same for all particles. Figure 5 shows the results of calculating I_v as a function of *r*. The dependences show that the intensity ~$5{\cdot}10^4$ W/cm^2 is required to evaporate an AlN particle ~50 μm in diameter. This intensity is lower than $I_0 \approx 7{\cdot}10^5$ W/cm^2, and we can expect that such particles will be evaporated in the radiation field with intensity I_0, which corresponds to the results of measuring the temperature of AlN particles.

3.3.2. Criteria of optical breakdown. It is known that compound AlN decomposes at the evaporation temperature. Al vapors have a low ionization potential 5.98 eV, and their presence in the interaction region can favor the optical breakdown in the region. The thermal explosion model [7,8] gives a good agreement of the breakdown thresholds near a solid target with the measured results. We use the main statements of the model to determine the possibility of the optical breakdown in our conditions.

To estimate the threshold coefficient of laser radiation absorption α_{th} and the time of transition of gas to a high-temperature state τ_{tr}, we use the following expressions [9]

$$\alpha_{th} = \frac{\lambda_1 T_1}{I r_0^2 (1+R)\alpha^0}, \tag{6}$$

$$\tau_{tr} \approx \frac{c_1 \rho_1 T_1}{(1+R) I \alpha_{th}}, \tag{7}$$

where T_1 is the vapor temperature, λ_1 is the thermal conductivity coefficient of vapor, c_1 and ρ_1 are the heat capacity and the density of vapor, respectively, r_0 is the size that determines the heat loss, $R = 1 - A$ is the reflectivity, $\alpha^0 = \mathrm{dln}\alpha/\mathrm{dln}T$. At $T_1 = 3000$ K, $I = I_0 \approx 7 \cdot 10^5$ W/cm^2, $r_0 = w(z) = 0.23$ mm, $A = 0.3$, $\alpha^0 \sim 10$, taking for Al vapors $\lambda_1 \approx 10^{-3}$ W/(cm·K), $c_1 = 0.78$ J/(g·K) and $\rho_1 = 1.2 \cdot 10^{-4}$ g/cm^3, we have $\alpha_{th} \approx 5 \cdot 10^{-4}$ cm^{-1} and $\tau_{tr} \approx 4.7 \cdot 10^{-4}$ s.

The breakdown is possible as two conditions are fulfilled as follows. First, time τ_{tr} must be smaller than the flying time of a particle through the interaction region $\delta t = 2w(z)/v$. According to figure 2, $\delta t = 2 \cdot 10^{-3}$ s for particles 50 μm in size, and inequality $\tau_{tr} < \delta t$ obeys. In this case, the vapor temperature T_1 is close to particle temperature T. Second, the coefficient of absorption of laser radiation α must be higher than the threshold value. The calculations by formulas from [10] showed that the coefficient of absorption of laser radiation with wavelength 1.07 μm in a thermal plasma of Al vapors at the atmosphere pressure and a temperature of 3000 K is $\alpha \approx 10^{-7}$ cm^{-1} << $\alpha_{th} \approx 5 \cdot 10^{-4}$ cm^{-1}, which indicates the absence of the optical breakdown.

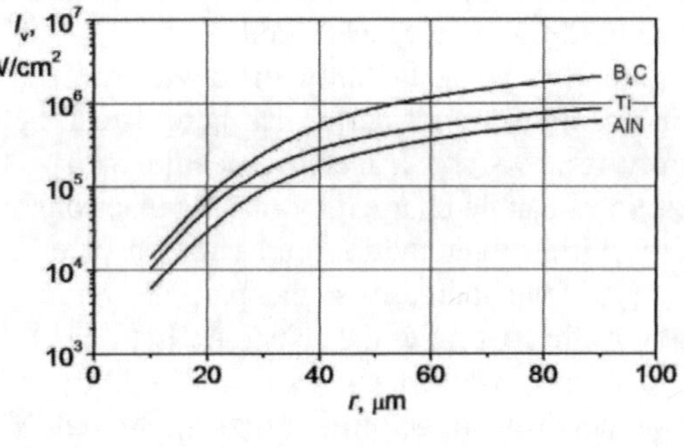

Figure 5. Dependences of I_v on r at $w(z) = 0.23$ mm, $h = 3.5$ cm, $p_{N2} = 750$ mmHg and $A = 0.3$.

3.3.3. Reactive motion of particles. Reactive acceleration of the particles under action of laser radiation was theoretically studied in detail for the first time in [11]. Standard geometry of coating deposition from the particle jet formed by a coaxial nozzle was considered. The particle collisions were not taken into account in the calculations. It was shown that stainless steel particles 45 μm in diameter can take additional velocity of ~100 m/s due to the action of the vapor recoil momentum at the mean radiation intensity ~10^5 W/cm^2.

In the case of our experiments when the jet is irradiated in a crossed geometry, the collisions must be taken into account, which will influence the mean particles velocity. A qualitative justification of the effect of reactive particle acceleration can be the mentioned above shift of the center of the illuminating region with respect to the jet axis along the laser beam.

Journal of Physics: Conference Series **751** (2016) 012026 doi:10.1088/1742-6596/751/1/012026

4. Formation of coatings and their properties

In all our experiments, the laser radiation output power *W*, lens focus length *F*, focal spot radius $w(z)$, initial intensity averaged over the spot I_0, radiation pulse duration τ_{in}, and distance *h* were unchanged and were 2 kW, 53 cm, 0.23 mm, $7 \cdot 10^5$ W/cm^2, 50 ms, and 3.5 cm, respectively

4.1. AlN coating

Figure 6 shows the photographs of coating on etched substrates of St.3 obtained by irradiation of AlN particles at p_{N2} = 1.8 mmHg and pulse repetition rate f = 0.5 Hz. The substrates shown in figures 6*a* and 6*b* were the anode and the cathode, respectively (discharge current I_d = 10 mA, distance between electrodes d = 20 mm, distance between the substrate and the jet axis d_1 = 10 mm); the substrate shown in figure 6*c* was used without discharge at d_1 = 10 mm. The substrates were preliminarily (without a jet) heated by laser, and the substrate-cathode was also additionally heated in a discharge for the same time 100 s. The weigher was immediately turned on after the preliminary heating, the substrate was additionally heated during time t = 60 s. After the complete heating cycle, the total increases in the substrate temperature ΔT_s with respect to room temperature (20°C) slightly differed from the heating at the end of the preliminary stages and were ΔT_s = 350, 385 and 350°C for the substrates shown in figures 6*a*, 6*b* and 6*c*, respectively.

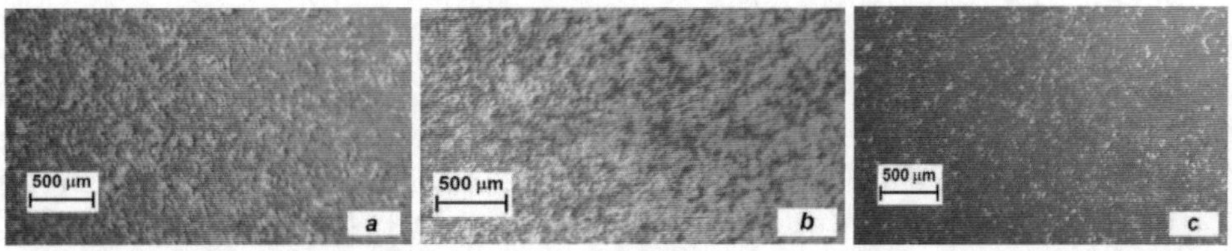

Figure 6. The photographs of coating on etched substrates of St.3 for substrate-anode (*a*), substrate-cathode (*b*) and substrate without discharge (*c*).

The X-ray diffraction spectra of the coatings showed only lines groups of AlN, α-Fe and Fe_2O_3; the latter two groups seem to belong to the substrate base metal.

The coatings shown in figure 6 are substantially different from each other in their adhesion properties. Quantitative measurements were not performed here. It can be only noted that the coating on the substrate-anode is hardly removed by a metallic needle with a small rounding radius. The reason can be that the dusty particles can be charged by plasma electrons in the discharge plasma [12] and can take additional velocity during their motion under action of an accelerating potential to the substrate-anode. According to [13], the increase in the particle velocity during depositing on the substrate improves the adhesion of the coating to the substrate. Let us perform a numerical estimation. Neglecting collisions between particles, we have the particle velocity $v_u = 0.9[qU/(\rho r^3)]^{1/2}$ cm/s (here, q is the particle charge in the elementary charge unit, U is the potential, V; density ρ, g/cm^3; radius *r*, μm). For AlN particles 1-5 μm in radius, we find that v_u = 22-2 m/s at U = 200 V and $q \sim 10^5$.

Note, that we did not obtain good adhesion when non-etched substrate-anode was used at the same p_{N2}, f, I_d and d, even we changed temperature and distance d_1 (the experiments were performed at ΔT_s = 510°C, d_1 = 10 mm and ΔT_s = 310°C and d_1 = 3 mm).

Another mechanism of accelerating particles is the acceleration under action of vapor recoil momentum. To determine the possibility of action of this mechanism, the substrate-cathode was mounted in the place of the auxiliary electrode (position *8*, figure 1) and the electrode-anode that becomes the auxiliary electrode is placed in position *7*. The distance from the substrate to the jet center is 5 mm, d = 20 mm and other parameters being unchanged. We used a non-etched substrate; its heating and the preliminary cleaning its surface were performed in a discharge. The value $\Delta T_s \approx 400$°C is approximately the same as that for the substrate-cathode in figure 6*b* (385°C). Unlike the substrate shown in figure 6*b*, these experiments demonstrated good adhesion of the coating to the substrate, despite the slowing-down of negatively charged particles in the discharge by the cathode electric field.

It can be noted that the results presented in this Section completely correspond to well-known facts [13,14]: a preliminary preparation of the substrate surface and quite high velocity of particle motion to the substrate are necessary conditions of obtaining required adhesion of a coating to a substrate.

4.2. Ti coating

Figure 7 shows the photographs with different magnification of a coating obtained by deposition of Ti particles irradiated by laser on an etched steel St.3 substrate-anode. The coating was deposited at the following conditions: $p_{N2} = 1.5$ mmHg, $f = 1$ Hz, $d = 20$ mm, $d_1 = 10$ mm, deposition time $t = 50$ s, $\Delta T_s = 300^{\circ}$C. The X-ray diffraction spectra were contained only α-Ti lines and also α-Fe lines belonging to the substrate base metal. It is likely that the TiN compound did not form because of high (3280°C) evaporation temperature of titanium and insufficient intensity of laser radiation for titanium to be evaporated. In figure 7*a*, bright points are like to solidified titanium droplets. The droplet can have a form of rings or droplets with a valley in the center surrounded with finer rings (figure 7*b*). The coating has a strong adhesion with the substrate, and it can be considered as the titanium facing on steel St.3 obtained in our conditions.

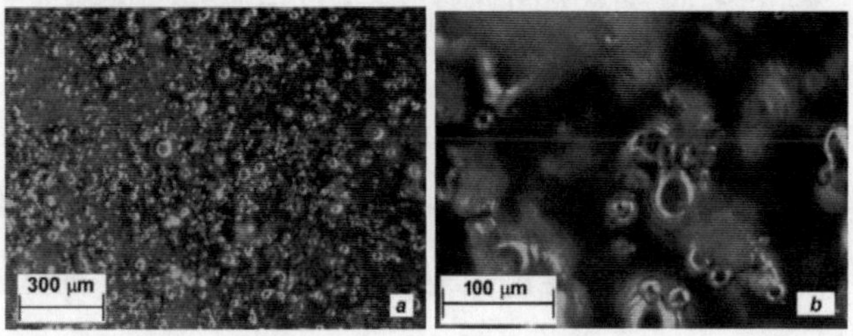

Figure 7. The photographs with different magnification of Ti coating on an etched steel St.3 substrate-anode.

Similar experiments were performed with a substrate that was subjected to sand-blasting. The difference for the conditions of deposition on the etched substrate was that we changed nitrogen pressure and turned on or did not turn on a discharge. In the case as $p_{N2} = 1.5$ mmHg and the discharge was turned on, the adhesion improved as compared to the adhesion obtained without discharge, but it was weaker than that on an etched substrate deposited with discharge. The increase in the nitrogen pressure to 15 mmHg during deposition without discharge insignificantly influences the coating adhesion and does not change the X-ray diffraction spectrum. At both nitrogen pressures, the X-ray diffraction spectra contained α-SiO_2 lines, along with α-Ti and α-Fe lines. It seems likely that the α-SiO_2 lines are due to quartz-sand particles that on the substrate surface during its sand-blasting. It seems likely that the existence of a sublayer of sand particles decreases the coating adhesion.

Thus, the experiments on deposition of Ti coating confirmed the positive influence of an auxiliary discharge on the adhesion properties of the coating and the influence of the method of preliminary preparation of a substrate on these properties.

4.3. Coating deposited by laser irradiation of a Ti and AlN particle mixture

At present, composite coatings attract significant interest, since such coatings can have high technological properties. Composite coatings are usually prepared from composite powders in which each of composite particles contains all initial components with desired properties [14].

An Al/AlN composite coating was prepared in [15] by plasma sputtering of a mixture of Al and AlN powders. The coating was a solidified layer of molten Al with embedded AlN particles. The coating combines two desired properties: high adhesion of Al melt to the substrate and high hardness due to the presence of AlN particles.

A jet containing a mixture of Ti and AlN particles with volume proportions Ti:AlN=3:1 or 1:1 was irradiated by laser beam at nitrogen pressure 1.5 and 4.5 mmHg. The substrate temperature was varied within 290-340°C, and auxiliary gas discharge was not used. We used substrates subjected and not subjected to sand-blasting. No adhesion was observed in the case of untreated substrates; in the case of the sand-blasted substrates washed by ethanol, the adhesion was better than that for unwashed substrates, but it was insufficient for practical applications.

Figure 8 shows the photographs with various magnification of the coating on the substrate of steel St.3 treated by sand-blasting obtained by laser irradiation of the Ti:AlN=3:1 particle mixture at p_{N2} = 1.5 mmHg, $t \approx 100$ s and $f = 1$ Hz. At figure 8, as compared to figure 7, the density of bright points and their size are smaller. The results of measurements of the X-ray diffraction spectra of the coating were ambiguous. While α-Fe and α-SiO_2 lines exist in all the spectra, the Ti and AlN lines are absent in the spectrum of the substrate, shown in figure 8, but the lines appeared, e. g., as the deposition time decreases and the pressure increases ($t = 30$ s and $p_{N2} = 4.5$ mmHg, respectively) at the constant ratio Ti:AlN=3:1. The absence of Ti and AlN lines in the spectra can be due to formation of the Ti/AlN solid solution on the surface or its part by a mechanism similar to that proposed in [15]. The element analysis of the solid solution was not carried out.

Figure 8. The photographs with various magnification of the coating on the steel St.3 substrate treated by sand-blasting without discharge.

5. Conclusions

As a result of this work, we designed the experimental unit for producing coatings on structural materials using irradiation of a dusty gas medium in the form of free vertical jet of microparticles in gas by horizontal laser beam. The main parameters of the medium were determined, and the temperature of irradiated particles was measured. We prepared AlN, Ti and Ti/AlN coating on substrates of steel St.3, determined the role of auxiliary gas discharge and the method of preliminary preparation of substrates.

The results can be used for development of basis of the technology of producing coating by this method.

Acknowledgments

This work was supported by the State Corporation “Rosatom” (State contract No. N.4h.44.90.13.1090 from 02.04.2013). We are thankful to collaborators of CA “SRC RF TRINITI” V A Barsuk and N T Jigailo for assistance in preparation of the experiments.

References

[1] Szekeres A, Cziraki A, Huhn G, Havancsak K, Vlaikova E and Socol G 2012 *J. Phys.: Conf. Ser.* **356** 012003
[2] Jagannadham K, Sharma A K, Wei Q, Kalyanraman R and Narayan J 1998 *J. Vac. Sci. Technol. A* **16** 2804
[3] Sun J, Wu J D, Ying Z F, Shi W, Zhou Z Y, wang K L, Ding X M and Li F M 2001 *Appl. Phys. A* **73** 91
[4] Kikoin I K 1976 *Tables of Physical Quantities* (Moscow: Atomizdat)
[5] Lebedeva V 1977 *Optical Spectroscopy Technique* (Moscow: Moscow State University Press)
[6] Kutateladze S S, Borishanskii V M 1958 *Reference Book on Heat Transfer* (Leningrad–Moscow: Gosenergoizdat)
[7] Nastoyashchii A F 1980 *Sov. J. Quant. Electron.* **10** 95

[8] Vedenov A A, Gladush G G, Yavokhin A N 1981 *Sov. J. Quant. Electron.* **11** 896
[9] Bondarenko A V, Golubev V S, Dan'shchikov E V et al 1980 *SU Academy of Sciences Reports* **253** (4) 867
[10] Raizer Y P 1987 *Physics of Gas Discharge* (Moscow: Nauka)
[11] Kovaleva I O, Kovalev O B 2012 *Opt. Las. Technol.* **44** 714
[12] Fortov V E, Khrapak A G, Khrapak S A, Molotkov V I, Petrov O F 2004 *Phys. Usp.* **47** 447
[13] Khasui A, Morigaki O 1985 *Surfacing and Spraying* (Moscow: Mashinostroyeniye)
[14] Puzryakov A F 2008 *Theoretical Basics of Plasma Spraying Technology* (Moscow: Bauman Moscow State Technical University Press)
[15] Shahien M, Yamada M, Yasui T, Fukumoto M 2011 *Coatings* **1** 88-107

Amorphous hydrogenated carbon films. Synthesis, structure and properties

A P Ryaguzov[1,3], R R Nemkayeva[1,2], B E Alpysbaeva[2] and R K Aliaskarov[2]

[1] National Nanotechnological Laboratory Open Type "al-Farabi KazNU", Kazakhstan
[2] Laboratory of engineering profile "al-Farabi KazNU", Kazakhstan
[3] National Research Nuclear University MEPhI (Moscow Engineering Physics Institute), Kashirskoe highway 31, Moscow, 115409, Russia

Abstract. Method of high-frequency ion-plasma (RF) sputtering of graphite target in the atmosphere of argon, methane and hydrogen with various ratios was used to synthesize amorphous hydrogenated carbon films (a-C:H) at low power of RF discharge in wide range of temperature and pressure of gas mixture from 0.5 to 0.9 Pa. Structural analysis by methods of atomic force microscopy and Raman spectroscopy showed significant dependence of films structure on thermodynamic and kinetic parameters of synthesis. Change of optical band gap in the range from 1.4 eV to 2.7 eV confirms considerable structural difference of amorphous carbon films obtained at different synthesis conditions. The influence of hydrogen on formation of structure is shown.

1. Introduction

All crystal forms of carbon have very different physical, mechanical and electronic properties. At the same time, the promising material for science and industry is amorphous carbon. Depending on the synthesis methods and conditions, taking a varying degree of hybridization, the carbon atoms in the amorphous phase may form graphite (GLC) and diamond-like (DLC) nanostructures. The amorphous DLC films are widely used as antireflective, chemically inert and durable antifriction coatings for various technical products [1-5].

In the synthesis of thin films, the passivation of carbon dangling bonds by hydrogen atoms takes an important position. Introduction of hydrogen reduces the density of π electrons, which leads to changes of optical, photoelectric and electrical properties.

In the paper it was used radio-frequency (RF) ion-plasma synthesis method using a magnetron at a frequency of 13.56 MHz and a gas mixture of Ar + CH_4 + H_2 in the ratios of 92/4/4 and 84/8/8. Specific power (P_d) of RF discharge varied from 1 to 1.5 W/cm^2. The average growth rate of the carbon film was ~ 3 Å / min. Purity of graphite target is 99.999 at.%, the purity of the gases Ar and CH4> 99,99at.%. The synthesis of films was carried out on quartz substrates.

2. Study of the surface morphology of the a-C: H films synthesized at different thermodynamic and kinetic parameters

The structure and surface morphology of a-C:H films prepared in gas mixture 92Ar + $4CH_4$ + $4H_2$, were studied by atomic force microscopy (AFM) (NT-MDT, Russia) using contact method with force modulation, which is useful for studying of high-resistance samples with a certain degree of electrostatic surface potential. In our case the resistivity of a-C:H films was $10^{14} \div 10^{15}$ ($\Omega \cdot$m). Figure 1 shows AFM images of the surface of a-C:H films prepared at 1.25W/cm^2 of RF discharge specific power and at a gas pressure of 0.7 Pa for three different substrate temperatures.

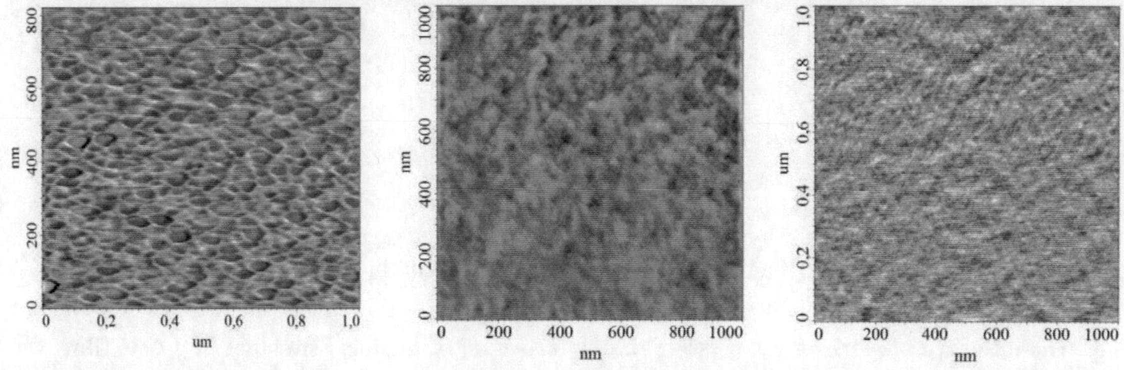

a) b) c)

Figure 1. AFM images of a-C:H films synthesized at atmosphere of gas mixture $92Ar+4CH_4+4H_2$ under P_d =1.25 W/cm^2, p=0.7 Pa: a) T_{sub} = 50 °C, b) T_{sub} = 150 °C, c) T_{sub} = 250 °C.

From figure 1 (a) it is seen that the film's structure consists of tiles of various shapes with sizes of less than 100 nm. The increase of synthesis temperature (fig. 1 b, c) leads to erosion of tile boundaries and at a substrate temperature of 250 °C surface morphology of the film has not any characteristic structural units. Usually the surface of DLC films consists of globules [6]. Therefore, the basis of tiles is not graphite or diamond crystals, but polymer structure, which formation is energetically favorable under the given conditions.

In addition, it was revealed that, the particle size, which determines the structure of the carbon film, increases with increasing gas pressure in the plasma of RF discharge. The increase of specific power leads to the increase of energy of condensing atoms and molecules and therefore to decrease of particles' size. In addition, there is an increase of the role of carbon and hydrogen ions during the structure formation.

Similar picture is observed for the gas mixture with doubles concentration of methane and hydrogen, i.e. $84Ar+8CH_4+8H_2$. AFM study of the films show, that surface morphology and structure are not uniform. Synthesis conditions significantly affect grain size, i.e. structural unit of carbon film. Depending on the constituent structural units and formation of polymeric chains, structure and form of the grain can be different. AFM data confirm defined in [7] condition of carbon films as nanoheteromorphous structures.

3. Study of the local structure of short-range order in the atomic structure of a-C:H films by Raman spectroscopy (RS)

Study of the local structure by Raman spectroscopy is performed on NTegra Spectra spectrometer (NT-MDT, Russia) with laser excitation of 473 nm (2.62 eV) and power of 1.5 mW.

Let us consider RS of a-C:H films synthesized in atmosphere of gas mixture $92Ar + 4CH_4 + 4H_2$ at specific power of RF discharge 1.5 W/cm^2. Figure 2 indicates that at T_{sub} = 50°C the position of G ~ 1545 cm^{-1}, therefore according to the three-stage model [8], the atomic structure of a-C:H films can be attributed to a diamond-like one, with a percentage of sp^3 hybridized bonds about 70-80%. Due to the increase of sp^2 site, G peak shifts towards the high-frequency region and takes value of 1590 cm^{-1}. At the same time, the appearance of D peak at 1360 cm^{-1} indicates increasing of C_6 rings.

It was revealed that with increasing specific power of RF discharge, G peak position considerably shifts towards the low frequency region and the amorphous phase of the a-C:H films is partially getting structured, passing from PLC state to the amorphous GLC and DLC structures.

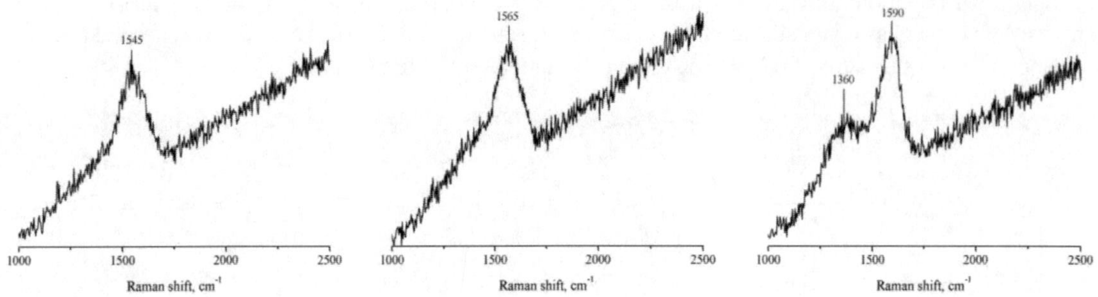

a) T_{sub} = 50 °C b) T_{sub} = 150 °C c) T_{sub} = 250 °C

Figure 2. Raman spectra of a-C:H films, synthesized in gas mixture $92Ar+4CH_4+4H_2$ at P_d = 1.5 W/cm^2 and p= 0.5Pa

The influence of working gas pressure on the formation of the local structure of a-C:H films was revealed. With increasing pressure of gas mixture 92Ar + $4H_2$ + $4CH_4$ we observed the rise of photoluminescence and shift of G peak towards the high-frequency region. Position of G peak is attributed [8-10] to the change of C-C sp^2 bonds amount per unit of volume. The increase of the gas pressure increases the number of hydrogen atoms per unit of volume, which in turn increases the formation of C-H bonds. This leads to a reduction of unbounded π electrons and localized states within the band gap. It can be seen from figure 3 that the photoluminescence differs for three different pressures. The rise of gas pressure leads to the increase in photoluminescence. Thus, with increasing pressure the part of sp^2 C-C sites increases as well. At the power of 1 W/cm^2 and the temperature T_{sub} = 150°C there is a sufficient concentration of π electrons associated with sp^2 C-C and sp^2 C-H sites which enough to provide a photoluminescence intensity greater than that of the Raman signal.

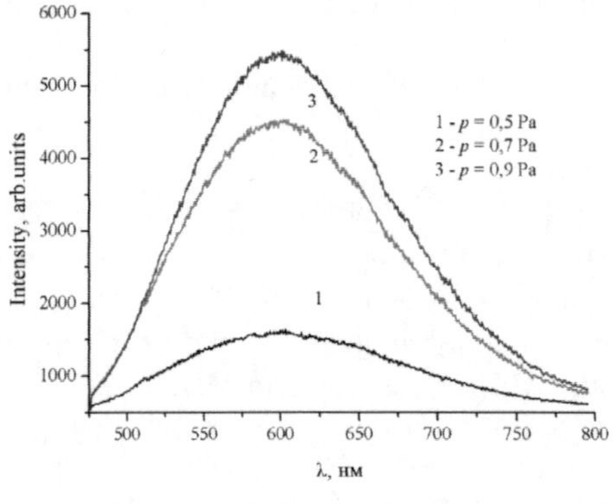

Figure 3. Photoluminescence spectra of a-C:H films, synthesized in gas mixture $84Ar+8CH_4+8H_2$ at P_d=1 W/cm^2 and T_{sub}=150 °C

At the same time, atomic force microscopy showed that, the increase in pressure leads to the increase in the size of grains forming the structure of the carbon film. Thus, the growth of the grains can be due to the increase in the number of sp^2 nanoclusters, which in turn leads to increase of the G peak intensity. Besides, shift of G peak position is due to increase in synthesis temperature.

Comparing the results of Raman spectroscopy with AFM data one can claim that if the structure is formed by particles with sizes of less than 20 nm, then the probability of the formation of nanoclusters with graphite or diamond-like phase is high enough at low specific power of RF discharge. The main criterion for their formation is the synthesis temperature T_{sub}, namely, $T_{sub} \leq$ 50°C is the favorable condition to form DLC nanoparticle, while T_{sub} ~250°C and Pd = 1.5 W/cm^2 promote formation of GLC structures. We have performed additional study about the influence of specific power of RF discharge on the structure of the films. It should be noted that at the specific power ranging from 2 to 3

W/cm^2, p = 0.5 Pa and T_{sub} = 50°C the position of G peak is 1545 cm^{-1}, and with the increase of P_d the trend of Raman spectrum, characterizing the photoluminescence, decreases. This work aims to identify the structural peculiarities of a-C:H films formed at low specific power of RF discharge, and therefore let us focus only on the brief information.

4. The optical properties of the carbon films synthesized by RF method

The bandgap (E_g) is a structurally sensitive parameter and its value depends on the atomic structure of the film. As it defined in [8], if the band gap of a-C film is less than 1 eV, then the structure of the film is formed mainly of sp^2 bonded structural units and the case of E_g =0 eV corresponds to the structure of crystalline graphite. The value of E_g ranging from 1 eV to 2 eV defines an amorphous diamond-like structure formed of sp^3 sites with its content from 40% to 85%.

We performed the study of optical transmission and reflection spectra in the range from 300 nm to 700 nm. As shown in Figure 4, with the increase in the pressure of gas mixture 92Ar + $4CH_4$ + $4H_2$ in the plasma of RF discharge the E_g value decreases. Reduction of E_g in the range from 2.5 eV to 1.45 eV is due to the increase of π electrons.

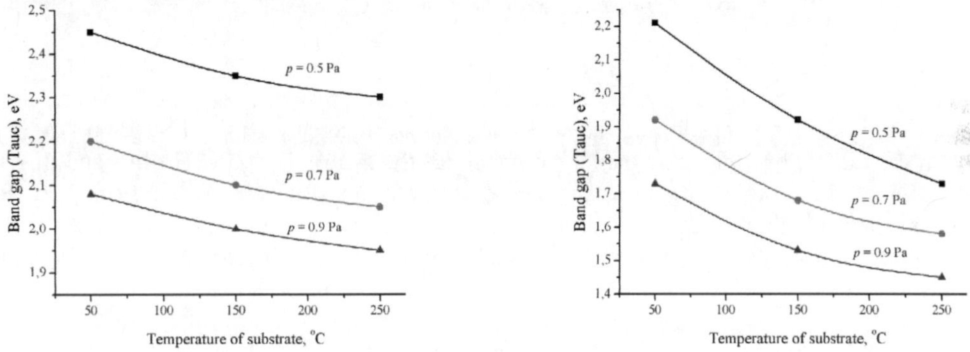

a) b)

Figure 4. Dependence of band gap E_g on the substrate temperature T_{sub} of a-C:H films synthesized at a) P_d= 1 W/cm^2; b) Pd = 1.5 W/cm^2

Figure 5 shows the plot of E_g versus synthesis temperature for a-C:H films prepared at different values of pressure 84Ar+$8CH_4$+$8H_2$ and specific power of discharge. E_g depends on the pressure of the working gas 84Ar + $8CH_4$ + $8H_2$ and the temperature at P_d = 1 W/cm^2. It can be noticed, that changes in Eg values correlate with photoluminescence data. Figure 5 (b) shows how much the power of discharge affect the band gap at different pressures and at synthesis temperature of 50°C synthesis. The increase in the gas pressure leads to a decrease in the slope of $E_g(P_d)$ function. With increasing the gas pressure from 0.5 to 0.9 Pa and the change of P_d from 1 to 1.5 W/cm^2, the value of ΔEg varies from 1 to 0.5 eV for three temperatures of synthesis. Thus, according to these data, this change of E_g values can be attributed to the decrease of possibility of carbon atoms to form the various structural units, which determine the atomic structure of the amorphous carbon film.

Comparing the data of Raman spectroscopy and optical band gap, it can be concluded that the G peak position at 1590 cm^{-1} (fig.2) at the value of the band gap of about 1.5 eV indicates the graphite-like structure, with a small fraction of sp^3 bonds. Hydrogen, as one of the structural elements of the amorphous matrix, incorporates into the structure to form it in such a way that there is a significant change in density of π electrons near the Fermi level. This process widens the band gap in graphite structures.

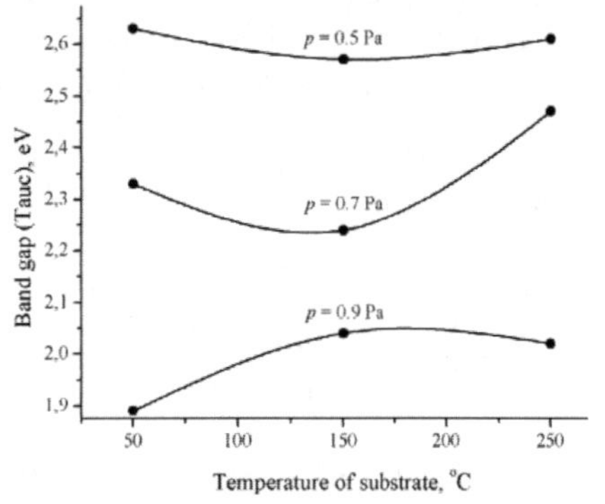

a)

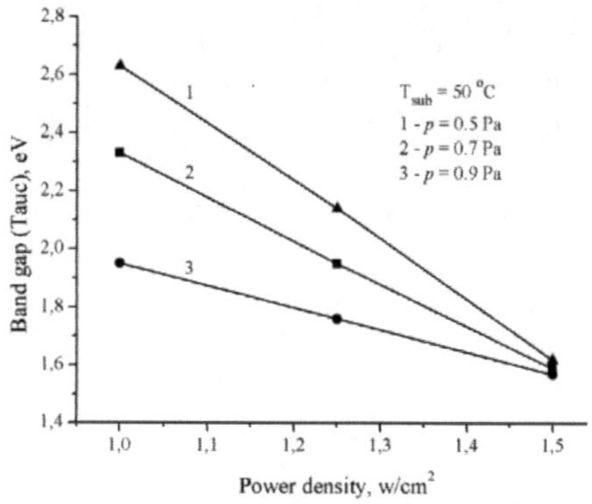

b)

Figure 5. Dependence of optical band gap on (a) substrate temperature at $P_d = 1$ W/cm^2 and (b) specific power of RF discharge at $T_{sub} = 50$ °C.

Conclusion

The results of atomic force microscopy and band gap values correlate with each other. Thus, the decrease of the grain size, constituting the structure, leads to decrease of E_g. The grain size depends strongly on the thermodynamic and kinetic parameters of the synthesis of a-C:H films and this affects the formation of the atomic structure. Changes of the atomic structure are confirmed by Raman spectroscopy data. Using two concentrations of gas mixture of argon, hydrogen and methane it is shown that hydrogen significantly affects the optical properties and structure formation of synthesized a-C:H films.

References

[1] Gangopadhyay A 1998 *Tribology Lett.* **5** 25
[2] Shum P W, Zhou Z F, Li K Y, Chan C Y 2004 *Thin Solid Films* **458** 203
[3] Yongjun W, Hongxuan L, Li J, Fei Zh, Qinghua K, Yongxia W, Xiaohong L, Weilong Q, Huidi Zh, Jianmin C 2011 *Surf. Coat. Technol.* **205** 3058
[4] Kartsova L A, Makarov A A 2002 *Rus. J. Appl. Chem.* **75** 1725
[5] P'yanova L G, Baklanova O N, Likholobov V A, Drozdov V A, Sedanova A V, Drozdetskaya M S 2015 *Protection of Metals and Phys. Chem. Surf.* **51** 407
[6] Liu C, Li G-Q, Gou W, Mu Z-X, Zhang Ch-W 2004 *Chin. Phys. Lett.* **21** 2249
[7] Ryaguzov A P, Yermekov G A, Nurmamytov T E, Nemkayeva R R, Guseinov N R, Aliaskarov R K 2016 *J. Mater. Res.* **31** 127
[8] Robertson J 2002 *Mater. Sci. and Eng R.* **37** 129
[9] Ferrari A C, Robertson J 2000 *Phys. Rev. B* **61** 14095
[10] Ferrari A C, Robertson J 2001 *Phys. Rev. B* **64** 075414

Journal of Physics: Conference Series **751** (2016) 012028 doi:10.1088/1742-6596/751/1/012028

Modification of polysulfone porous hollow fiber membranes by air plasma treatment

V V Volkov[1,2], **R G Ibragimov**[3], **I Sh Abdullin**[3], **R T Gallyamov**[3], **A A Ovcharova**[1] **and A V Bildyukevich**[4]

[1] Topchiev Institute of Petrochemical Synthesis, Russian Academy of Sciences, Moscow, Leninskii pr. 29, Moscow, Russia
[2] National Research Nuclear University MEPhI (Moscow Engineering Physics Institute), Kashirskoe highway 31, Moscow, 115409, Russia
[3] Kazan National Research Technological University, Karl Marx st. 68, Kazan, 420015 Russia
[4] Institute of Physical Organic Chemistry of the National Academy of Sciences of Belarus, Surganova st. 13, Minsk, 220072 Belarus

E-mail: vvvolkov@ips.ac.ru

Abstract. Air plasma treatment was used to enhance the surface hydrophilic properties of the polysulfone porous hollow fiber membranes prepared via a dry-wet phase invertion technique in the free spinning mode in air. Membranes prepared had porous asymmetric structure with macroporous support on the shell side and fine-porous selective layer on the lumen side. The wettability of the inner membrane surfaces were checked by contact angle measurements and FTIR was used to compare the surfaces before and after plasma treatment. Membrane morphology was examined with confocal scanning laser microscopy (CSLM). Contact angle measurements confirm that air plasma treatment affords improvement in the wettability of polysulfone membranes and FTIR results show that air plasmas chemically modify the lumen side membrane surface, however, there is no significant change in membranes chemical structure after modification. CSLM data obtained, as well as gas permeability (He and CO2) measurements show that after plasma treatment pore etching occurs.

1. Introduction

Polysulfone (PSf) is an important material in the field of polymeric membranes because of its mechanical, thermal and chemical stability as well as its excellent film forming properties. It is used in different morphologies, such as flat sheet or hollow fiber membranes, primarily for microfiltration and ultrafiltration separation applications [1].

Modified PSf membranes can be obtained in different ways. One applicable option is adding modifying macromolecules [2] or carbon nanotubes [3], in the polymer dope solution as an additive. By using interfacial polymerization method, polymers could be coated onto PSf membranes [4, 5], and the polymerization occurs at the membrane surface and pore surface.

Low-temperature plasma treatment with non-polymer-forming gases has been used in the last years as a useful tool to modify the surface properties of different materials, including PSf porous membranes [6-14]. This technique is more useful and effective than most chemical and thermal

methods. The active species generated in plasma can activate the upper molecular layers on the surface, thus improving wettability or adhesion without affecting the bulk of the polymer.
Three basic phenomena affecting porous membrane properties can take place in plasma using non-polymer-forming gases [14]: etching and reactions giving volatile products result in increase of pore diameter and porosity; modification of the chemical structure of the surface layer, giving hydrophilization or hydrophobization, depending on plasma conditions; deposition of polymer film made of volatile products that come from the etched surface, which can result in lower porosity. In this study, the air plasma modification of porous asymmetric PSf hollow fiber membrane surfaces has been investigated using confocal scanning laser microscopy (CSLM), IR spectroscopy, contact angle and gas permeability measurements.

2. Experimental

Porous asymmetric PSf hollow fiber membranes were the object of research. Membranes were prepared via a dry-wet phase invertion technique in the free spinning mode in air when bore fluid was brought into liquid polymer solution orifice. Membranes prepared had porous asymmetric structure with macroporous support on the shell side and fine selective layer on the lumen side.
For hollow fiber low-temperature non-equilibrium plasma modification the following experimental setup was used. It includes high frequency generator (HF generator), vacuum chamber, filling gas feed system, high voltage rectifier, high frequency plasma torch (a device for high frequency capacitive discharge generation) and control equipment.

Device for high frequency gas discharge (Fig. 1) is a dielectric gas-filled camera close to dielectric stand-by with two parallel pad electrodes. Pad electrodes are connected with gas-filled camera by loop electrodes with some of them plugged in the HF generator while the rest of them being grounded. Electrodes plugged in the HF generator as well as grounded electrodes are connected in parallel. Electrodes plugged in the HF generator are next but one interspersed with grounded electrodes. All loop electrodes are placed perimeter-wise outside of the dielectric gas-filled camera providing plasma space distribution in the camera. Such electrode configuration when loop electrodes are in-phase plugged in the HF generator and grounded stabilizes electromagnetic component of the field and doesn't produce zero potential points in the bulk of dielectric gas-filled camera which results in plasma burning uniformly and stable in the camera space.

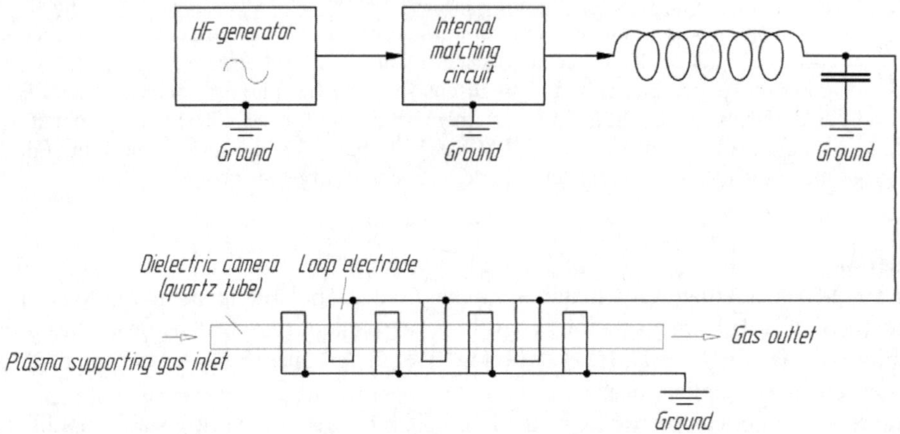

Figure 1. Gas discharge camera electrode connection diagram

Hollow fiber membranes plasma modification was carried out as follows: first, dielectric gas-filled camera was vacuumized; after that, 10 specimens of the membranes were placed into the camera followed by plasma supporting gas transmission at 110 Pa. Air was used as the plasma supporting gas with mass flux 0 to 0.24 g/s.

Turning the HF generator on lets discharging plasma fill the dielectric camera providing spatial uniformity and combustion stability. The duration of plasma treatment was 3 minutes. Dielectric gas-filled camera is made of quartz tube having high optical clarity, mechanical and thermal stability and also small dielectric quartz loss in operating band.

Wetting contact angle measurements of inner surface of the membranes were carried out via a conventional sessile drop method using the LK-1 goniometer. For surface hydrophilicity properties characterization of unmodified and plasma treated membranes and determining their surface energy values water and ethylene glycol were used as testing liquids couple. Such system is widely used for membrane surface energy determination, as surface energy components of both water and ethylene glycol are well-known. For image capture and digital processing of drops images, DropShape software was used providing Laplace-Young contact wetting angle calculation. Measurement error was ± 2°. Experiments were carried out at room temperature (23 ± 2 °C).

Hollow fiber gas permeability was measured via the manometric technique. Carbon dioxide and helium were chosen as operating fluids, as their molecular mass difference provides reliable way to determine Knudsen gas flow on the ideal selectivity value (permeability coefficients of individual gases ratio).

Membrane surface chemical composition analysis was performed via IR spectroscopy technique. IR Fourier spectrometer IRAffinity-1 (Shimadzu, Japan) was used for absorption spectra analysis.

Membrane morphology for both unmodified and plasma treated membranes was examined with confocal scanning laser microscopy (CSLM Olympus Lext OLS-4000, Japan). After scanning, surface images of different points with various magnifications were obtained for each membrane specimen. Based on scanning data, surface micro relief characteristics were calculated. Surface roughness was the key characteristic. When determining the surface roughness main criteria were arithmetic average of the roughness profile (R_a) and roughness height (R_z). Roughness characteristics were evaluated based on all the scan points. For each specimen surface scans were obtained in 10 different points and then R_a and R_z values were averaged.

3. Results and discussion

Active components generated in plasma may activate upper molecular layers of membrane surface, therefore increasing their hydrophilicity without producing changes in the membrane bulk. Under bombardment by plasma ions radicals are produced which may interact with gas molecules. Hollow fiber membranes surface properties analysis shows that under air plasma treatment oxygen-containing functional groups appear on the membrane surface, in particular, carbonyl and carboxyl groups, which formation is connected with terminal groups oxidation under chemical bonds breaking. As can be seen from Table 1, average contact wetting angle value both for water and ethylene glycol decreases after plasma treatment, while surface energy value increases. It indicates that plasma treatment enhances hydrophilic surface properties of both inner and outer surfaces of hollow fiber PSf membranes.

Table 1. Wetting contact angle θ and surface energy γ_s values for unmodified and plasma treated membranes (for lumen hollow fiber surface)

Parameter	Unmodified membrane	Plasma treated membrane
θ_{av} (H_2O), °	81,3	72,5
θ_{av} (ethylene glycol), °	69,7	65,3
γ_s^{av}, mJ/m^2	24,0	33,2

Table 2 gives gas permeability data for unmodified and plasma treated membranes. Mean pore size (d_{av}) values are derived from calculation based on Dusty Gas Model (DGM), considering both Knudsen and Poiselle flows contribution.

Table 2. Gas permeability and mean pore size values for unmodified and plasma treated membranes

Parameter	Unmodified membrane	Plasma treated membrane
P/l (CO_2), (m^3/m^2*h*bar)	560	880
P/l (He), (m^3/m^2*h*bar)	1300	1300
Ideal selectivity α (He/CO_2)	2,3	1,5
d_{av}, nm	24	34

As can be seen from table 2, for plasma treated membranes ideal selectivity value is 1.5, which corresponds to mixed flow mode with Poiselle flow mode prevailing, while for untreated membranes α = 2,3 corresponds to Knudsen flow mode prevailing. Membrane selectivity decrease after plasma treatment indicates that average membrane pore size increases because of etching phenomena; such conclusion is in the good agreement with data calculated by DGM.

Fig. 2 depicts photomicrographs for selective layer dimensions evaluation for both unmodified (a) and plasma treated (b) membranes. Thickness variation of the membrane selective layer is observed in the cross-section: for untreated membrane the thickness value was ~28 μm while for plasma treated membrane it decreased to ~24 μm.

As separation selectivity (under constant external conditions) depends exclusively on inner membrane material structure, membrane permeability directly depends on selective layer thickness. Therefore, membrane selective layer thickness decrease increases overall membrane performance as well as the membrane units performance, allowing to reduce overall dimensions of separation plants. Membrane selective layer thickness decrease proves the fact that plasma treatment influences on the inner membrane surface.

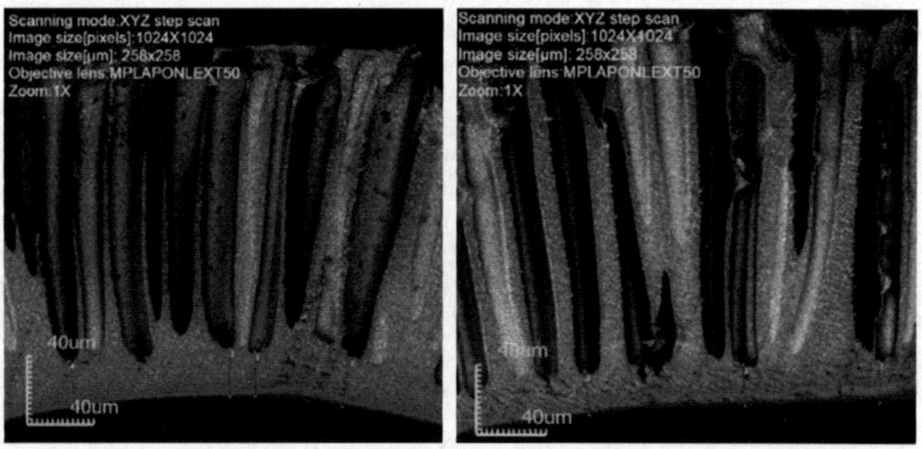

a *b*

Figure 2. Hollow fiber membrane selective layer dimensions: a – unmodified specimen, b – plasma treated specimen.

The result of low-temperature plasma treatment on the membranes is a change in their surface morphology, which influences on the separation processes characteristics. Data obtained show that after plasma treatment pore etching occurs. Membrane surface topography changes after plasma treatment. For unmodified membrane, arithmetic average of the roughness profile (R_a) is 0.15 μm and roughness height (R_z) is 0.68 μm. Plasma treatment causes minor morphology changes and results in

Journal of Physics: Conference Series **751** (2016) 012028 doi:10.1088/1742-6596/751/1/012028

R_a and R_z values decrease to 0.11 and 0.49 μm, respectively. So, smoothing of surface heterogeneity is observed, and for plasma treated membranes, their surface becomes less rough.

Results obtained by mechanical properties change analysis show that strength properties of membranes don't change significantly after plasma treatment, keeping initial values.

IR spectroscopy data show that there is no significant change in membranes chemical structure after modification. It should be noted that slightly increased absorption is observed near the 1650 and 1685 cm^{-1} absorption bands, which corresponds to carbonyl stretching vibrations, and near the 1041 cm^{-1}, which corresponds to C-O bond stretching vibrations. It proves that oxygen content in the upper membrane layer slightly increases, and hydrophilic surface properties enhance. IR-Fourier chemical structure analysis for plasma treated membranes shows that plasma treatment leads to the surface layer oxidation and oxygen-containing functional groups formation.

4. Conclusion

Hollow fiber porous polysulfone membranes low temperature air plasma treatment was carried out. It is shown that plasma treatment allows membrane surface to get enhanced hydrophilic properties due to the surface layer oxidation. Plasma treatment also reduces membrane selective layer thickness and smoothens its roughness without decline in mechanical properties. Low temperature air plasma treatment can be successfully used to obtain modified polysulfone hollow fiber membranes with enhanced surface hydrophilic properties.

Acknowledgements

This work was performed at the Topchiev Institute of Petrochemical Synthesis and supported by the Russian Science Foundation, project no. 14-49-00101. V.V. Volkov, A.A. Ovcharova and V.A. Bildyukevich acknowledge the financial support of RSF.

References

[1] Schott K 1997 *Handbook of Industrial Membranes* (Elsevier Science) p 904.
[2] Korminouri F, Rahbari-Sisakht M, Matsuura M and Ismail A F 2015 *Chem. Eng. J.* **264** 453–61
[3] Plisko T V, Bildyukevich A V, Volkov V V and Osipov N N 2015 *Petrol. Chem.* **55** 318-32
[4] Chu L Y, Wang S and Chen W M 2005 *Macromol. Chem. Phys.* **206** 1934–40
[5] Liubimova E S, Bildyukevich A V, Melnikova G B and Volkov V V 2015 *Petrol. Chem.* **55** 795–802.
[6] Steen M L, Jordan A C and Fisher E R 2002 *J. Membr. Sci.* **204** 341
[7] Gancarz I, Pozniak G and Bryjak M 1999 *Europ. Polym. J.* **35** 1419
[8] Wavhal D S, Fisher E R 2005 *Desalination* **172** 189-205
[9] Gancarz I, Pozniak G and Bryjak M 2000 *Europ. Polym. J.* **36** 1563
[10] Castro Vidaurre E F, Achete C A, Gallo F, Garciac D, Simão R and Habert A C 2002 *Mater. Res.* **5** 37
[11] Kim K S, Lee K H, Cho K and Park C E 2002 *J. Membr. Sci.* **199** 135
[12] Asfardjani K, Secui Y, Aurelle Y and Abidene N 1991 *J. Appl. Polym. Sci.* **43** 271
[13] Hopkins J and Badyal J P S 1996 *J. Polym. Sci. A* **34** 1385
[14] Gancarz I, Pozniak G and Bryjak M 1999 *Europ. Polym. J.* **35** 1419

Multiscale simulation of ion beam impacts on a graphene surface

K B Dybyspayeva[1], A Zhuldassov[1], A Ainabayev[1], A F Vyatkin[2], K Alekseev[3] and Z Insepov[1,3,4*]

[1] Nazarbayev University, Astana, 010000, Kazakhstan
[2] Institute of Miscroelectronics Technology, Chernogolovka, Moscow region, Russia
[3] Moscow Engineering Physics Institute, Moscow Russia
[4] Purdue University, West Lafayette, IN USA

E-mail: *zinsepov@purdue.edu

Abstract. Multiscale study of single and multilayer graphene irradiation is presented in this paper. Ab-initio density-functional theory (DFT) was used to study point defects, and a large scale parallel molecular-dynamics (MD) simulations were used for studying formation of gas cluster ion impacts. Moreover, Raman spectra of pure and defect graphene samples were studied from DFT calculations. Threshold energies for creating craters on the surface of graphene were obtained from MD and compared with published papers. The results of simulations were also compared with experimental craters and surface shape.

1. Introduction

Since its first discovery in 2004 graphene has been one of the favored topics in the scientific area [1]. Researchers have published numerous works about the studies of graphene properties and their applications in different fields [2-5]. Despite the fact that graphene is a 2-dimensional thin material it has proven to be mechanically strong and chemically stable [2]. Moreover, it has excellent electrical conductivity properties that are widely used to design graphene-based electronics. It is also applicable for creating graphene-based biomedical devices [3].

Performing and experiment of graphene bombardment is a challenging task. Firstly, experiment is expensive. Secondly, there are several circumstances that prevent from conducting a quality work. Samples of graphene are not perfect: most of the time, the coverage percentage is low. Moreover, during the experiment, bombardment can make graphene flakes merely fly off. Therefore, to study this process there is a need for simulations which can provide some insight on favorable conditions for bombardment, such as, threshold energy, cluster size etc.

Recently, defect formation on graphene surface has been addressed in a considerable number of articles. For instance, Hopster discusses the damages in graphene in the result of irradiation by slow highly-charged ions [4]. However, mechanism of defect formation during HCI interaction with target material is not clear and it requires further investigations. For carbon materials such as HOPG, carbon nanomembranes (CMN) and graphene it is shown that high potential energy (charge state ~ $q > 25$) and low kinetic energy (several keV) required to form induced friction type defect in graphene, hillocks in HOPG and nanopores in CNM [5-6]. In [7] authors propose bombardment by electron and oxygen plasma treatment for the purposes of building nanopores on the surface of graphene. Plasma treatment can create nanopores in graphene and membranes made of plasma treated graphene shows good results for salt rejection, but plasma treatment do not allow fully control size of nanopores and theirs distribution. Celebi, on the other hand, proposes a method of drilling nanopores on graphene with focused ion beam (FIB) [2]. Although, using FIB drilling allows to create controlled size and uniformly distributed nanopores, but drilling millions nanopores might be formidable task in terms of consuming energy and time. Irradiation of graphene by ions creates single and double vacancies. However, these vacancies tend to reconstruct over time due to the self-healing property of graphene [8]. And this reconstructed graphene lattice still cannot serve as membrane because it still remains impermeable for atoms and molecules as shown in simulations [9]. In the present work, we propose a new method to make nanofilters. We state that irradiation by gas cluster ion beams (GCIB) is a very effective way for building controllable nanopores on the surface of graphene. Such membranes can be used to design water desalination filter with an efficient water transport. We present our preliminary theoretical analysis on irradiation of samples of graphene by GCIB both using density functional theory (DFT) and molecular dynamics (MD).

Evolutions in methods of calculating electronic structure, especially in density functional theory (DFT) [10], combine with the steady increase of available computing capacities; give new opportunities to calculate large systems. Accuracy of DFT technique makes this method an alternative in assisting the design and development of the generation of new devices for nanoscale applications. Unique functional of the electron density is the total energy of the system main principle of DFT. O. Lehtinen, J. Kotakoski used DFT to study the impacts of ion irradiation on production of defects in graphene [8]. In present study DFT was used to investigate graphene lattice reconstruction when single vacancy is introduced in graphene lattice. In addition, we considering DFT as method for calculation Raman spectra of graphene and in present work we show that G peak in graphene changes in position, width and shape with number of layers. It has been shown experimentally [12, 13, 14] and theoretically [15] that the G peak in graphene changes in position, width and shape with number of layers. In [12] the authors investigated variations in the Raman spectrum as a function of the number of graphene layers by using micro Raman spectroscopy and atomic force microscopy (AFM). The variations in the Raman G peak (1580 cm^{-1}), G* peak (2450 cm^{-1}), and 2D peak (2700 cm^{-1}) were observed as functions of the number of graphene layers.

Graf et al. [13] reported on Raman mapping of single- and few-layer graphene flakes resting on a silicon oxide substrate. Lateral resolution of 400 nm allows one to address neighboring sections with various layers of graphene down to a single graphene sheet, previously determined with the scanning force microscope (SFM). Gupta [14] presented the first formal report of Raman scattering studies on nGLs (Graphene Layers), where n is in the range $1 < n < 20$, as determined by AFM. An empirical force-constant method has been used by Wang et al [15] to investigate phonon dispersions of graphite and graphene. Their experimental and theoretical results clearly show that certain Raman peaks change with the number of graphene layers (n).

To study electronic and vibrational properties of crystals or molecules Raman spectroscopy is widely used. CASTEP (Cambridge Serial Total Energy Package) [16] is used to compute a Raman intensity. CASTEP delivers a correct symmetry assignment of Raman active vibrational modes and a quantitative measure of the Raman intensity change around the transition.

2. Simulation and modeling

2.1 DFT simulation

2.1.1 The model. Figure 1 shows 5x5x1 graphene supercell and four layer graphene which were used to calculate Raman spectrum of perfect graphene film before ion irradiation. A finite displacements technique was used to calculate the dependence of the graphene G peak position on the layer numbers of graphene. It is known that the G peak position shifted to lower energies if the number of graphene layers increases.

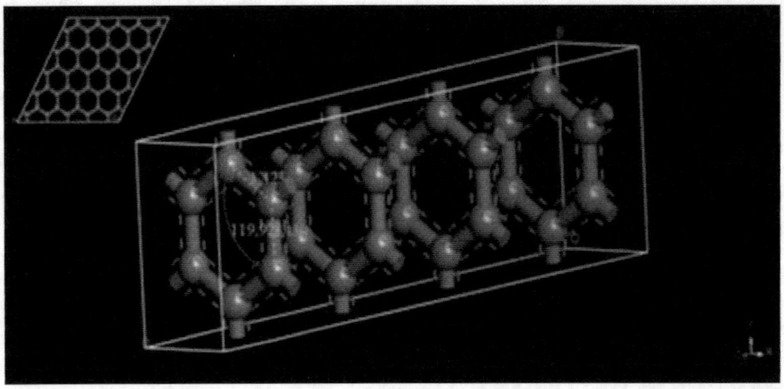

Figure 1. a) - supercell 5x5x1 lattice structure of graphene and b)- four layer grapheme.

2.1.2 Calculation details. We used a first-principles Raman calculation technique based on density functional theory. We used CASTEP which is a first-principles calculation code with using the plane-wave pseudopotential approach. We adopted GGA (PBEsol) [17] as the term exchange correlation with a cutoff energy from 750 to 940 eV and all calculations was performed by norm-conserving pseudopotential (2s22p2). The k-point separation was 0.04 Å-1 Monkhorst-Pack grid is 7x7x1, represented 25 irreducible k-points.

For graphene layers we considered the elementary cell, the distance between layers 0.335 nm, from 1 layer to 10 layers. Layers stacked directly on top of each other (AA stacking) [18].

2.1.3 Results and discussions. Irradiation by single charged ions, in general, produces single or double vacancies in graphene lattice. For high energy ions (1000 keV) graphene is almost transparent. Although single charged ions create defects as vacancies, graphene reconstructs its lattice and remaining flat with sp2 hybridized structure [11].

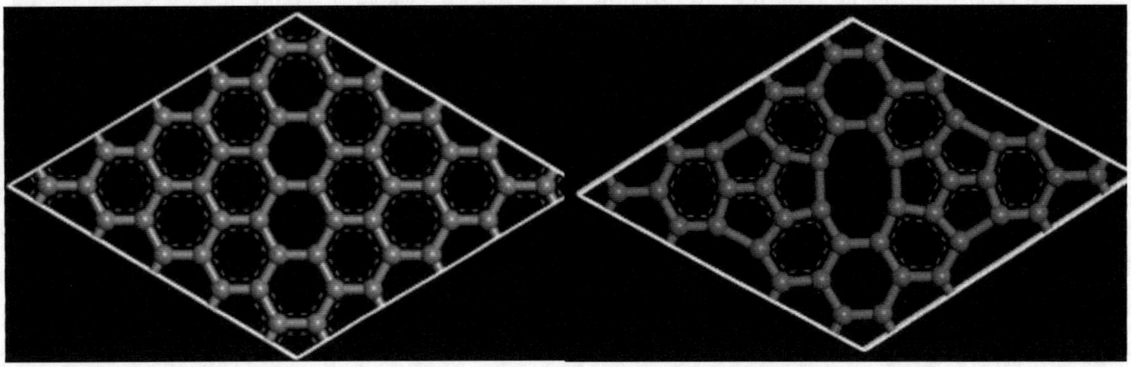

a) b)

Figure 2. a) -pristine graphene lattice and b) –reconstructed lattice of graphene mainly in the form of a network of pentagons and heptagons after introducing single vacancyin pristine graphene lattice.

In Figure 2, it is shown DFT optimized graphene lattice after removing one atom from its lattice. Reconstructed graphene lattice were obtained by a first-principles technique based on DFT. We used a first-principles calculation code with using the plane-wave pseudopotential approach. We adopted GGA (PBEsol) as the term exchange correlation with a cutoff energy from 750 to 940 eV and all calculations was performed by norm-conserving pseudopotential (2s22p2). The k-point separation was

0.04 Å-1 Monkhorst-Pack grid is 5x5x1, represented 25 irreducible k-points. As can be from Fig. seen lattice of graphene relaxed mainly in the form of a network of pentagons and heptagons with lower density and remaining flat with sp2 hybridized structure. This kind of lattice is still able to hold volatile substances and can serve as volatile substance storage for conduction ion beam analysis [11]. It means that graphene irradiation by single charged ions are not appropriate for creation membranes with few nanometer defects for filtration purposes, for example, for water desalination.

The shift of Raman spectrum G-peak was obtained in our DFT calculations of a multilayer graphene stucture with 1-10 layers (Figure 3) by using CASTEP software. The results of these calculations were in accordance with the statement that as number of graphene layers increases G peak position shifted to lower energies. Figure 2 shows comparison of the calculated (this work) and experimental (this work) Raman G-peaks with other experimental and calculated G-peaks means [12-15]. It can be seen from the comparison that CASTEP result in a good agreement with other experimental and theoretical works which means that CASTEP are reliable tool to conduct Raman shift related calculations.

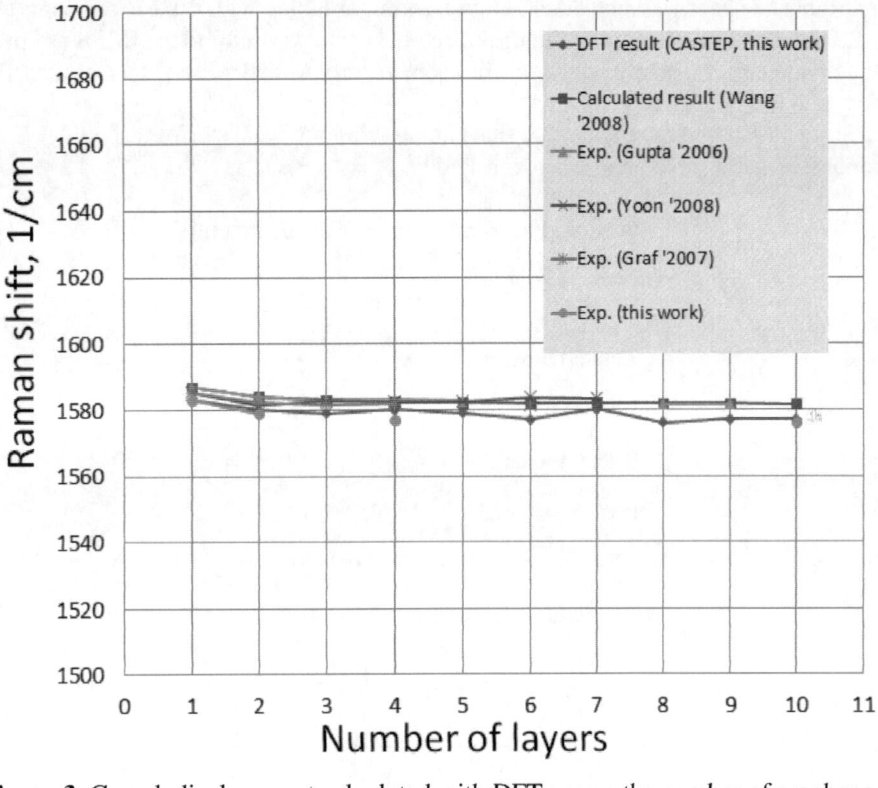

Figure 3. G-peak displacement calculated with DFT versus the number of graphene layers.

One of the main future goals of using CASTEP for calculation Raman spectra of a defect containing graphene sample is to determine different types of defects (vacancies, craters, holes) created by ion irradiation by evaluation of Raman spectra of irradiated graphene. Thus, by comparing experimental Raman spectra of GCIB irradiated graphene samples with calculated defect containing graphene, one can understand the kind of defects GCIB produces in the graphene structure. This study is still underway and the results will be published elsewhere.

2.2 MD Simulation

Many papers have been dedicated to the MD study of graphene irradiation with ions. In their models, scientists examine bombardment processes of single layer [6] and multi-layer graphene sheets [7] as well as graphene on various substrates such as SiC [8]. In the present work, we developed the simulation model of a single layer graphene irradiated with argon cluster in the result of which we observe formation of nanopores on the surface of the material. By controlling the fluency of irradiation and the size of the cluster, it is possible to create pores of a desired diameter. We believe that such technology can be useful to create energy efficient filters for water desalination. It has also been tested and discussed in [9] where the experimental material was MoS2.

2.2.1 The model. Irradiation of graphene with argon cluster was studied using large-scale molecular massively parallel simulator (LAMMPS) package [19]. The simulation model consisted of an individual argon cluster and a single layer graphene sheet. Incident energy of the cluster was varied between 2.8 eV/atom to 28 eV/atom in –z-direction, and size of the graphene sheet was chosen between about 100×100 Å^2 and 1000×1000 Å^2. Size of the cluster was chosen to be between 10 and 1070 atoms in the cluster. Tersoff potential is applied to describe carbon-carbon interactions of the graphene sheet, placed in the *xy*-plane at $z = 0$. Lennard-Jones and Buckingham potentials were used for Ar-Ar and Ar-carbon interactions, respectively.

Modeling graphene sheet was implemented by building a lattice with 4 basis atoms in a rectangle crystal cell. Lattice parameters and coordinates of the basis atoms used in LAMMPS are presented in the Table 1. Parameters of Tersoff potential are taken from LAMMPS BNC.tersoff data file, which was then converted to LAMMPS “real” units.

Initially, cluster temperature was set to 0K. The simulations were performed in NVE ensemble. Boundary conditions were set to periodic in *x, y,* and *z*.

Table 1. The parameters of Tersoff potential used in MD simulations [13].

Parameter	Value/Coordinates
Lattice constant	1.421 A
Lattice vector a1	(3, 0, 0)
Lattice vector a2	(0, 1.732, 0)
Lattice vector a3	(0, 0, 2.357)
Basis atom 1	(0, 0, 0)
Basis atom 2	(0.333, 0, 0)
Basis atom 3	(0.5, 0.5, 0)
Basis atom 4	(0.833, 0.5, 0)

2.2.2 Results and discussions. Upon irradiation, one can observe defect formation on graphene sheet, when the incident energy is higher than the threshold one. The purpose of this computational method is to explain cluster ion-solid interaction on a molecular level and to provide guidelines in determining optimal conditions to achieve desired material processing results in further cluster ion beam irradiation experiments.

Threshold energy was obtained from this simulation study, which was compared with the experimental data obtained by irradiation of graphene and graphene oxide films with Ar and Ar+O_2 mixture cluster ion beams at Exogenesis Corp., USA. The simulation data was also compared with the similar research study by Zabihi et al [6].

It was observed that threshold penetration energy of the cluster depends on the cluster radius as well as on the graphene sheet area. Figure 4 shows snapshots of our molecular dynamics simulations of argon cluster bombardment on a graphene sheet with the cluster energy of 10.12 eV/atom for graphene area of 100x100Å^2 and 1000x1000Å^2 at various time instants.

Figure 6 shows comparison of our simulation results for threshold values with the simulation data in [6]. Both set of simulation values have the same orders of magnitude for the 100xa100Å^2 graphene area for small cluster sizes. However, the present study shows that the threshold energies per atom are higher for Ar_{50}, and lower for less number of atoms in the cluster.

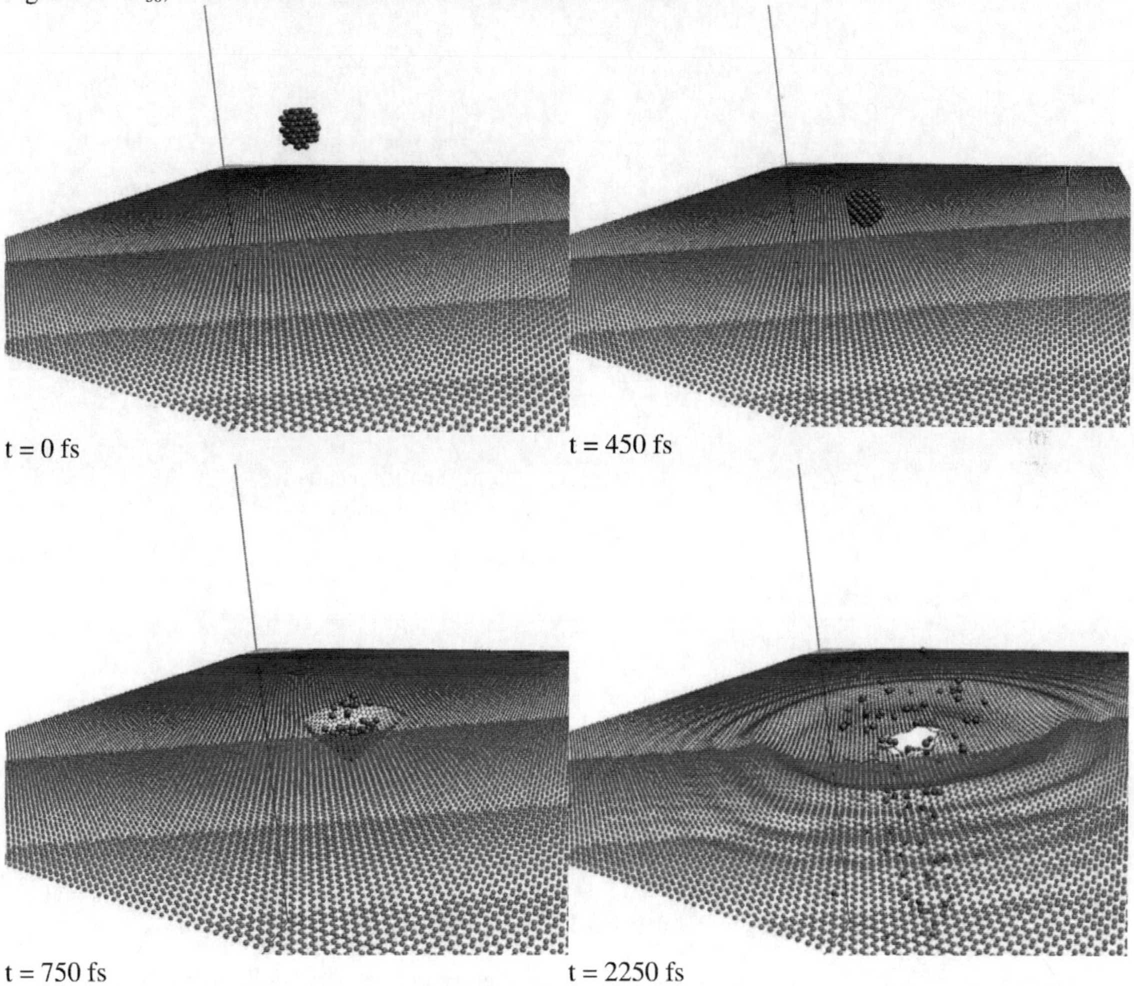

Figure 4. Simulation snapshots of argon cluster bombardment on a graphene sheet with the cluster at different timesteps. Images are processed using Atomeye.

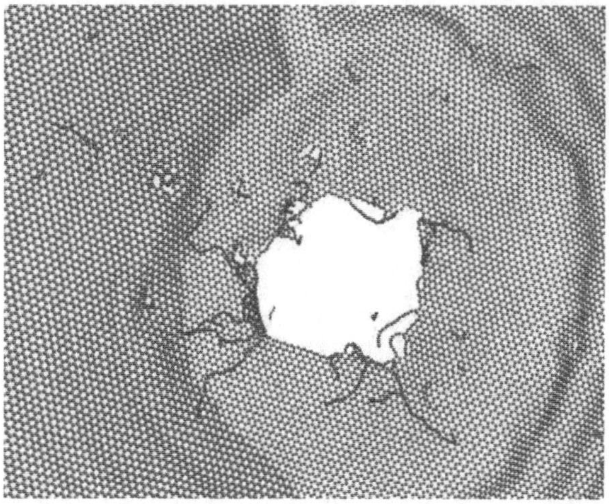

Figure 5. Top view of the crater formed in the result of GCIB bombardment at t = 3000 fs.

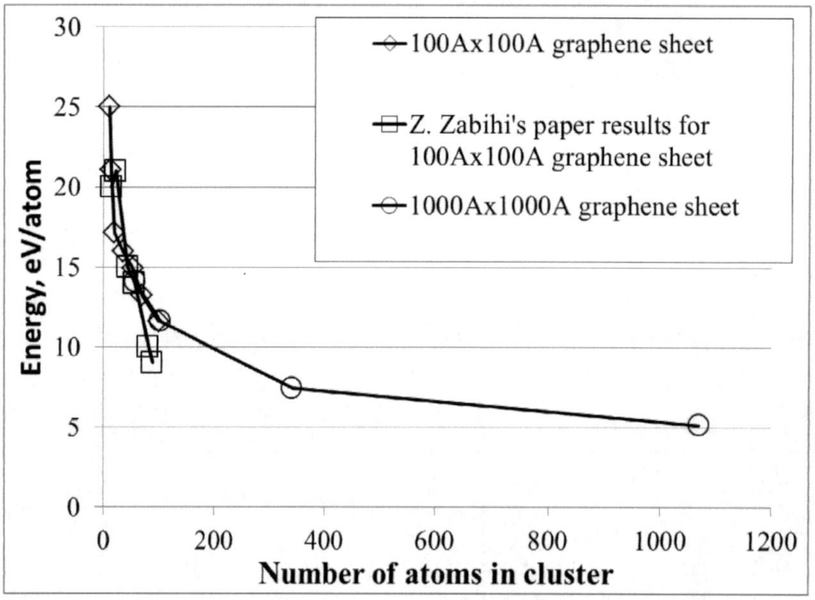

Figure 6. Threshold penetration energy *versus* cluster size, NVE ensemble.

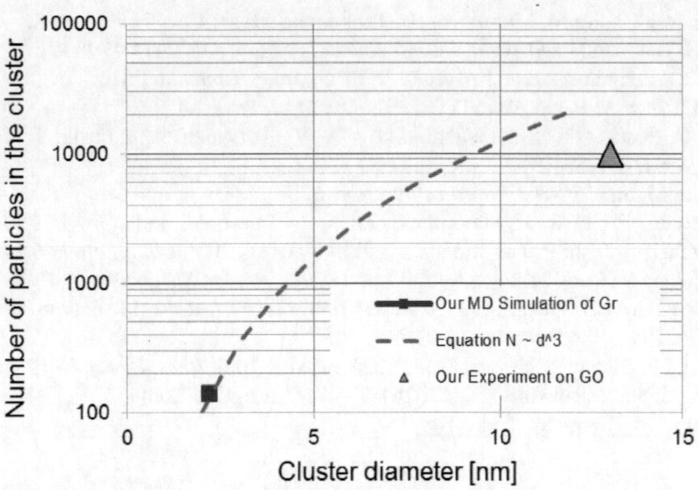

Figure 7. Cluster diameter versus number of particles.

Figure 7. shows how cluster diameter changes with number of particles in the cluster which is described by equation N ~d3. Our simulation results for holes diameter formed in graphene after cluster collision in agreement with graph line described by equation N ~ d3.

3. Conclusion

DFT calculation was conducted for obtaining the dependence of Raman spectra on the number of graphene layers: single layer and multilayer graphene were studied. DFT calculations are in good agreement with the Raman spectra measurements for graphene.

Large-scale atomistic MD simulations were used for obtaining threshold energies for pore formation in graphene with Ar cluster ions. The obtained threshold energies are in good agreement with the previous MD simulations by Zabihi et al [6] of small cluster irradiation on graphene. However, the threshold energies depend on the cluster size and for realistic experimental gas cluster sizes of 1070 atoms in cluster, the threshold energies are higher. It was also shown that small substrate sizes may lower threshold energies. Our MD simulations showed a high stability, strength and resistance of an ideal graphene film against making holes with cluster ion bombardments. Small holes were able to self-healing after the cluster atoms leave the graphene surface.

Acknowlegments

This work was funded in part by the Nazarbayev University World Science Stars program, under Grant No. 031-2013 of 12/3/2013, by the Ministry of Education and Science of the Republic of Kazakhstan contract № 265 of 12.02.2015 and in part by the Ministry of Education and Science of Russia, the agreement No.14.607.21.0047 (RFMEFI60714X0047).

References

[1] Novoselov K S, Geim A K, Morozov S V, Jiang D, Zhang Y, Dubonos S V, Grigorieva I V and Firsov A A 2004 *Science* **306** 666.

[2] Celebi K, Buchheim J, Wyss R M, Droudian A, Gasser P, Shorubalko I, Kye J-I, Lee C and Park H G 2014 *Science* **344** 289-92.

[3] Yang Y, Asiri A M, Tang Z, Du D and Lin Y 2013 *Mater. Today* **16** 365-73.

[4] Hopster J, Kozubek R, Ban-d'Etat B, Guillous S, Lebius H and M Schleberger 2014 *2D Mater.* **1** 1011011.

[5] Sumedh P S, Sergei N S, Ivan V, Vlassiouk R R and Unocic G M, Veith S D and Shannon M M 2015 *Nature Nanotech.* **10** 459–64.

[6] Zabihi Z and Araghi H 2015 *Nucl. Instr. Meth. Phys. Res. B* **343** 48-51.

Journal of Physics: Conference Series **751** (2016) 012029 doi:10.1088/1742-6596/751/1/012029

[7] Wu X, Zhao H, Zhong M, Murakawa H , Tsukamoto M 2014 *Carbon* **66** 31-8.
[8] Wang Z, Bi K, Guan H and Wang J 2014 *Journal of Materials* **2014** 479808 5.
[9] Heiranian M, Farimani A B and Aluru N R 2015 *Nature Comm.* **6** 8616.
[10] Payne M C, Teter M P and Ailan D C 1992 *Rev. Mod. Phys.* **64** 4.
[11] Lehtinen O, Kotakoski J, Krasheninnikov A V, Tolvanen A, Nordlund K and Keinonen J 2010 *Phys. Rev. B* **81** 153401.
[12] Yoon D and Moon H 2009 *J Korean Phys. Soc.* **55**.
[13] Graf D, Molitor F, Ensslin K, Stampfer C, Jungen A, Hierold C and Wirtz L. 2007 *Nano Lett.* **7** 238-42.
[14] Gupta A, Chen G, Joshi P, Tadigadapa S and Eklund P C 2006 *Nano Letters* **6** 12 2667-73.
[15] Wang H, Wang Y, Cao X J, Feng M and Lan G 2009 *Raman Spectrosc.* **40** 1791–6.
[16] Clark S J, Segall M D, Pickard C J, Hasnip P J, Probert M J, Refson K, Payne M C 2005 *Journal Z. Kristall* **220**.
[17] Haas P, Tran F, Blaha P, Schwarz K and Laskowski R 2009 *Phys. Rev B* **80** 195109.
[18] URL: http://digital.csic.es/bitstream/10261/2757/1/AAdigCSIC.pdf
[19] URL: https://www.lammps.sandia.gov

THE DISTRIBUTION OF CAPTURED NON-WETTING LIQUID DISPERSED IN NANOPOROUS MEDIUM RECOVERY METHOD

A A Belogorlov, V D Borman, S A Bortnikova and V N Tronin

National Research Nuclear University MEPhI, Molecular Physics Department, Kashirskoe highway, 31, Moscow, 115409, Russian Federation

E-mail: svetlana.sweta7@yandex.ru, AABelogorlov@mephi.ru

Abstract. In this paper the recovery method of a distribution function of non-wetting liquid captured in porous medium presented. Computation algorithm described. The recovered distribution functions for the system nanoporous medium (Libersorb 23) - non-wetting liquid (water) at the temperature of experiments 6^0C presented. The used for calculation experimental data obtained in the time relaxation of metastable states of the non-wetting liquid in a porous medium experiment [Borman at al. 2016 *Phys Rev* E **93** 022142].

1. Introduction

The study of non-wetting liquid entrapment phenomenon continuous more than a twenty years [1, 2]. This phenomenon is observed not only for disordered porous media [2], but also for porous media with an ordered structure, such as MCM and SBA type materials [2, 3]. Basic models for describing the process of intrusion - extrusion non-wetting liquid from the pores are models based on the Laplace equation and its modifications [1, 2]. In many models to describe entrapment phenomenon proposed a change in contact angle of wetting or assumption of pore size distribution spatial heterogeneity. Also in recent years have been found such effects during intrusion - extrusion non-wetting liquid process, as the existence of metastable states [4] and their power law relaxation [5]. The effect of pores spatial distribution and it's correlation in space can have a significant contribution on the non-outflow phenomenon.

The aim of this work was to develop a method to determine the pore size distribution for pores from which non-wetting liquid does not outflow after reducing the excess pressure to zero. The proposed method allowed us to determine the characteristic size of the pores in assumption of cylindrical pores. Estimates are provided for the relaxation experiments to investigate metastable states Libersorb 23 - water system at a temperature of 6^0C and observation time 0, 1, 10 and 100 minutes. The calculated pore size distribution functions demonstrate that there is a bimodal distribution of pores from which liquid does not outflow after the decrease excess pressure to zero.

2. The Experiment

The method has been tested on the experimental results of the metastable states relaxation study presented in [5]. Grafted silica gel Libersorb 23 (L23) as nanoporous medium and water as non-wetting liquid have been used for study. The characteristics of nanoporous medium and method for obtaining the experimental data described in [5, 6]. The initial experimental data of pressure dependences of a change in the volume of the L23 water system for analysis presented on Fig 1. This is a typical intrusion - extrusion cycles with time delay after the first cycle ($t = 0$ min) 1, 10 and 100 min. Increasing of time delay increases quantity of non-wetting liquid outflowed from pores and each intrusion curve presents new state of empty pores after extrusion. The data for calculation presented on Fig 2. This data normalized on volume of pores and compressibility of empty porous medium.

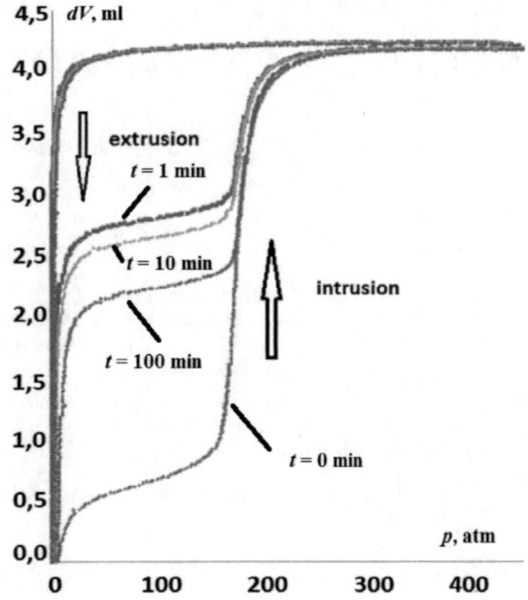

Figure 1. Dependences of internal camera volume changes with applied pressure for water intrusion - extrusion cycles in nanoporous medium L23 at 6 with time delay 0, 1, 10 and 100 minutes.

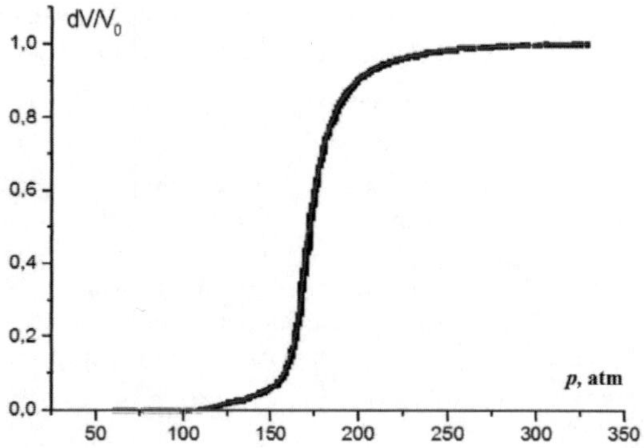

Figure 2. Normalized experimental data for $t = 0$ min.

It is well known method pore size distribution function recovery based on intrusion non-wetting liquid in porous medium [7]. Examining of intrusion experiments as independent and the comparison of distribution functions will provide information about the distribution of non-wetting liquid captured in a porous medium.

3. The Calculation method

According to the classical method of determining the pore size distribution function $f(R)$ by mercury porosimetry [7], the recovery of distribution function in the approximation of cylindrical pores of a porous medium, where the relationship between pressure (p) and a pore radius (R) defined by the Laplace equation $p = -\frac{2 \cdot \sigma \cos\theta}{R}$, where σ - liquid surface tension, θ - contact wetting angle. Calculation of $f(R)$ is carried out according to:

$$f(R) = \frac{p^2}{2 \cdot \sigma \cos\theta} \cdot \frac{dV}{dp}, R = -\frac{2 \cdot \sigma \cos\theta}{p}, \tag{1}$$

where dV/dp - derivative of the experimental dependence of $dV(p)/V_0$. Since the experimental methods of obtaining dependencies $dV(p)/V_0$ in the mercury and water porosimetry are close, the task of restoring the distribution of functions in the approximation of cylindrical pores is reduced to the computation of the derivative dV/dp, and the definition of the value of θ. For derivative dV/dp initial experimental dependences previously been converted, taking into account the deformation of porous medium, is not filled with the non-wetting liquid, and normalized to the full pores volume V_0, determined in the first intrusion - extrusion cycle (see fig.2). It was then carried out smoothing the experimental curves. New values of the experimental points ($V_j = \frac{1}{n}\sum_{i=m}^{i=m+n-1} V_i; p_j = \frac{1}{n}\sum_{i=m}^{i=m+n-1} p_i$) at $n = 5$, $m = 1 + (j-1)\cdot n$, $j = 1, 2, 3, ..., k$. Averaging over five points, as subsequent processing steps is optimal. The results before and after the smoothing procedure the same, within measurement error. The next stage of calculation consisted numerical differentiation smoothed experimental dependence. The result of numerical differentiation of the experimental dependence is not a smooth function and as a consequence, the further processing of the results is complicated, because it contains numerous peaks that are artifacts of the numerical differentiation. Also, the presence of a large array of data requires additional software packages for further analysis. Therefore, to optimize the processing techniques, depending on the numerical differentiation obtained were approximated by an analytical function. Preliminary analysis showed that the function:

$$F(x) = \sum_{i=1}^{5} \frac{A_i}{w_i \cdot \sqrt{\pi/2}} \exp\left[-2(\frac{x-\bar{x}_i}{w_i})^2\right], \tag{2}$$

with reliability $R^2 \geq 0.998$ approximates the obtained numerical dependence. Increasing the number of functions more than five ($i > 5$) does not increase reliability R^2, but fewer functions ($i < 5$) can not achieve $R^2 \approx 1$. The correctness of approximating function choice (2)has been proofed by the integration and comparison to the original experimental data according to:

$$\int_{50}^{a} F(x)dx, \tag{3}$$

where the upper limit of integration $a = 50 \div 350$ atm, and x is replaced by the pressure p. The obtained calculation of the approximation function according to (3) describes the experimental data within measurement error. Thus, this method of calculation allows high accuracy to determine the analytical form of the derivative dV/dp and use this function in subsequent calculation. Figure 3 shows two pore size distribution functions calculated by the above-mentioned methods in accordance to the equation (1) there $dV/dp = F(x)$ for the system L23 water at a temperature 6^0C. The contact Angle is determined from the work [3] and is equal to 124^0.

As it seen from figure 3, the distribution function of pore size after a time delay of $t = 1$ min $f_1(R)$ is below the distribution function obtained from the experimental data in terms of the empty porous medium $f_0(R)$ in the whole range of radii. Thus, the pore size distribution function of a porous medium filled non-wetting liquid can be defined as the difference between the two distribution functions:

$$f_1^c(R) = f_0(R) - f_1(R). \tag{4}$$

The result of subtraction according to 4 for pore size distribution function $f_1^c(R)$ at 1 minute time delay and pore size distribution function of the empty porous medium $f_0(R)$ are presented on figure 4. As an additional check, we calculated the definite integral of a function $f_1^c(R)$. The value obtained within the measurement errors consistent with the part of liquid captured in the porous medium.

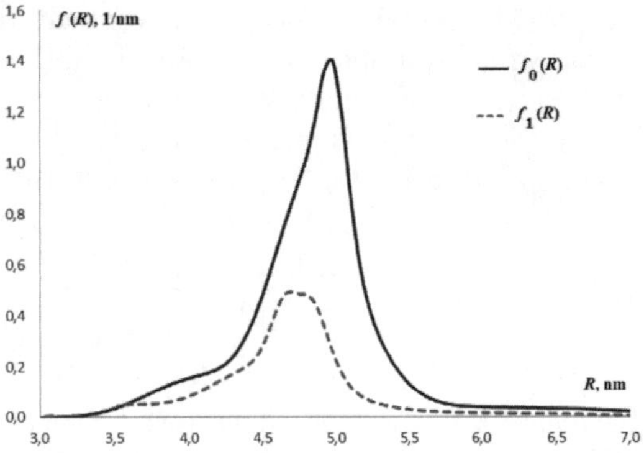

Figure 3. The distribution function obtained from the experimental data in terms of the empty porous medium $f_0(R)$ and $f_1(R)$ the distribution function of pore size after a time delay of $t = 1$ min .

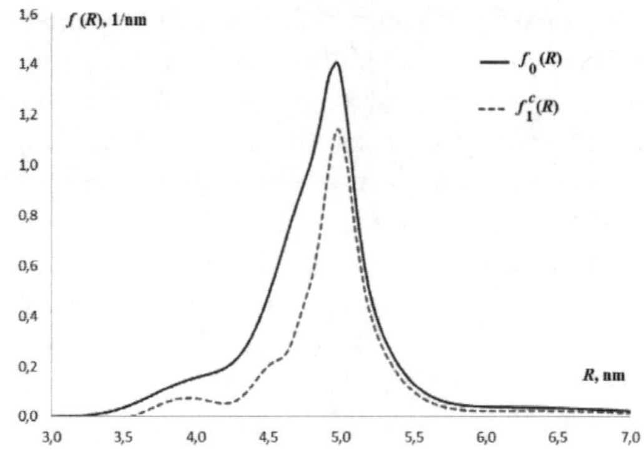

Figure 4. The pore size distribution function of pores filled by non-wetting liquid $f_1^c(R)$ with 1 minute time delay and pore size distribution function of the empty porous medium $f_0(R)$.

4. Results

In accordance to the procedure described above were also obtained pore size distribution functions of pores filled by non-wetting liquid at time delay 10 and 100 minutes, respectively. The results are shown on Figures 5 and 6. Also on the figures, for comparison, the pore size distribution function $f_0(R)$ presented.

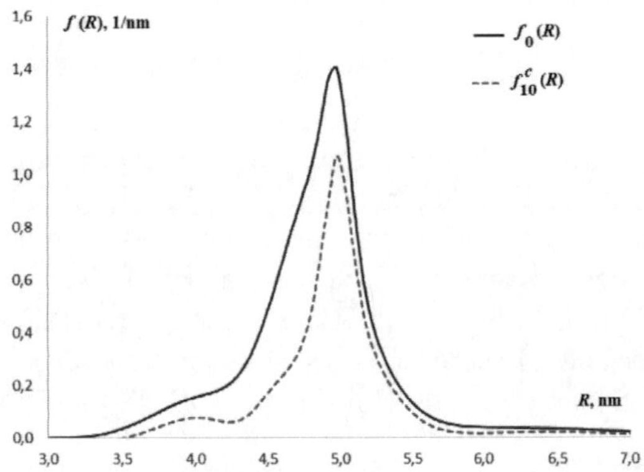

Figure 5. The pore size distribution function of pores filled by non-wetting liquid $f_{10}^c(R)$ at 10 minutes time delay and pore size distribution function of the empty porous medium $f_0(R)$.

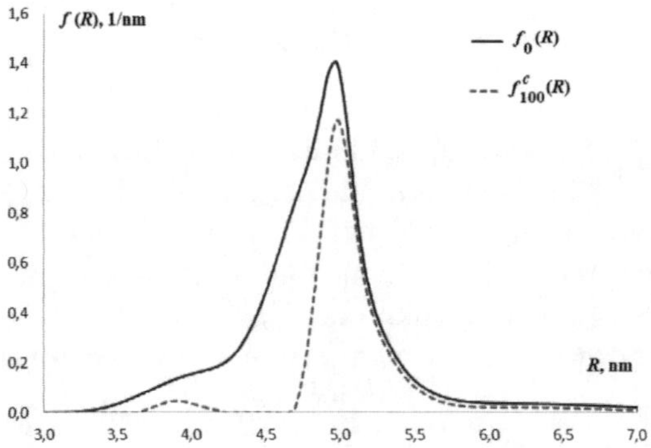

Figure 6. The pore size distribution function of pores filled by non-wetting liquid $f_{100}^c(R)$ at 100 minutes time delay and pore size distribution function of the empty porous medium $f_0(R)$.

As seen from Figures 4 - 6, the amount of liquid in the porous medium decreases with increasing time between cycles. The distribution function then the liquid is not outflow after reducing the excess pressure to zero is changing. The fluid does not flow out of the pores of large and small size and bimodal distribution occurred. The magnitude of peaks with increasing time delay decreases, which corresponds to an increase of the amount of liquid flowing out of the pores at time delay increase. Also, it can be assumed on the spatial heterogeneity of the pore distribution. Bimodal distribution may be indicative of the spatial configurations existence, where a large pores are surrounded by a smaller ones.

5. Conclusion

We propose a method of recovering pore distribution function of a porous medium filled with the non-wetting liquid. We obtain the distribution function of captured liquid for experiments to investigate the relaxation of non-wetting liquid (water) in a porous medium (L23) at a temperature of 6^0C. It is shown that the distribution function is bimodal. Assumed that this type of function may be due to the spatial heterogeneity of pores and in the existence of spatial configurations, where large pores are surrounded by small pores.

Acknowledgments

This work was performed within the framework of the Center of Physics of Nonequilibrium Atomic Systems and Composites supported by MEPhI Academic Excellence Project (contract 02.03.21.0005). The authors also express their gratitude to the Ministry of Education and Science of Russian Federation for the financial support (task No. 3.720.2014/K).

References

[1] Sean P R and Karen J E 2002 *J. Coll. Int. Sci.* **250** 175
[2] Lefevre B, Saugey A, Barrat J L, Bocquet L, Charlaix E, Gobin P F, and Vigier G 2004 *Coll. Surf. A: Physicochem. Eng. Aspects* **241** 265
[3] Belogorlov A A, Bortnikova S A and Mingalev P G 2015 *Physics Procedia* **72** 33
[4] Borman V D, Belogorlov A A, Grekhov A M and Tronin V N 2015 *Int. J. Mod. Phys.* B **29** 1550097
[5] Borman V D, Belogorlov A A and Tronin V N 2016 *Phys. Rev.* E **93** 022142
[6] Borman V D, Belogorlov A A and Tronin V N 2016 *Coll. Surf. A: Physicochem. Eng. Aspects* **496** 63
[7] Lowell S, Shields J E, Thomas M A and Thommes M 2004 *Characterization of Porous Solids and Powders: Surface Area, Pore Size and Density* (Springer Netherlands) p. 350

Journal of Physics: Conference Series **751** (2016) 012031 doi:10.1088/1742-6596/751/1/012031

System nanoporous media – non-wetting liquid, as a basis for the development of shock damper

A A Belogorlov, V D Borman, V A Byrkin, D A Paryohin and V N Tronin

National Research Nuclear University MEPhI (Moscow Engineering Physics Institute), Kashirskoe highway 31, Moscow, 115409, Russia

E-mail: AABelogorlov@mephi.ru

Abstract. In this work the method of system nanoporous medium – non-wetting liquid testing for damping devices presented. The method was tested for four systems nanoporous medium – non-wetting liquid. As a nanoporous medium was used commercial samples produced by Sigma-Aldrich hydrophobic silica Fluka 60 C8, Fluka 100 C8, Fluka 100 C18 and Fluka 90 C18. As a non-wetting liquid was used distilled water. These systems have been investigated in the temperature range 20 ÷ 60° C, the maximum energy of impact was 100 J. According to the method, time, power and energy characteristics of the process of non-wetting liquid intrusion - extrusion was determined. Also presented the dependences of impact processes characteristics on temperature.

1. Introduction

Recent studies have shown that systems “nanoporous medium – non-wetting liquid” are promising for use in damping devices and storage of mechanical energy [1-5]. One of the basic requirements for such devices it is stable work in a given temperature range without critical settings changing. At the same time, the requirements must be carried out to limit impacts on the protected object. For devices that can be used system “nanoporous medium – non-wetting liquids” power values, energy and time characteristics of the system depends on the choice of nanoporous media and non-wetting liquid [6-11]. It is also shown that such systems tend to change performance parameters under the influence of external action [12,13].

The aim of this study was to test promising methods of analysis using different porous media for absorbing mechanical energy devices. Also as part of this work we investigated the effect of surface modification and pore size of porous media on the process of filling nanoporous media by non-wetting liquid under pulsed pressure changes at a given temperature and energy of impact.

To determine the time, energy and power characteristics of the process of filling nanoporous medium by non-wetting liquid was used developed in MEPhI technique [14]. Distilled water was used as a non-wetting liquid, and industrially manufactured hydrophobized silica Fluka 60 C8, Fluka 100 C8, Fluka 100 90 Fluka C18 and C18 were used as the nanoporous media, differ an average pore size and surface modification.

2. Experiment

2.1 Nanoporous media

The nanoporous media under study were commercially available silica gels with a random structure of pores obtained in the sol–gel process Fluka 60 C8 - Silica gel 60 C_8, Reversed phase (Sigma-Aldrich catalogue # 18948), Fluka 100 C8 - Silica gel 100 C_8, Reversed phase (Sigma-Aldrich catalogue # 60755), Fluka 90 C18 - Silica gel 90 C_{18}, Reversed phase (Sigma-Aldrich catalogue # 60757) and Fluka 100 C18 - Silica gel 100 C_{18}, Reversed phase (Sigma-Aldrich catalogue # 60756). Using the pycnometry (Ultrapyc 1200e, Quantachrome Instruments) and porometry (Autosorb IQ, Quantachrome Instruments) methods [15], we determined the density (ρ), specific surface area (S_p), specific volume of pores (V_p), and average pore size within the cylindrical model of pores (<*R*>).The results are presented in Table 1.

Table 1. Characteristics of the porous media under study

Nanoporous medium	ρ, g/cm^3	V_p, cm^3/g	S_p, m^2/g (BET)	<*R*>, nm	Modificator's length [16], nm
Fluka 60 C8	1,9685	0,38	230	1,9	1,2
Fluka 90 C18	1,5975	0,36	148	2,7	2,45
Fluka 100 C18	1,6125	0,46	183	3,9	2,45
Fluka 100 C8	1,7603	0,56	267	3,9	1,2

2.2 Experimental bench and measurement procedure

The experimental stand was a base plate on which the strain gauge load cell (*F*) with a measuring range of 10 N ÷ 10 kN with measurement error is less than 0.5% and the upper plate. The base plate and top plate are connected to the uprights. Steel cables, on which load of 10 kg can freely move, is tensioned between the plates. On one of the pillars slide-wire gauge displacement (*l*) is fixed with a measuring range of 0÷140 mm with a margin of error of less than 0.05%. Assembled pressure chamber was placed on the force sensor. Inside the chamber, the porous medium is located a given weight, and the remaining volume of the chamber is filled with liquid (distilled water). The chamber was sealed through a sealing system of moving rod. This rod was connected by the plate with rod of displacement sensor. After the fall of the load at a predetermined height relative to the rod chamber, a rod inside the chamber starts to move, creating an overpressure therein by changing the internal volume of the chamber. When this strain gauge recorded the force acting on the support, and, consequently, the pressure in the chamber $p = F / S$, where S - the rod area. Changing the internal volume of the chamber respectively defined by a movement sensor as $V = l \cdot S$. Data from the force and displacement sensors were recorded every 20 microseconds in the file and subsequently processed. To investigate the process at different temperatures chamber was equipped with temperature control system. The temperature inside of chamber was maintained by the thermostat LOIP FT-316-40 with accuracy ± 0.2 ° C. More information about measuring system and the measurement procedure are described in [14].

Experimental results of pressure change in the chamber *p(t)* and the internal chamber volume *V(t)* from the time shown in Fig. 1. Power, time and energy characteristics of the systems is also determined according to the procedure described in [14].

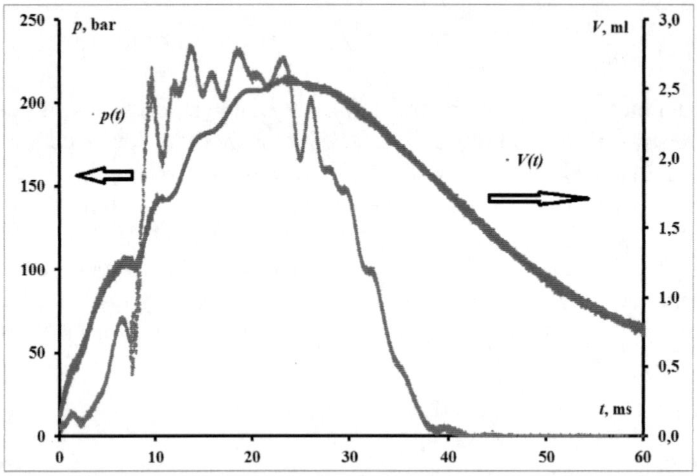

Figure 1. The cell pressure and the internal volume change in time
(Sample Fluka 100 C18, impact energy 30 J and T = 40 ° C)

Intrusion time of the porous medium by non-wetting liquid, characteristic filling pressure and the specific dissipated energy were identified by the experimental. The experiments were performed at temperatures of 20 ÷ 60 ° C and the energy of impact 20 ÷ 100 J.

3. Results and discussion
Figures 2-7 summarizes the results of investigated systems.

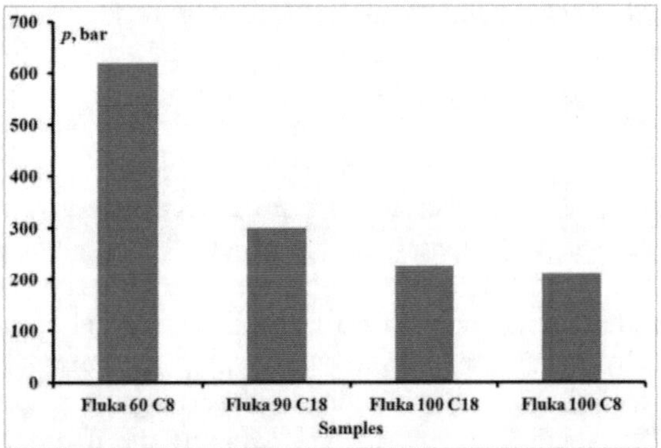

Figure 2. Typical filling pressure at T = 40 °C and impact energy = 40 J for systems Fluka 90 C18, 100 C18, C8 100, and 100 J for the system Fluka 60 C8

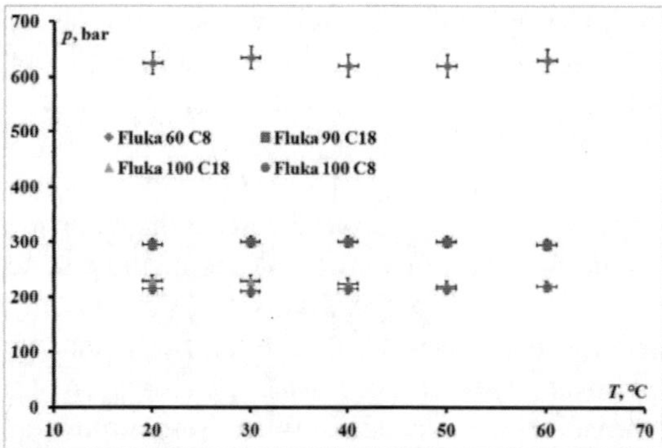

Figure 3. Typical filling pressure of investigated systems. Impact energy 40 J for systems Fluka 90 C18, 100 C18, C8 100, and 100 J for Fluka 60 C8 system at the temperature range 20 ÷ 60 °C

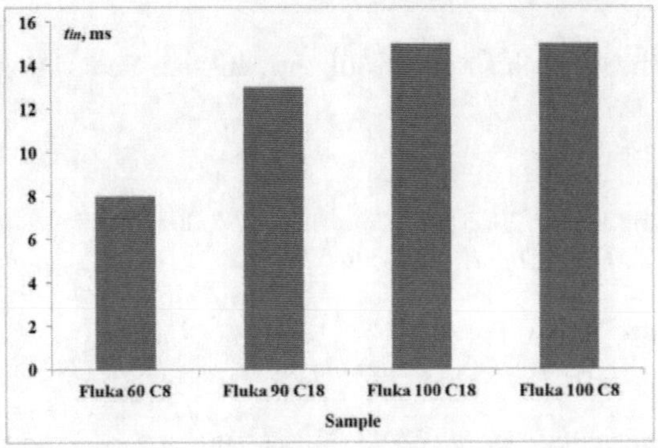

Figure 4. Typical intrusion time at T = 40 °C and impact energy = 40 J for investigated systems

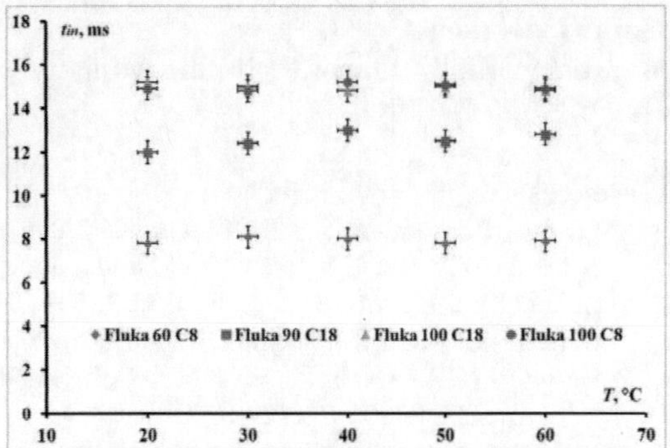

Figure 5. Typical investigated systems intrusion time. Impact energy 40 J for investigated systems at the temperature range 20 ÷ 60 °C

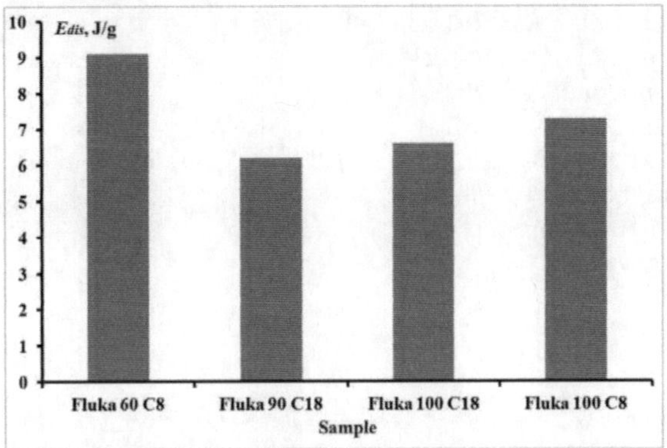

Figure 6. Typical dissipated energy at T = 40 °C and impact energy 40 J for systems Fluka 90 C18, 100 C18, C8 100, and 100 J for the system Fluka 60 C8

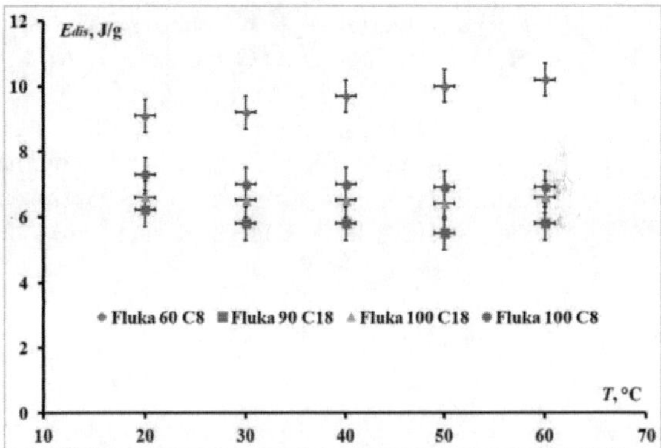

Figure 7. Typical dissipated energy of investigated systems. Impact energy 40 J for systems Fluka 90 C18, 100 C18, C8 100, and 100 J for Fluka 60 C8 system at the temperature range 20 ÷ 60 °C

Typical pressure decrease (Fig. 2) with radius increase (Table 1) and modificator chain length decrease (in case of similar radii (Table 1). But typical pressure (Fig. 3), within the error limits, does not depend on temperature in temperature range 20 ÷ 60 °C.

Typical liquid intrusion time increase (Fig. 4) with radius increase (Table 1) and modificator chain length decrease (in case of similar radii (Table 1). But typical liquid intrusion time (Fig. 5) also does not depend on temperature, within the error limits, in temperature range 20 ÷ 60 °C.

Typical dissipated energy (Fig. 6) does not have straight dependence on radius and modificator chain length (Table 1). Typical dissipated energy (Fig. 7) also does not depend on temperature, within the error limits, in temperature range 20 ÷ 60 °C.

Increasing the temperature from 20 to 60 °C has no significant effect on the absorption characteristics of the process of mechanical energy, but fluid characteristics such as surface tension and viscosity are reduced from 72 to 66 mN/m and 1.002 to 0.467 10^{-3} Pa s, respectively [26]. However, the typical filling pressure and intrusion time does not change within an error. Therefore, we can conclude non-viscous liquid transport in porous media. This confirms, on independent samples, the earlier results on the test specimen Libersorb 23 [14].

Acknowledgements
This work partially supported by the Ministry of Education and Science of the Russian Federation (task no. 3.720.2014/K).

References
[1] Qiao Y, Punyamurtula V K, Xian G, Karbhari V M and Han A 2008 *Appl. Phys. Lett.* **92** 063109.
[2] Iwatsubo T, Suciu C V, Ikenagao M and Yaguchio K 2007 *J. Sound Vibr.* **308** 579-90.
[3] Suciu C V 2010 *Adv. Mater. Res.* **123-125** 987-90.
[4] Surani F B, Kong X, Panchal D B and Qiao Y 2005 *Appl. Phys. Lett.* **87** 163111.
[5] Belogorlov A A, Grekhov A M, Grobanov E V, Konukov V V, Lisichkin G V, Tronin V N and Troyan V I 2006 *RU Patent* 2309307.
[6] Fadeev A Y and Eroshenko V A 1995 *Rossiiskii Khimicheskii Zhurnal* **39** 93 (in Russian).
[7] Eroshenko V A and Fadeev A Y 1996 *Zhurnal Fizicheskoi Khimii* **70** 1482 (in Russian).
[8] Fadeev A Y and Eroshenko V A 1997 *J. Coll. Int. Sci.* **187** 275.
[9] Lefevre B, Saugey A and Barrat J L 2004 *Coll. Surf. A* **241** 265.
[10] Han A, Kong X and Qiao Y 2006 *J. Appl. Phys.* **100** 014308.
[11] Lu W, Punyamurtula V K and Qiao Y 2011 *Inter. J. Mater. Res.* **2011** 587-90.
[12] Xu B, Qiao Y, Li Y, Zhou Q and Chen X 2011 *Appl. Phys. Lett.* **98** 221909.
[13] Kim T, Lu W, Lim H, Han A and Qiao Y 2011 *Appl. Phys. Lett.* **98** 053106.
[14] Borman V D et al 2009 J. Exp. Theor. Phys. 108 389.
[15] Lowell S, Shields J E, Thomas M A and Thommes M 2004 *Characterization of Porous Solids and Powders: Surface Area, Pore Size and Density* (Springer Netherlands) p 350.
[16] Suciu C V, Iwatsubo T, Yaguchi K, Ikenaga M 2005 *J. Coll. Int. Sci.* **283** 196–214.

Journal of Physics: Conference Series **751** (2016) 012032 doi:10.1088/1742-6596/751/1/012032

Applying a cellular automaton model to describing the anomalous relaxation of the metastable states in the disordered porous media

V D Borman, A N Mikheev and V N Tronin

National Research Nuclear University MEPhI (Moscow Engineering Physics Institute), Kashirskoe highway 31, Moscow, 115409, Russia

E-mail: al.mikheev84@gmail.com

Abstract. We propose a numerical model of outflow of the nonwetting liquid from previously completely filled nanoporous media based on cellular automata theory. It was shown that the relaxation of this system has an avalanche-like behaviour, which is inherent to phenomenon of self-organized criticality (SOC). Avalanche size and energy distributions were obtained. It was found that spectra have a typical for SOC theory power–law dependence $N(s) \sim s^{-\beta}$ with $\beta = 2.73$, and $N(E) \sim E^{-\beta_{en}}$ with $\beta_{en} = 2.25$.

1. Introduction

In paper [1, 2] anomalously slow (power-law) relaxation of the system "disordered nanoporous media – nonwetting liquid", when $\theta \sim t^{-\alpha}$ $(0 < \alpha < 0.2)$, was observed. Based on that, in [3, 4] it was assumed that in this kind of systems self-organized criticality, where power-law relaxation is typical, is possible. This model first proposed by Bak to describe relaxation of "sandpile" system [5, 6]. It was shown in [5, 6] that after critical angel is reached further relaxation happens through avalanche-like transitions between different metastable states of a system, moreover avlanches have a power-law size distribution $N(s) \sim s^{-\beta}$. Later it became clear that this mechanism is characteristic of the whole class of systems in the critical state with common features: they are all non-linear, open and dissipative [7]. The critical state is maintained in a wide region of the phase diagram through the feedback mechanism. Such properties make the mathematical description of the relaxation process a very difficult problem [8], there are only limited number of models, which can be exactly solved. One of the main tools in the study of systems that are capable of self-organized criticality is numerical simulation, using cellular automata theory [9]. In this model evolution of the system in time is described by a discrete way (iteratively), and the system is defined as cellular structure. Each cell on grid is on some state that corresponds to certain state of the element in simulated system. To use the model initial condition and rules of transition between states are required. The advantage of this model is that information about local mechanisms "is not lost" with this approach. It is especially important in modelling processes in disordered systems, where the dynamics of local configurations plays crucial role in relaxation of the metastable states [10].

In [3, 4] feedback mechanism, by which SOC can be implemented in such systems, was proposed. It was shown that the interaction between local configurations depends on the relative volume θ of the confined liquid. In other words, liquid volume change leads to shift of the energy barriers spectrum to lower values, resulting formation of critical conditions for other metastable states. According to the above, decay of the metastable states in systems of nonwetting liquid – disordered nanoporous media is considered as cooperative process with long-range correlations. Besides, there was also mentioned a local mechanism, whereby maintaining the system in critical state occurs due to change of the energy barriers in local configuration of pore after its emptying.

Numerical experiment could help to verify the idea of SOC scenario for relaxation in system of nonwetting liquid in disordered nanoporous media. This approach allows to look at the process “from the inside”, based on the basic mechanisms of the system relaxation. Moreover, numerical simulation allows to get data about the characteristics, which are unmeasurable in real experiment, such as avalanche size distribution or energy barriers spectrum of the metastable configurations in the system. Namely, power-law distribution of the avalanches in relaxation process could be a substantial argument in favour of presence of self-organized criticality in such systems. This paper is dedicated to consideration of using cellular automata model in simulating “nonwetting liquid – nanoporous media” systems' relaxation process.

2. Main part

Herein, porous material is considered within the framework of the model of randomly distributed polydisperse overlapping spheres [11]. Let a system has a cubic form with side L. Then, if $L \gg \overline{R}$ [12]:

$$\varphi = 1 - exp(-\eta) \tag{1}$$

Here, φ is the porosity of system, $\eta = \dfrac{N}{L^3} 4\pi \displaystyle\int f(R) R^2 \, dR$, $f(R)$ is the probability density function of the radii.

Offered results were obtained for the system with $L = 50\,\overline{R}$, so the condition $L \gg \overline{R}$ can be regarded as fulfilled. Then, setting $f(R)$, L and φ and using (1), we can get number of pores N corresponding to given system.

In this paper, outflow of the nonwetting liquid from previously completely filled nanoporous media was studied. Besides, initially filled with liquid only the part of the system that belongs to percolation cluster of empty pores. Relaxation of this subsystem will be discussed further. The external pressure is removed linearly from p_{max} to p_{min} by the time T_p.

Conditions of the outflow of a liquid from a pore is determined based on statistical theory of fluctuations. We can write isothermal work for liquid outflow as:

$$\delta A = p\,V + \sigma_{lg}\,(S_{neut} - S_{lg}) - (\sigma_{ls} - \sigma_{gs})\,S_{ls} = p\,V + \sigma\,\delta S - \delta\sigma\,S, \tag{2}$$

here, p – external pressure, V – volume, available for liquid outflow from pore; σ_{lg} – relative surface energy of the liquid–gas; σ_{gs} – relative surface energy of the gas–solid; σ_{ls} – relative surface energy of the liquid–solid; S_{neut} – area of the “neutral surface” that changes from “liquid–liquid” to “liquid–gas” after the outflow of liquid occurs; S_{lg} – area of the “liquid–gas” menisci; S_{ls} – area of the “liquid–solid” surface.

The probability of “filled pore – empty pore” transition, according to the statistical theory of fluctuations, is then given by:

$$\omega \sim exp\left(-\frac{\delta A}{T}\right) \tag{3}$$

If we now assume that only two states of pore are possible – filled and empty, – then from normalization condition it follows that outflow probability is:

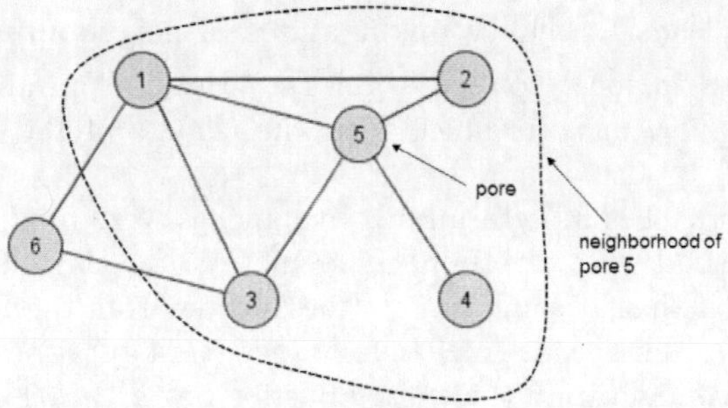

Figure 1. The system as a graph

$$\omega = \frac{exp\left(-\dfrac{\delta A}{T}\right)}{1 + exp\left(-\dfrac{\delta A}{T}\right)} = \frac{1}{1 + exp\left(\dfrac{\delta A}{T}\right)} \tag{4}$$

It is obvious that after the liquid is extruded from the pore the "liquid – gas" surface area increases for pores of its local configuration, therefore their energy barriers for outflow are decreased. This causes a feedback mechanism, by which the system is maintained in critical state, and avalanche-like evolution becomes possible, when outflow from one pore leads to chain of decays of metastable states in some region.

Assuming that $\overline{R} \sim 1\,\mathrm{nm}$, $\sigma \sim 10^2\,\dfrac{\mathrm{mJ}}{\mathrm{m}^2}$, we can estimate the value of the energy barrier for outflow:

$$\delta A \sim \sigma R^2 \sim 10^{-18} \cdot 10^2 \cdot 10^{-3}\,\mathrm{J} \sim 1\,\mathrm{eV} \sim 10^4\,\mathrm{K} \tag{5}$$

It follows from (5) that for systems with a pores of nanometre size characteristic value of the energy barrier is much bigger than the temperature $T \sim 300\,\mathrm{K}$. Thus, if $\delta A < 0$, then the outflow probability $\omega \approx 1$, therefore it can be assumed that the outflow occurs guaranteed in this case.

In accordance with (2), for the calculation of work for liquid outflow the values of V, S_{neut}, S_{lg} and S_{ls} for pore are needed. In this paper, intersections of more than three spheres are not regarded, so every local configuration was considered as a set of double and triple intersections. All these cases were completely analysed in [13, 14, 15].

It is now possible to turn to a discussion of a cellular automaton for the "nonwetting liquid – nanoporous media" system. Evolution of the system in this model is described in a discrete way, therefore for an adequate simulation of process it is necessary to select as the time unit the minimum characteristic time, for which there is a change in system. In this case it would be the characteristic hydrodynamic time.

Basically, any cellular automaton boils down to following points:

1. Finite number of states for cells.
2. Definition of a neighbourhood for each cell.
3. Rules of transition between the cell states.

The system considered here is an undirected graph, where each pore is corresponded to some vertex with the number i, and the edge $i - j$ means that there is an intersection between i-th and j-th pores. Let us denote by D_i the neighbourhood of i-th pore. State of each cell (pore) is defined by the energy barrier of outflow, which depends on the values of V, S_{neut}, S_{lg} and S_{ls}:

$$\delta A = \delta A(V, S_{neut}, S_{lg}, S_{ls})$$

Outflow from the pore i leads to change of the states of neighbouring pores, in other words $\delta A_k \xrightarrow{P} \delta A'_k$ for all pores in local configuration D_i. Here, P means the transformation of parameters V, S_{neut}, S_{lg}, S_{ls} in accordance with the change of the local configuration after emptying of i-th pore.

In this paper, two approaches in determining avalanches were used: the first one was called "Bak's approach", the second was corresponded to energy of avalanche. Let us explain what is meant by the first. Suggest that at some time t_k the outflow from the i-th pore occurs, and this pore has avalanche label n. This leads to change of local configuration of the i-th pore. Let for some pore j in this local configuration the energy barrier was $\delta A_j > 0$ before, and $\delta A'_j < 0$ after the outflow from the i-th pore. Then for this pore j label of avalanche n is assigned, and the power of the avalanche n is increased by 1. Otherwise, if the i-th pore initially has no avalanche label, then for i label of avalanche $M+1$ is assigned, where M is the total number of avalanches at the moment. The energy of the avalanche is considered as the sum of energy changes for the local configurations in the avalanche:

$$E = \sum_{i=1}^{s} \left\{ \sum_{k=1}^{z_i} \delta E_k \right\}_i , \quad (6)$$

here, s – power of avalanche according to "Bak's approach", z_i – the number of filled neighbours of i-th emptied pore, δE_k – change of the surface energy of the k-th neighbour.

It is expected that both spectra should have power dependence, that is $N(s) \sim s^{-\beta}$ and $N(E) \sim E^{-\beta_{en}}$.

As it was mentioned above, in [3, 4] was made a guess about the self-organized criticality scenario of relaxation in "nonwetting liquid – nanoporous media" systems, which means the presence of some feedback mechanism. In this paper, the feedback mechanism is local: the ouflow of liquid from pore leads to decrease of the barriers of pores in local configuration due to increase of "liqid – gas" surface, and avalanche-like evolution becomes possible, when outflow from one pore leads to outflow of liquid in some area. Analysis of avalanche size distribution could be an argument in favour or against of presence of self-organized criticality in such systems.

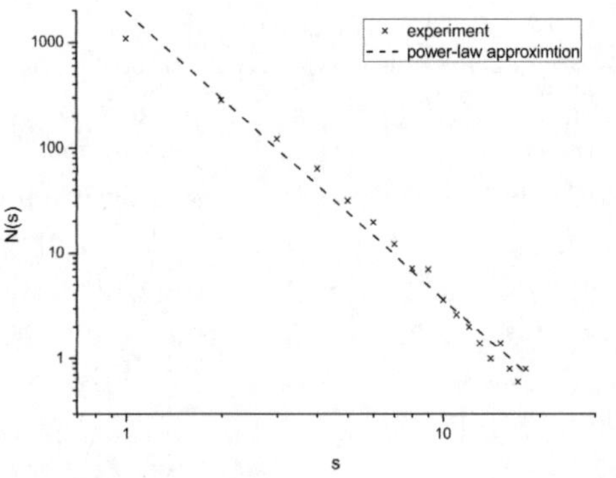

Figure 2. "Bak's approach" avalanche size distribution in log-log scale.

Accordingly, avalanche size distributions were obtained for the system with Gaussian radii distribution (1, 0.28), average radius of pores $\overline{R} = 3\,\text{nm}$, porosity $\varphi = 0.5$, $T = 293\,\text{K}$, $\sigma = 72\,\dfrac{\text{mJ}}{\text{m}^2}$, $\delta\sigma = 23\,\dfrac{\text{mJ}}{\text{m}^2}$. In Fig. 2 avalanche size distribution, according to "Bak's approach", is shown. As it was expected, avalanches have a power-law distribution $N(s) \sim s^{-\beta}$, and $\beta = 2.73 \pm 0.08$. Coefficient of determination $R^2 = 0.9844$. It may be noted that the power of avalanches does not reach an arbitrarily large value, and that there is some "downturn" in the end. This can be explained by the following factors:

1. Limited size of the system. Spreading, avalanche can reach the boundary of the system, after which the avalanche "movement" in this direction becomes impossible. This is especially important for pores near the border.

2. In this process the situation, when at one time several avalanches occur in different part of the system, is quite common. At the same time in calculating the power of avalanches it is assumed that they cannot merge or intersect. Thus, in reality not all of the volume available for the development of avalanches (or to be more precise, not all of the pores in this volume), but only the part limited by the "fronts" of other avalanches spreading at that moment.

3. Relatively small observation time and number of experiments. The probability of the occurrence of the avalanche of great power is really small, so to observe the avalanche with sizes of the order of system size the data sample must be much bigger.

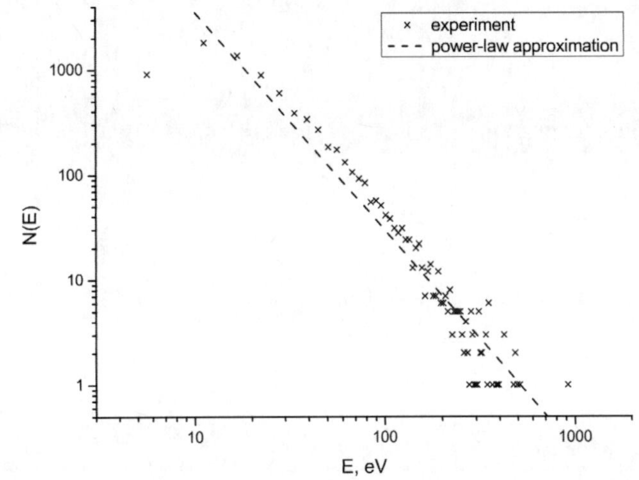

Figure 3. Distribution of avalanches' energies in log-log scale.

Similarly, the energy spectrum of avalanches was obtained for the same parameters of the system. By assumption, it should also have power-law form. In Fig. 3 distribution of avalanches' energies is shown in log-log scale. After excluding the first point, approximation gives a power dependence $N(E) \sim E^{-\beta_{en}}$ with $\beta_{en} = 2.25 \pm 0.16$ and coefficient of determination $R^2 = 0.9395$. Thus, typical for SOC phenomena power-law function provides a good approximation for both spectra.

3. Conclusion

Proposed cellular automaton allows to simulate relaxation process of "nonwetting liquid – nanoporous media" system with variety values of parameters σ, $\delta\sigma$, T, $\overline{R}$, ΔR, φ. In this paper, based on the described above local feedback mechanism, possibility of self-organized criticality in such systems was studied. For this purpose avalanche size and energy distributions were obtained. These spectra have a power-law form, which can be the evidence that proposed local feedback mechanism may indeed lead to self-organized criticality in this type of systems.

References

[1] Borman V D, Belogorlov A A, Zhuromskii V M and Tronin V N 2015 *J. Exp. Theor. Phys.* **121** 1027–1041

Journal of Physics: Conference Series **751** (2016) 012032 doi:10.1088/1742-6596/751/1/012032

[2] Borman V, Belogorlov A A and Tronin V 2016 *Coll. Surf. A* **496** 63–68
[3] Borman V, Tronin V and Byrkin V 2016 *Phys. Lett. A* **380** 1615–1620
[4] Borman V and Tronin V 2016 *Phys. A* **457** 391405
[5] Bak P, Tang C and Wiesenfeld K 1988 *Phys. Rev. A* **38**(1) 364–374
[6] Bak P, Tang C and Wiesenfeld K 1987 *Phys. Rev. Lett.* **59**(4) 381–384
[7] Turcotte D L 1999 *Rep. Prog. Phys.* **62** 1377
[8] Dhar D 1999 *arXiv:cond-mat/9909009*
[9] Kari J 2005 *Theor. Comput. Sci.* **334** 3–33
[10] Langer J S 2014 *Rep. Prog. Phys.* **77** 042501
[11] Byrkin V A 2015 *J. Porous Media* **18** 463–475
[12] Dhar D 1997 *Phys. A* **242** 341 – 346
[13] Gibson K and Scheraga H 1988 *Molecular Physics* **64** 641–644
[14] Gibson K D and Scheraga H A 1987 *J. Phys. Chem.* **91** 4121–4122
[15] Rychkov G and Petukhov M 2007 *J. Comput. Chem.* **28** 1974–1989

Journal of Physics: Conference Series **751** (2016) 012033 doi:10.1088/1742-6596/751/1/012033

Monitoring local configuration and anomalously slow relaxation of a nonergodic system of interacting liquid nanoclusters in a disordered confinement of a random porous medium

V D Borman, A A Belogorlov, S A Bortnikova and V N Tronin

National Research Nuclear University MEPhI (Moscow Engineering Physics Institute), Kashirskoe highway 31, Moscow, 115409, Russia

E-mail: VNTronin@mephi.ru

Abstract. The relaxation of a confined nonwetting liquid dispersed in a disordered nanoporous medium has been experimentally studied in the system consisting of water and the L23 hydrophobized silica gel. Discovered that the relaxation of these states to study the system occurs as a result of the relaxation of local metastable configurations of filled and empty pores of the porous medium. Such relaxation occurs abnormally slowly back to the power law with the exponent . $\alpha < 0.16$ The observed anomalously slow relaxation of such a system and comparison with the time dependence of the volume of the confined liquid obtained in confirm the correctness of the description of disordered media on the basis of the notion of local metastable structures.

1. Introduction

The disordered system of pores of a random nanoporous medium can be filled with an incompressible liquid at a pressure that can be estimated by the Laplace formula. It was established in [1, 2] that, when a certain critical degree of filling is achieved and the excess pressure is then removed, a dispersion transition occurs and a part of the liquid can remain in the disordered porous medium. This means that the confined nonwetting liquid is effectively in a wetting state in the form of an ensemble of liquid nanoclusters in the disordered confinement of the porous medium. By varying the degree of preliminary filling of the system of pores, it is possible to change ensembles of pores in the bulk of the random medium in which the liquid is confined [1, 3]. Confinement of the nonwetting liquid is observed as a part of the liquid remaining in the porous medium after its filling [4–13]. As a result of the relaxation process, a transition occurs from the nonergodic metastable state of the ensemble of liquid nanoclusters in pores of the disordered medium to the ergodic state of the empty porous medium immersed in the liquid. The porous medium studied in [1, 2] is a silica gel that is obtained in the sol–gel process of formation of the random medium.

The states and properties of disordered media such as glasses, colloids, polymers, and loose media have been actively studied in recent years [14–45]. In the absence of the consistent inclusion of metastable multiparticle correlations, numerical studies were performed and phenomenological models such as shear transformation zone (STZ), dynamic heterogeneity (DH), and random first order transition theory (RFOT) were introduced and discussed (see, e.g., [17, 21, 22]). These models involve the notion of local structures and are used to describe states and relaxation of glasses, colloids, polymers, and loose media, as well as liquid–glass transitions and sol–gel process, which result in a random order. The state of these media is nonergodic; these media are characterized by the anomalously slow relaxation of nonequilibrium states, which is usually described by a stretched exponential (stretched-exponential relaxation) [18, 28, 34]. In the case of metastable states,typical of the average relaxation time becomes konechnym, while remaining large.

Anomalously slow relaxation means that a system cannot reach any point of the phase space in any large observation time [17,45]. Anomalously slow relaxation is attributed in phenomenological models to the assumed existence and decay of metastable states of random local structures in a disordered medium and to the energy distribution of these states.

Within the consistent statistical description and analytical theory of percolation [41], the escape-time distribution of clusters of the confined liquid in pores for the ground state with a fractal percolation cluster was calculated in [3], where a power law of a decrease in the volume of the confined liquid with time was predicted. According to [3], relaxation is a discrete equilibrium process with the overcoming of numerous local maxima appearing because of random local configurations of filled and empty pores with various sizes in the disordered medium. The condition of the trapped non-wetting fluid (water) in the pores was studied by various methods, such as high-resolution quasielastic neutron scattering[36].The relaxation of the metastable states of the confined liquid has not yet been studied experimentally, but the anomalously slow escape of a nonwetting liquid from a porous medium, e.g., mercury from porous glass, was sometimes observed [37,38]. It was established that the volume of confined liquid depends on the temperature, observation time, and sizes of granules of the porous medium [38–40]. The characteristic relaxation (escape) time of a part of the liquid (water in the L23 porous medium) that is not in the confined (metastable) state after filling and removal of excess pressure is 0.1 s at temperatures from 279 to 323 K [42].

In this work, the relaxation of the confined dispersed liquid is observed in the system consisting of water and the L23 hydrophobized silica gel, which is a disordered nanoporous medium. Two metastable states of the dispersed liquid in this system were revealed in [1, 2]. According to [2, 3], one of them, quasi-ergodic, appears through the formation of local configurations of filled pores belonging to the percolation cluster and empty pores separated by a finite potential barrier from the state of the empty medium. The other state, nonergodic, is formed owing to the decay of the fractal percolation cluster of filled pores into individual clusters of filled pores; this decay results in the disappearance of paths for the escape of the liquid through this percolation cluster.

In the present work found that the relaxation of these states to study the system occurs as a result of the relaxation of local metastable configurations of filled and empty pores of the porous medium. Such relaxation occurs abnormally slowly back to the power law with the exponent $\alpha < 0.16$. The observed anomalously slow relaxation of such a system and comparison with the time dependence of the volume of the confined liquid obtained in [3] confirm the correctness of the description of disordered media on the basis of the notion of local metastable structures.

2. Main part

The nanoporous medium under study was the commercially available KSK-G silica gel with a random structure of pores obtained in the sol–gel process. Using the porometry and pycnometry methods, we determined the parameters of the resulting Libersorb 23 (L23) porous medium: density ρ = 1.7798±0.0016 g/cm^3, specific surface area of pores S = 212±7 m^2/g, and specific volume of pores v = 0.62±0.02 cm^3/g. The pore size distribution appeared to be close to a Gaussian distribution with the mean radius $\bar{R}$ = 5.0±0.2 nm and the relative FWHM $\Delta R / \bar{R}$ = 0.07.

The aim of the performed measurements was to determine the time dependence of the volume fraction of pores θ filled with the confined liquid. The confined liquid is defined as the fraction of the liquid that remains in pores after the complete filling of pores at an increase in the pressure and the subsequent removal of excess pressure. The other fraction of the liquid in pores flows from pores rapidly in the pressure decreasing time. The extrusion time of this fraction of the liquid from the L23–water system is smaller than 1 s [35,42]

In our experiments, we used deaerated distilled water (28 ml) and 5 g of this porous medium, which were placed in a high-pressure chamber described in [6,43]. To maintain and measure excess pressure p, to measure change in the volume ΔV of the system at the intrusion–escape of the liquid from pores, and to maintain and measure the temperature, we used the bench described in [6,43]. Measurements of θ(t) have been performed at each point and each temperature several times.

The measurements showed that the relative volume of liquid $\theta_2 = \frac{\Delta V_2}{V_{pore}}$ remaining after the preliminary filling of the porous medium the relative volume $\theta_1 = \frac{\Delta V_1}{V_{pore}}$ of the liquid depends on the size θ_1, These dependences are obtained when increasing the system pressure for 15 seconds before filling fraction of the pores θ_1, the subsequent reduction of the overpressure to zero for 15 seconds and after a timeout of 1 min according to the methods described in [2, 43]

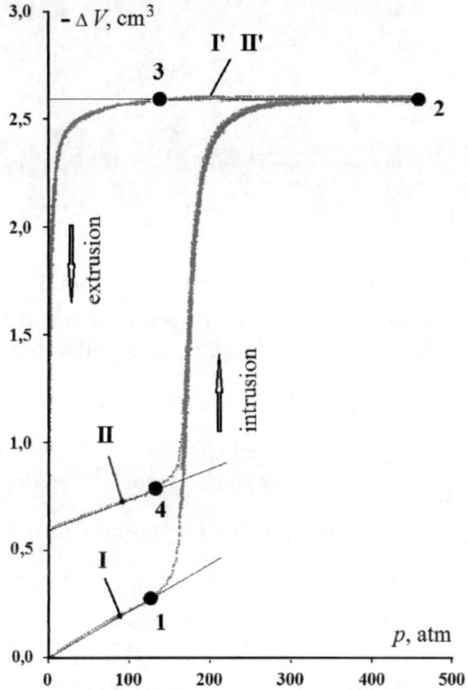

Figure 1. The dependence of the filled volume of the pressure when filling porous (I) and after spilling part of the liquid (II) when reducing excess pressure to zero within 15 seconds and after a timeout of 1 min according to the methods described in [2, 43]

The experiments showed that the volume fraction θ_2 of the liquid remaining in pores depends not only on the degree of preliminary filling θ_1, but also on the escape expectation time $t_{\exp}$. For this reason, to reduce errors and to improve the reproducibility of the measurements, the dependence of θ_2 on the expectation time $t_{\exp}$ was determined under the condition of the complete preliminary filling $\theta_1 = 1$. The experiments were performed as follows. The L23 porous medium was filled with water at an increase in the excess pressure to p = 40 MPa in time t_{in} = 15 s until complete filling. Then, excess pressure was reduced to zero in time t_{out} = 15 s. The times t_{in} and t_{out} were much larger than the known characteristic filling–escape times for this system, which are < 0.1 s [42]. After that, the volume fraction θ_2 of the confined liquid was determined after the expectation time t_{exp} = 60–10000 s. On figure 1 shown the dependence of the filled volume of the pressure when filling porous (I) and after spilling part of the liquid (II) when reducing excess pressure to zero within 15 seconds and after a timeout of 1 min .To this end, we performed the aforementioned cycle of an increase and a subsequent decrease in the pressure. The dependence $\theta_2(t_{exp})$ for various temperatures is shown in Fig. 2 in a log–log scale.

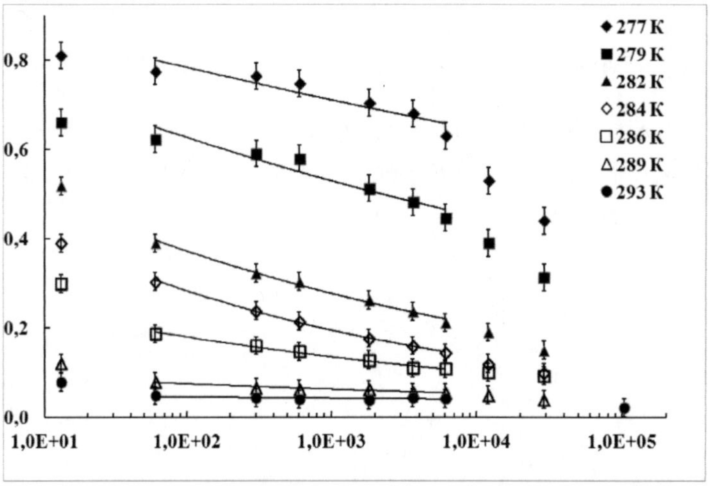

Figure 2. Dependences θ(*t*) at seven temperatures from 277 to 293 K in the time interval from 10 to $1\cdot10^5$ s. Experimental points in comparison with the power-law approximations $\theta \sim t^{-\alpha}$ shown by solid lines in the time interval from 60 to $6\cdot10^3$ s

It can be seen that the dependences of the volume fraction of the confined liquid on the time t_{exp} in this scale are straight lines within the measurement error. This indicates a power-law character of the relaxation of the volume fraction of the confined liquid θ_2 in the time t_{exp}. Correspondingly, the resulting dependences $\theta_2(t_{exp})$ were approximated by an inverse power law $\theta_2 = At^{-\alpha}$. According to Fig. 2, the slope α of the dependences $\theta_2 = At^{-\alpha}$ first increases with the temperature from α = 0.05 to the maximum α = 0.16 at T = 283 K and then decreases at T > 283 K to the value α =0.04 at T = 293 K (see Fig.). Such a dependence can be explained by the acceleration of the decay of the metastable state of the system at an increase in the temperature from T = 277 K to T = 283 K and by a transition at $T \geq 293$ K to a slower decay of the state of the confined liquid in which the volume fraction of the liquid θ_2 is independent of the temperature.

The experiments allow to recover the distribution function of the filled pore size for different relaxation times of the system obtained in the study of the temporal relaxation of metastable States of a nonwetting liquid in a porous medium. Figure 4 shows the recovered distribution function of the filled pore size for the system Liberar-23 – water at a temperature of 279K experiments and the waiting times 10 and 100 minutes, respectively.. Also in the figures for comparison, are the distribution function of the pores of the porous medium .

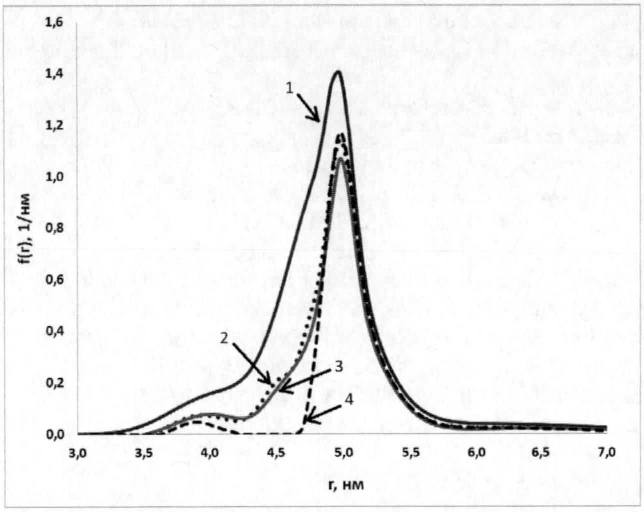

Figure 3. Distribution function of the liquid-filled pores corresponding to different relaxation times 1 - initial distribution,2 - distribution of the pore with the remaining liquid after 1 minute 3 - the distribution of the pore with the remaining liquid after 10 minutes,4 - the distribution of the pore with the remaining liquid after 100 minutes.

On fig. 3 presents the distribution function of the liquid-filled pores corresponding to different relaxation times 1 - initial distribution,2 - distribution of the pore with the remaining liquid after 1 minute 3 - the distribution of the pore with the remaining liquid after 10 minutes,4 - the distribution of the pore with the remaining liquid after 100 minutes. From figure 4 it is seen that in the relaxation process there is a change in the number of filled pores with sizes of 3.5-4.2 nm at maximum pore size distribution of 5 nm

Conclusion

From fig. 3 shows that relaxation process on the 0-100 times h occurs preferential leakage of fluid from local configurations of the pore sizes of 4.1-4.6 nm(figure 4.4) while discharge from the pores of the other radii in the distribution of the pore size occurs much more slowly(figure 4.2,4.3). This confirms the idea of the existence in the studied disordered local environment configuration While the system under consideration occurs, the range of times corresponding to the relaxation time of the cluster of filled pores of different radius. This means that not all of then, the liquid flows for the same time and the number of pores involved in the leakage, determined by the distribution function of the pore size depends on the time of observation. This looks disconcerting nature of the relaxation to a power law with $\alpha \sim 0.1$.

Acknowledgments

This work was partially supported by the Russian Foundation for Basic Research (project nos. 14-08-00895 a and 14-08-00805 a). The work of V.D. Borman and V.N. Tronin was supported by the Ministry of Education and Science of the Russian Federation (task no. 3.720.2014/K). This work was performed within the framework of the Center "Physics of nonequilibrium atomic systems and composites" supported by MEPhI Academic Excellence Project (contracts No. 02.a03.21.0005, 27.08.2013).

References

[1] Borman V D, Belogorlov A A, Byrkin V A, Tronin V N and Troyan V I 2012 *Jetp Lett* **95** 511.

[2] Borman V D, Belogorlov A A, Byrkin V A, Lisichkin G V, Tronin V N and Troyan V I 2013 *J. Exp. Theor. Phys.* **117** 1139.

[3] Borman V D, Belogorlov A A, Byrkin V A and Tronin V N 2013 *Phys. Rev. E* **88** 052116.

[4] Eroshenko V, Regis R C, Soulard M and Patarin J 2002 *C. R. Phys.* **3** 111.
[5] Lefevre B, Saugey A, Barrat J L, Bocquet L, Charlaix E, Gobin P F and Vigier G 2004 *J. Chem. Phys.* **120** 4927.
[6] Borman V D, Belogorlov A A, Grekhov A M, Lisichkin G V, Tronin V N and Troyan V I 2005 *J. Exp. Theor. Phys.* **100** 385.
[7] Kong X and Qiao Y 2005 *Appl. Phys. Lett.* **86** 151919.
[8] Han A, Kong X and Qiao Y 2006 *J. Appl. Phys.* **100** 014308
[9] Surani F B and Qiao Y 2006 *J. Appl. Phys.* **100** 034311.
[10] Qiao Y, Cao G and Chen X 2007 *J. Am. Chem. Soc.* **129** 2355.
[11] Han A, Lu W, Kim T, Chen X and Qiao Y 2008 *Phys. Rev. E* **78** 031408.
[12] Han A, Lu W, Punyamurtula V K, Kim T and Qiao Y 2009 *J. Appl. Phys.* **105** 024309.
[13] Liu L, Chen X, Lu W, Han A and Qiao Y 2009 *Phys. Rev. Lett.* **102** 184501.
[14] Falk M L and Langer J S 2011 *Ann. Rev. Cond. Matt. Phys.* **2** 353.
[15] Rycroft C H and Bouchbinder E 2012 *Phys. Rev. Lett.* **109** 194301.
[16] Bouchbinder E and Langer J S 2009 *Phys. Rev. E* **80** 031138.
[17] Langer J S 2013 *arXiv:1308,6544v2*.
[18] Langer J S 2012 *Phys. Rev. E* **85** 051507.
[19] Bouchbinder E and Langer J S 2011 *Phys. Rev. Lett.* **106** 148301.
[20] Borman V D, Belogorlov A A, Grekhov A M and Tronin V N 2014 *Eur. Phys. J. B* **87** 249.
[21] Biroli G and Garrahan J P 2013 *J. Chem. Phys.* **138** 12A301.
[22] *Dynamical Heterogeneities in Glasses, Colloids, and Granular Media* 2011 ed L Berthier, G Biroli, J P Bouchaud, L Cipelletti and W van Saarloos *Int. Ser. Monogr. Phys.* **150** (Oxford Univ. Press, Oxford).
[23] Andersen H C 2005 *Proc. Natl. Acad. Sci.* **102** 6686.
[24] Kirkpatrick T R, Thirumalai D and Wolyynes P G 1989 *Phys. Rev. A* **40** 1045.
[25] Berthier L and Biroli G 2011 *Rev. Mod. Phys.* **83** 587.
[26] Chander D and Garrham J P 2010 *Ann. Rev. Phys. Chem.* **61** 191.
[27] Debenedetti P G and Stilliger F H 2001 *Nature* **410** 259.
[28] Potuzak M, Welch R C and Mauro J 2011 *J. Chem. Phys.* **135** 214502.
[29] Luo A M and Ottinger M Ch 2014 *Phys. Rev. E* **89** 022137.
[30] Tanaka H, Jabbari-Farouji S, Mennier J and Bonn D 2005 *Phys. Rev. E* **71** 021402.
[31] Tanaka H, Kawasaki T, Shintani H and Watanabe K 2010 *Nat. Mater.* **9** 324.
[32] Tanaka H 2012 *Eur. Phys. J. E* **35** 113.
[33] Langer J S 2013 *Phys. Rev. E* **88** 012122.
[34] Phillips J S 1996 *Rep. Prog. Phys.* **59** 1133.
[35] Borman V D, Belogorlov A A and Tronin V N 2015 *Phys. Proc.* **72** 4-9.
[36] Mamontov E, Kumzerov Y A and Vakhrushev S B 2005 *Phys. Rev. E* **71** 061502.
[37] Kumzerov Y A, Nabereznov A A, Vakhrushev S B and Savenko B N 1995 *Phys. Rev. B* **52** 4772.
[38] Rigby S P and Edler K J 2002 *J. Coll. Int. Sci.* **250** 175.
[39] Porcheron F, Thommes M, Ahmad R and Monson P A 2007 *Langmuir* **23** 3372.
[40] Kong X and Qiao Y 2005 *Phil. Mag. Lett.* **85** 331.
[41] Grinchuk P and Rabinovich O 2003 *J. Exp. Theor. Phys.* **96** 301.
[42] Borman V D et al 2009 *J. Exp. Theor. Phys.* **108** 389.
[43] Borman V D et al 2011 *J. Exp. Theor. Phys.* **112** 385.
[44] Valluri S R, Jeffrey D J and Corless M R 2000 *Can. J. Phys.* **78** 823.
[45] Vasin M 2011 *J. Stat. Mech.: Theory Exp.* **2011** P05009.

Scenario SOC abnormally slow relaxation of the system clusters the fluid in disordered nanoporous confinement environment

V D Borman and V N Tronin

National Research Nuclear University MEPhI (Moscow Engineering Physics Institute), Kashirskoe highway 31, Moscow, 115409, Russia

E-mail: VNTronin@mephi.ru

Abstact. The relaxation of the system of the nanoporous medium with the nonwetting liquid is a self-organized criticality process characterized by waiting for fluctuation necessary for overcoming a barrier of local metastable state with the subsequent avalanche decay of local metastable configurations of pores. The dependence of the interaction between local configurations on the number of filled pores belonging to the infinite percolation cluster of filled pores serves as an internal feedback initiating the SOC process.

1. Introduction

Anomalously slow relaxation and hysteresis properties are a characteristic phenomena of many disordered media [1-4]. These media include atomic, molecular and polymeric glass of atomic particle size, colloid media, nanocomposites with nanoscale particles and loose material with a micron scale particles, as well as spin glasses.

The phenomenon of hysteresis is associated with the existence of long-lived metastable states and related spatial inhomogeneity of a disordered medium. The assumption of the existence of heterogeneous inhomogeneities and local configurations is central in phenomenological models of description of anomalously slow relaxation in disordered media. Analysis of these models performed in several recent reviews [1,2,5,6]. From this analysis it follows that the cause of the inhomogeneous of disordered media, and therefore the mechanism of anomalously slow relaxation are unclear until today.

A system of liquid clusters is a nonlinear dissipative systems. The relaxation of such system can be considered in the frameworks of model of self-organized criticality (SOC) [7,8]. This model was proposed in [9,10] to describe relaxation as an avalanche process of falling of a pile of sand. It is commonly accepted [7,8] that the SOC state appears in nonlinear dissipative systems in the critical state that relax without the external control parameter through rapid avalanche transitions between different metastable states of a system. The critical state is maintained not at a point, but in a wide region of the phase diagram of states through a nonlinear feedback mechanism because of the existence of the internal mechanism leading to the dynamic self-organization of transitions between intermediate metastable states of the system. Self-organized criticality is characteristic of the systems with fractal objects. However, physical mechanisms responsible for avalanche transitions between metastable states, as well as the reason for the appearance of the feedback mechanism leading to the dynamic self-organization of transitions between such states, are not yet understood. In particular, numerical studies [11] indicate that the properties of SOC with avalanche relaxation for spin-glass systems are not manifested at a finite number of neighbors and a divergent number of neighboring interacting spins are necessary.

In this work, we propose a new mechanism of anomalous relaxation of states of a liquid confined in a random nanoporous medium, which can be characterized as the SOC mechanism in revealed properties. The interaction energy of a system of clusters is calculated in the quasiparticle approach [12] as the sum of the energies of the local liquid cluster configurations interacting with clusters in neighboring pores. The process of relaxation involves overcoming the metastable state of the local configurations and the subsequent rapid hydrodynamic extrusion of the liquid. The dependence of the interaction between local configurations on the number of filled pores belonging to the infinite percolation cluster of filled pores serves as an internal feedback initiating the SOC process.

The calculations of the density of states and extrusion time distribution functions of liquid clusters give a power-law time dependence of the relative volume θ of the confined liquid $\theta \sim t^{-\nu}$ ($\nu \sim 0.1$). Such a dependence is characteristic of known self-organized criticality phenomena.

2. Main part

We assume that the disordered nanoporous medium includes pores with different sizes and the size of the porous medium is much larger than the maximum size of pores, so that the porous medium can be considered as infinite and the infinite percolation cluster of empty pores is formed in it.

According to [13,14], the metastable state of filled pores in the porous medium is formed at times $t > \tau_0$, relaxes slowly at times much larger than τ_0, and then decays at times $t > t_0$. It was shown in [35] that the relaxation of the metastable state in the Libersorb23–water system with a narrow pore size distribution with the relative width $\Delta R / \overline{R} < 0.1$ occurs at times $t_0 \sim 10^5$ s much larger than the thermal equilibrium establishment time $\tau_e < \tau_0$ at the extrusion of the liquid from the pore $t_0 >> \tau_0, \tau_e$. Therefore, such a state of the disordered porous medium with the confined nonwetting liquid can be described by local equilibrium distribution functions and can be considered thermodynamically with distribution functions in $5N$-dimensional phase space of pores coordinates, pore radii and their occupation numbers n_i of all N pores in the porous medium [13-15]. For the filling of pores in the porous medium, it should contain an infinite cluster through which the transport of the liquid is possible [13,14]. Consequently, the percolation cluster of filled pores is formed inside this percolation cluster of pores at filling [13]. It follows that the ground state of partially filled porous medium is a state containing a percolation cluster of pores with the percolation cluster of filled pores the inside of it.

The state of the system containing the infinite cluster of filled pores can have many realizations; i.e., this state is degenerate. Consequently, this state can be characterized by the probability $P(\theta)$ for the pore to belong to the infinite cluster of filled pores.

The liquid is removed from pores of porous medium only if at least one of its neighboring pores belongs to the infinite cluster of filled pores through which the extrusion of the liquid from a granule of the porous medium is possible. According to [13-15] a change in the complete thermodynamic potential, i.e., the energy of the porous medium filled to the degree of filling θ, at extrusion of liquid from pores in the case of small changes of the degree of filling θ can be represented in the form of the sum of the energies $\delta\varepsilon\ (\vec{r}_1, R_1, \theta)$ of local metastable configurations of a pore and its environment. This energy is the sum of the change of energy $\delta\varepsilon\ (r_1, R_1, \theta)$ of the interface between the porous medium and liquid and the energy $\delta\varepsilon_{int}(\vec{r}_1, R_1, \theta)$ necessary for the formation of menisci in the throats of the neighboring pores, which is proportional to the difference between the amount of meniscus before and after emptying pores per nearest neighbor. This quantity can be treated as the energy of the interaction $\delta\varepsilon_{int}$ of the liquid cluster in the pore with a liquid cluster in neighboring pores. Since the extrusion of liquid occurs through percolation cluster of filled pores, the energy $\delta\varepsilon_{int}(\vec{r}_1, R_1, \theta)$ can be characterized by the probability $P(\theta)$ for the pore to belong to the infinite cluster of filled pores and depends on the number of nearest neighbors z of depleted pore and the number of unfilled pores in its environment, composing the local configuration of depleted pore.

It was shown in [13-15] that a change in energy $\delta\varepsilon$ of pore at the extrusion of the liquid from this pore can be represented as a sum of changes in energy of various local configurations containing n empty pores $\delta\varepsilon_n$, divided by the number of nearest neighbors [13,14]

$$\delta\varepsilon(\theta\ ,R) = \frac{1}{z}\sum_{n=0}^{z-1}\delta\varepsilon_n, \tag{1}$$

Expression for the energy $\delta\varepsilon_n$ of the local configuration containing n empty pores was obtained in [13,14]:

$$\delta\varepsilon_n = -4\pi R^2\delta\sigma(1-q(\frac{R_0}{R})^\alpha) + 4\pi qR^{2-\alpha}R_0^\alpha\sigma(1-\theta\ )^n(P(\theta\ ))^{z-n}\frac{z-2n}{z}\frac{z!}{n!(z-n)!} \tag{2}$$

Here, σ is the surface energy of the liquid, $\delta\sigma = (\sigma_{ls} - \sigma_{sg})$ is the difference between the surface energies of the solid–liquid and solid gas interfaces, $P(\theta)$ is the probability for the pore to belong to the infinite cluster of filled pores (monotonic function of the degree of filling θ), $R_0 \sim \bar{R}/z$ is the minimum radius in the pore size distribution, $\alpha \approx 0.3$, and $q \sim 1$ [13,14]. The relative number of filled pores in the configuration can be estimated from the relation $z-n \sim \theta$. Therefore, as follows from Eq. (2) $\delta\varepsilon_n$ at a given degree of filling θ can be both positive and negative, depending on the number n. A negative $\delta\varepsilon_n$ value means that the extrusion of the liquid from pores in the n th state is energetically favorable. A positive $\delta\varepsilon_n$ value means the existence of the energy barrier for the decay of metastable configurations. Thus, Eq. (1,2) indicates the existence of local maxima and minima of the energy necessary for the extrusion of the liquid from the n th state. We now calculate the density of states $g(E,\theta)$ corresponding to the presence of the local configuration in a state with the energy E at the degree of filling θ:

$$g(E,\theta) = \int_0^\infty dR\sum_{n=0}^{z-1}\delta(E-\delta\varepsilon_n(R))f(R). \tag{3}$$

Figure 1 shows the energy dependence of the density of states of the local configuration for various degrees of filling θ. The calculations were performed with the results for the surface tension coefficient $\sigma(T)$ from [16]. The surface tension coefficient at $T = 293$ K is 75 mJ/m^2 [17]. The $\delta\sigma$ value at $T = 293$ K is 22 mJ/m^2. The quantity $R_0 \sim \frac{\bar{R}}{\bar{z}}$ was estimated within the model of randomly distrbuted spheres [17,18].

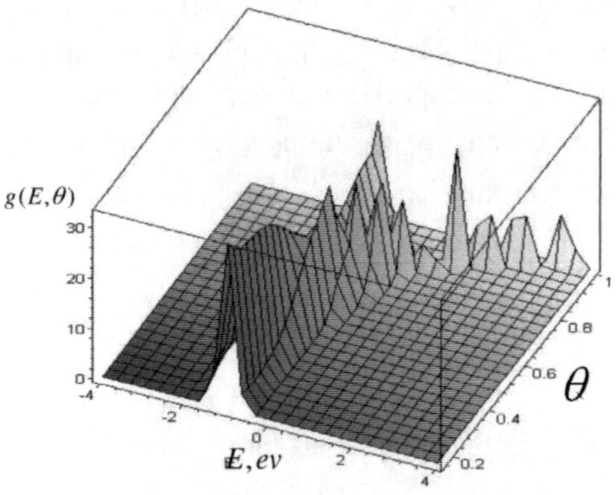

Figure 1. Density of states of the local configuration versus the energy and degree of filling of the medium θ.

It is seen in Fig. 1 that the states of the system at the degree of filling $\theta = 0.5$ are concentrated at negative energies of local configurations $E < 0$ or at energies below the temperature $E < T$. The characteristic liquid extrusion time τ for these states is determined by the hydrodynamic time τ_0 of the motion of the liquid in the porous medium. At the degree of filling $\theta = 0.7$, some local configurations are in states with a negative energy. The liquid flows out of these configurations in the hydrodynamic time. The other part of the liquid is in a state for which the potential barrier is positive and amounts $E \sim 1ev$. The characteristic liquid extrusion time for these states is determined by the potential barrier for extrusion and can significantly exceed the hydrodynamic time, $\tau >> \tau_0$, when the barrier is above the temperature, i.e., $\delta A > T$. These configurations correspond to a metastable state of the liquid with the decay time determined by the barrier. At the degree of filling $\theta = 0.8$, the number of local configurations with a negative energy $E < 0$ decreases, whereas the number of local configurations with a positive barrier increases. At the same time, states with a higher barrier, $E > 1.5ev$, appear in addition to states with $E \sim 1ev$. These configurations correspond to the appearance of additional metastable states of the liquid with a longer decay time, which is determied by the height of the barrier.

From the analysis made above follows that the extrusion of the liquid from different local configurations of pores can take different times, and the number of configurations of pores involved in extrusion depends on the observation time. As a result, the state of the system relaxes and the amount of the liquid remaining in the porous medium $\theta(t)$ depends on the observation time.

To calculate this dependence following [13,14] can be done by the distribution function $F^V(t)$ for the extrusion times from a pore with the radius R and volume $V = (4\pi/3)R^3$, using the relaxation time

$$\tau_n = \tau_0 exp(\delta\varepsilon_n(R,\theta,T)/T). \quad (4)$$

It follows from Eq. (6) that the relaxation time τ_n of the nth configuration is determined by the waiting time of fluctuation $\sim exp(\delta\varepsilon_n(R,\theta,T)/T)$ necessary for overcoming the local barrier $\delta\varepsilon_n(R,\theta,T)$ and by the time τ_0 of the subsequent barrierless hydrodynamic extrusion of the liquid from the local configuration. According to Eq. (6), the quantity $\delta\varepsilon_n(R,\theta,T)$ serving as the barrier for the nth local configuration depends on the radius of the pores, the probability of belonging to the percolation cluster $P(\theta)$, and the degree of filling θ of the porous medium with the liquid and, thereby, according to Eq. (1,2), it depends on the interaction between local configurations in the percolation cluster. At extrusion of liquid and the decay of the *n*th metastable state is accompanied by a decrease in the number of filled pores in the percolation cluster and, hence, in the number of neighboring pores. Therefore, the energy of the multiparticle interaction in Eq. (1,2), as well as thereby the lifetime of local configurations with different numbers *n*, decreases. As a result, the decay of the *n*th local metastable configuration reduces the decay time of local metastable configurations with different numbers *n* and with different, initially higher energy. In this case, the dependence of the interaction between local configurations on the number of filled pores belonging to the infinite percolation cluster of filled pores serves as an internal feedback at the decay of metastable states in a wide region of the degree of filling $\theta > \theta_c$, where θ_c is the percolation threshold.

As follows from Eq. (1,2), the volume fraction $\theta(t)$ of pores from which the liquid is not extruded can be represented in the form

$$\theta(t) = \sum \theta_n(t), \theta_n(t) \sim \theta_{np}\left(\frac{\tau_{qn}}{t}\right)^{a_n}, a_n = \frac{1}{1+(2-\alpha)\frac{\Delta R}{\bar{R}}\frac{\varepsilon_{n\max}}{T}}, \tag{5}$$

$$\tau_{qn} \sim \tau_0 \exp\left(\frac{\varepsilon_{n\max}}{T}\right), \theta_{np} \sim \theta \int_0^{\infty} \eta(\delta\varepsilon_n(R)) dR f^V(R).$$

where nd $\varepsilon_{n\max} = \max \delta\varepsilon_n$ is the maximum energy barrier for the decay of the local metastable configuration in the *n*th state.

As follows from Eqs. (5), the exponents a_n and times and specify the relaxation law of the state of the n th local configuration, are determined by the energy barrier $\varepsilon_{n\max}$. The volume fraction of the liquid remaining in the local metastable configuration in the *n*th state decreases according to a power law with the exponent a_n and charateristic time $\tau_{nq} \sim \tau_0 \exp(\frac{\varepsilon_{n\max}}{T})$. According to Eqs. (5), different states $\theta_n(t)$ relax with different charateristic times τ_{nq} and different exponents a_n. The slowest relaxing state of the local configuration corresponds to $\tau_{nq} = \max$ and $a_n = \min$ and, hence, $\varepsilon_{n\max} = \max$; whereas the slowest relaxing state to $\tau_{nq} = \min$ and $a_n = \max$ and, therefore, $\varepsilon_{n\max} = \min$.

It follows from Eq. (5) that the decay of the local metastable configuration in the *n*th state results in the disappearance of the corresponding term in sum (5) and, consequently, in a decrease in the degree of filling θ. As a result, according to Eq. (1,2), the energy of the interaction between local configurations decreases and, therefore, the energy barrier $\delta\varepsilon_m(\theta)$ for extrusion from other states changes. For this reason, according to Eqs. (1,2), the volume fraction of the liquid remaining in the metastable state in the porous medium should be determined from the self-consistent equation

$$\theta(t) = \sum_{n=0}^{z-1} \theta_{np}(\theta(t))\left(\frac{\tau_{qn}(\theta(t))}{t}\right)^{a_n(\theta(t))}. \tag{6}$$

To analyze this equation, we note that the terms in sum (9) include different exponents $a_n(\theta(t))$ and times $\tau_{qn}(\theta(t))$. As a result, the relaxation of the system determined by the dependence $\theta(t)$ at each time instant occurs as the «rapid» relaxation of the *n*th state, which is determined by the maximum exponent a_n and the minimum time τ_{qn}, and as the «slow» relaxation of the *n*th current state, which is determined by the minimum exponent a_n and the maximum time τ_{qn}. According to Eq. (6), the set of such modes contributes to the total relaxation of the metastable state at times $t \sim t_0$

>> τ_0. Therefore, the relaxation of the metastable state in times $\tau_0 << t \sim t_0$ corresponds to «the trajectory of motion» of the system in the (θ, n) space with the slowest relaxation of the local metastable configurations in the *n*th states. The relaxation of the state of the local metastable configurations on this trajectory occurs because of feedback with the minimum exponent a_n and maximum time τ_{qn}. Figure 2 shows the energy of the state of the local configuration $\delta\varepsilon_n$ corresponding to the *n*th mode for the configuration of the pore with the radius equal to the average radius $R = \overline{R}$ for a Gaussian pore volume distribution $f^V(R)$. The calculations were performed with the same parameters as for Fig. 1.

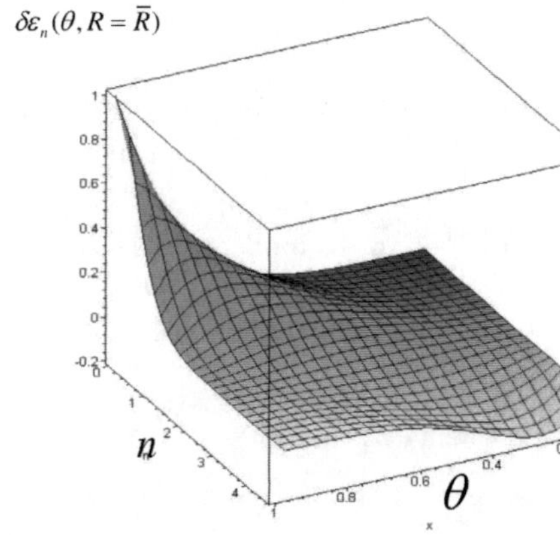

Figure 2. Energy of the state of the local configuration $\delta\varepsilon_n$ corresponding to the *n*th mode for the configuration of the pore with the radius equal to the average radius $R = \overline{R}$ for a Gaussian pore volume distribution $f^V(R)$. The states on the crest correspond to the critical state of the system.

It is seen in Fig. 2 that the energy of the state of the local configuration $\delta\varepsilon_n$ as a function of the variables θ, n has a maximum (crest) corresponding to the states and degrees of filling of local metastable configuration with the slowest relaxation. Consequently, the states on the crest determine the critical state of the system throughout the range $\theta > \theta_c$. According to Eq. (6), the time dependence of the degree of filling for the critical state is given by the expression

$$\theta(t) \approx \theta_{n(\theta(t))p}(\theta(t))\left(\frac{\tau_{qn(\theta(t))}(\theta(t))}{t}\right)^{a_{n(\theta(t))}(\theta(t))}. \quad (7)$$

Here, $n(\theta(t))$ is the number corresponding to the state of the local configuration on the slowest relaxing trajectory (critical state).

Expressions (7) describe the relaxation of the formed metastable state corresponding to the mean-field picture considered in [20]. According to [15], the formed metastable state corresponding to bound states of interacting local configurations of filled pores relaxes slowly according to a power law with the exponent a given in Eqs. (10) in times $t \sim \tau_q$. The formed metastable state decays at times $t > t_0$ because of a decrease in the degree of filling θ and, as a result, a decrease in the energy barrier for extrusion $\varepsilon(R, \theta_1)$. In this case, the ε_z value decreases and the exponent a increases according to Eqs. (7). Consequently, the decay rate of the metastable state increases with the relaxation of the system.

The final expressions indicate that the relaxation of the system occurs through the decay of the local metastable configuration responsible for the formation of critical conditions for the decay of other local metastable configurations with a different, initially higher energy. The decay of the metastable state of some local configurations reduces the number of filled pores in this percolation cluster and, therefore, the number of neighboring pores. As a result, the energy of the multiparticle interaction decreases and critical conditions appear for the decay of other metastable configurations. In this case, the system relaxes on the trajectory in the (configuration number, degree of filling) space corresponding to the critical state with the slowest relaxing mode. The time dependence of the relative volume θ of the confined liquid on this trajectory has the form of a power law given by Eq. (7)) (Figure 3) Figure 3 shows a relaxation picture in approach of an average field [20].

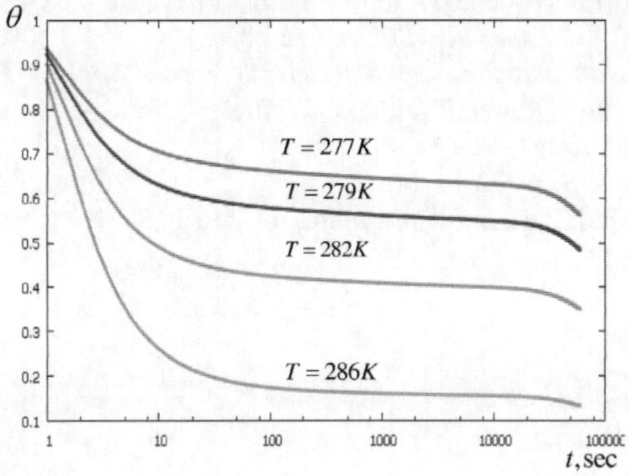

Figure 3. Qualitative depending of the volume fraction remaining liquid on time at different temperatures.

From Fig. 3 it is seen that the relaxation of the formed metastable state of confinement liquid occurs at times t = 100-6000s according to a power law $\theta(t) \sim t^{-\alpha}$ (18) with $\alpha \sim 0.1$. At times t>10^5 s metastable state decays.

Conclusion

To summarize, the relaxation of the system of the nanoporous medium with the nonwetting liquid is a self-organized criticality process characterized by waiting for fluctuation necessary for overcoming a barrier of local metastable state with the subsequent avalanche decay of local metastable configurations of pores. The dependence of the interaction between local configurations on the number of filled pores belonging to the infinite percolation cluster of filled pores serves as an internal feedback initiating the SOC process.

The model of the relaxation of the porous medium with the nonwetting liquid developed in this work demonstrates possible mechanisms and scenarios of SOC for disordered atomic systems.

Acknowledgements

The work was supported by the Ministry of Education and Science of the Russian Federation (task no. 3.720.2014/K). This work was performed within the framework of the Center "Physics of nonequilibrium atomic systems and composites" supported by MEPhI Academic Excellence Project (contracts No. 02.a03.21.0005, 27.08.2013).

References

[1] Biroli G and Garrahan J P 2013 *J. Chem. Phys.* **138** 12A301.
[2] Langer J S 2014 *Rep. Prog. Phys.* **77** 042501.
[3] Dotsenko V S 1993 *Phys. Usp.* **36** 455–85.
[4] Biroli G 1998 *Hysteresis in Magnetism* (Elsiver: Amsterdam)

[5] Berthier L and Biroli G 2011 *Rev. Mod. Phys.* **83** 587
[6] Stillinger F H and Debenedetti PG 2013 *Ann. Rev. Condens. Matter Phys.* **4** 23.
[7] Aschwanden M J 2013 *Self-organized criticality systems* (Open Academic press GmbH Co).
[8] Aschwanden M J et al 2016 *Space Sci. Rev.* **198** 47–166.
[9] Bak P, Tang C and Wiesenfeld K 1987 *Phys. Rev. Lett.* **59** 381.
[10] Bak P, Tang C and Wiesenfeld K 1988 *Phys. Rev. A* **38** 364.
[11] Andersen J C et al 2013 *Phys. Rev. Lett.* **111** 097203.
[12] Landau L D and Lifshitz E M 1985 *Statistical Physics* **9** (Pergamon: New York).
[13] Borman V D, Belogorlov A A, Byrkin V A and Tronin V N 2013 *Phys. Rev. E* **88** 052116.
[14] Borman V D, Belogorlov A A, Byrkin V A, Lisichkin G V, Tronin V N and Troyan V I 2013 *J. Exp. Theor. Phys.* **117** 1139.
[15] Borman V D, Belogorlov A A and Tronin V N 2015 *Phys. Proc.* **72** 4-9.
[16] Borman V D et al 2011 *J. Exp. Theor. Phys.* **112** 385.
[17] *Handbook of Chemistry and Physics: A Ready Reference Book of Chemical and Physical Data* 2013 ed W M Haynes (Boca Raton, Florida : CRC Press)
[18] Byrkin V A 2015 *J. Porous Media* **18** 463-75.
[19] Byrkin V A 2015 *Phys. Proc.* **72** 14-21.
[20] Rigby S P and Edler K J 2002 *J. Colloid Inter. Sci.* **250** 175.

Evolution of the filling pressure of the porous medium by non-wetting liquid at pulse pressure changes

A A Belogorlov, V A Byrkin, V D Borman and V N Tronin
National Research Nuclear University MEPhI (Moscow Engineering Physics Institute), Kashirskoe highway 31, Moscow, 115409, Russia
E-mail: VAByrkin@mephi.ru

Abstract. The paper presents the results of experimental studies of impact filling of nanoporous medium with non-wetting liquid. With increasing the impact energies nontrivial pattern of pressure changes in the system over time is observed. A physical explanation is proposed for the observed phenomena.

1. Introduction

The last few decades systems consisting of nanoporous medium dispersed in a non-wetting liquid cause an increased interest from both the practical and theoretical points of view. Non-wetting liquid can infiltrate into the porous medium only with an excess pressure. Liquid infiltration tends to increase the solid-liquid interfacial energy and thus the absorbed energy can be estimated as $E \sim \Delta\gamma S$, where $\Delta\gamma$ is a solid-liquid interfacial tension, and S is the specific surface area of the medium. Therefore this energy for nanoporous media can reach several orders of magnitude superior to traditional damping materials and shape-memory materials [1, 2]. As a consequence, the prospects of using devices based on systems consisting of a nanoporous medium immersed in a non-wetting liquid associated mainly with the absorption of mechanical energy of impact or explosion. Particularly there are considered the development of automotive dampers [3], the system of protection against strains and by the earthquake [4].

At impact or explosion a rapid growth of pressure is occured in system with rates of $\dot{p} = 10^4 \div 10^5$ atm/s. However, most of modern research is related to the quasi-static regime of infiltration with pressure rate not exceeding $\dot{p} < 10$ atm/s. Not numerous studies [5, 6], that the physics of the infiltration process at high rates of pressure is significantly differ from infiltration at low speeds. Nevertheless, a number of papers [7, 8] studied the dynamic filling by molecular modeling. In [6] shows that the observable system can absorb more energy density in comparison with the case of quasi-static pressure. A lot of papers [3] are devoted to experimentally investigation of the development of car damper. Despite this, a set of experimental data is presented rather poorly and full-fledged studies are missing. The answer is absent to the question of how the system behaves under shock impacts at different energies.

Journal of Physics: Conference Series **751** (2016) 012035 doi:10.1088/1742-6596/751/1/012035

2. Experiment

In the present study we investigated the dynamics of shock filling with water of commercially available silica gels with a random structure of pores obtained in the solgel process Libersorb 23 (L23). L23 was modified in the laboratory headed by Prof. G.V. Lisichkin for obtaining the hydrophobized surface of pores. Using the pycnometry (micro-Ultrapyc 1200e, Quantachrome Instruments) and porometry (Nova 1200e, Quantachrome Instruments) methods, we determined the density (ρ), specific surface area (S_p), specific volume of pores (V_p), and pore size distribution by the classical Barrett-Joyner-Halenda (BJH) method within the cylindrical model of pores and porosity (φ) of this material. The results are presented in Table 1.

A sample of the porous body weight $2 \div 10\,\mathrm{g}$ was placed in a liquid-permeable container in the pressure chamber with volume of $\sim 60\ \mathrm{cm}^3$. The chamber was filled with distilled water. A falling weight from certain height is designed to impart a load pulse. Impact energy is determined by the variation of height. The direct method of measurement is described in detail at [5].

Table 1. Characteristics of the studied porous media.

	L23 (C8)
ρ, g/cm^3	1.7798 ± 0.0016
V_p, cm^3/g	0.66 ± 0.02
S_p, m^2/g (BET)	212 ± 7
$\varphi = V_p/(V_p + 1/\rho)$	0.66 ± 0.02

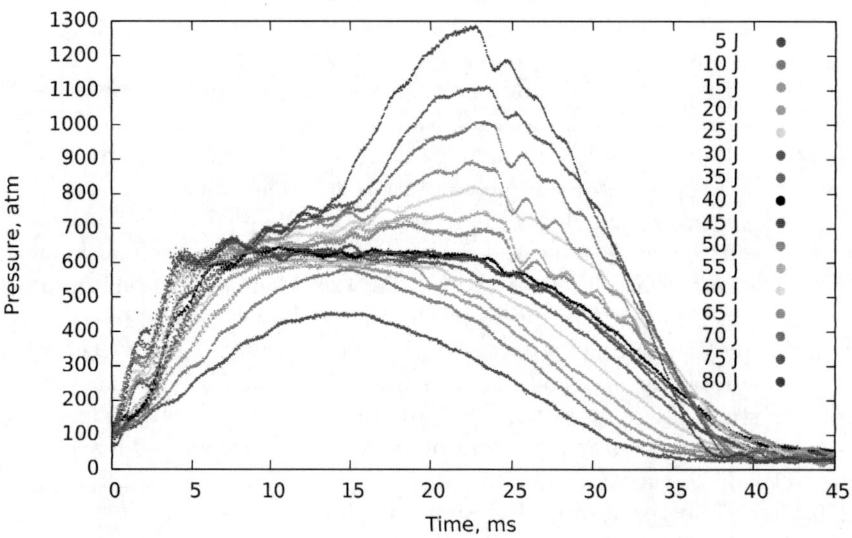

Figure 1. The dependence of the volume of the trapped liquid on the degree of filling for narrow pore size distribution $f_1(R)$: $\Delta R/R \ll 1$

The results of experimental measurements of pressure versus time with a porous medium L23 are shown in Fig. 1, the range of variation of energy ranged from 5 to 80 J. The analysis of the experimental data shows that the pressure does not behave in a trivial way with increasing impact energy. Under quasi-static compression system nanoporous medium – non-wetting liquid there is threshold pressure p_{c0}, at which the liquid infiltrate to the pores of the porous medium [9]. This forms an inhomogeneous front of porous media filling. This process is typical at filling macroscopic porous medium by non-wetting liquid. The threshold infiltration character was set for non-wetting liquids by [10], as well as for granular porous bodies - zeolites having a pore size ($R = 0.3 \div 1.4\,\mathrm{nm}$) and silochrome ($R = 4 \div 120\,\mathrm{nm}$) when filled with liquid (molten) metals, hydrophobized porous media with a framework of a silicon oxide ($R = 3 \div 50\,\mathrm{nm}$) when filled with water, a solution of ethylene glycol and salt solutions [4, 9, 11, 12]. At the same time, at the rapid compression $\dot{p} = 10^4 \div 10^5\,\mathrm{atm/s}$, as shown in the work [5, 6] for the systems "granular porous body Silohrom CX 1.5 - Wood's alloy" and "Fluka 100 - water", it was found that infiltraion occurs at the percolation threshold at a pressure far above the threshold pressure at quasi-static mode p_{c0}. So far the first system pressure threshold value of $p_0 = 1.6p_{c0}$ [5], and for the second — $p_0 = 2p_{c0}$ [6].

3. Disscussion

These experiments (Fig. 1) shows that at the energy of 5 J pressure is not reached value p_{c0}, corresponding to the beginning of the infiltration pressure at quasi-static mode [13], and at an energy of 10 J maximum pressure is in the interval $p_{c0} < p < p_c$, where the value p_c corresponds to the pressure of the beginning of infiltration at impact infiltration. The magnitude of the change in volume of the system within the measurement error coincides with a change in volume due to deformation, and this means that there is no filling of the porous body. With increasing impact energy is observed infiltration to the porous medium with a pronounced plateau in the pressure versus time. At higher energies, where the energy of compression exceeds the maximum value determined by the energy density of the porous body filling, fill mode at constant pressure is not realized. In these circumstances, as the filling of the porous body is not provided the necessary speed energy absorption characteristic growth time pressure and response system wetting liquids - nanoporous medium is elastic deformation, which is accompanied by the end of the filling pressure increases to a maximum and then only outflow.

Let us consider the characteristic times of the system: τ_p is the time of pressure changes; τ_z is the hydrodynamic time of clusters available barrier-free filling of pores; τ_v is the the time of change of volume; τ_d makes sense to the characteristic time of formation of the available pore pressure changes over time. The relations between these times in case of impact of filling should be arranged as follows: $\tau_z > \tau_p \gg \tau_d$.

In times t, which satisfy the inequality $\tau_v > \tau_z > t \geq \tau_p > \tau_d$, arising accessible pores do not have time to fill with liquid, causing the porous body is in a state above the percolation threshold at affordable pores. In times $\tau_v > t \geq \tau_z > \tau_p > \tau_d$ begin the process of filling the infinite cluster available now through clusters filled pores of finite size.

Conclusion

Thus, with the rapid change in pressure ($\tau_z > \tau_p \gg \tau_d$) filling the porous body occurs by rapid (taking place simultaneously throughout the volume of pellets filled clusters of finite size at times $t \sim \tau_z$) and slow (taking place at times $t \sim \tau_v \gg \tau_z$) the process of filling the infinite cluster of available pore fluid flowing through the pores filled with clusters of finite size. Consequently the dynamics of filling granules of a porous body can be represented as formation of the environment for filling, i.e. as the formation of clusters available since then filling some of these clusters. As in the experiments recorded by filling granules of porous medium occurs when the correlation length $\xi \approx \bar{R}/|\theta - \theta_{c0}|^v$, $v = 0.8$ [14] becomes comparable to the size of the granules L or higher than the ($\xi \geq L$), the process of filling the granules can be considered as a homogeneous and at the same time flowing throughout the space since the formation of clusters of granules filled pores.

Acknowledgments

The reported study was partially supported by RFBR, research project No. 14-08-00805a. The authors also express their gratitude to the Ministry of Education and Science of the Russian Federation under the grant to the President of the candidates No. MK-7793.2015.2.

References

[1] Chang K C, Soong T T, Lai M L and Nielsen E J 1993 *Earthquake Spectra***9** 371–87
[2] Eroshenko V, Grosu Y, Tsyrin N, Stoudenets V, Nedelec J M and Grolier J P E 2015 *J. Phys. Chem. C* **119** 10266–72
[3] Suciu C V and Buma S 2013 *Proc. of the FISITA 2012 World Automotive Congress* (*Lect. Notes Electr. Eng.* **198**) (Berlin: Springer) pp 21–32
[4] Chen X, Surani F B, Kong X, Punyamurtula V K and Qiao Y 2006 *Appl. Phys. Lett.* **89** 241918
[5] Borman V D, Belogorlov A A, Lisichkin G V, Tronin V N and Troyan V I 2009 *J. Exp. Theor. Phys.* **108** 389–410
[6] Surani F B, Kong X, Panchal D B and Qiao Y 2005 *Appl. Phys. Lett.* **87** 163111
[7] Liu H and Cao G 2013 *J. Chem. Phys.* **139** 114701
[8] Liu H and Cao G 2014 *J. Phys. Chem. C* **118** 25223–33
[9] Borman V D, Belogorlov A A, Grekhov A M, Lisichkin G V, Tronin V N and Troyan V I 2005 *J. Exp. Theor. Phys.* **100** 385–97
[10] Sahimi M 1993 *Rev. Mod. Phys.* **65** 1393–534
[11] Xu B, Qiao Y and Chen X 2013 *J. Mech.Phys. Solids*
[12] Kong X and Qiao Y 2006 *J. Appl. Phys.* **99** 64313
[13] Sahimi M 2011 *Flow and transport in porous media and fractured rock: From classical methods to modern approaches* 2nd ed (Weinheim: VCH)
[14] Isichenko M 1992 *Rev. Mod. Phys.* **64** 961–1043

Journal of Physics: Conference Series **751** (2016) 012036 doi:10.1088/1742-6596/751/1/012036

Percolation threshold of the permeable disks on the projective plane

V D Borman, A M Grekhov, I V Tronin and V N Tronin

National Research Nuclear University MEPhI (Moscow Engineering Physics Institute), Kashirskoe highway 31, Moscow, 115409, Russia

E-mail: IVTronin@mephi.ru

Abstract. The percolation threshold and wrapping probability for the two-dimensional problem of continuum percolation on the projecive plane have been calculated by the Monte Carlo method with the Newman–Ziff algorithm for completely permeable disks. It has been shown that the percolation threshold of disks on the projective plane coincides with the percolation threshold of disks on the surfaces of a torus and Klein bottle, indicating that this threshold is topologically invariant.

1. Introduction

Problems of the percolation theory receieving increasing attention during the last years both fundamental physics and in applications. From the fundamental point of view percolation problems are connected with the probability theory, conformal field theory and statistical physics. In applied problems, models of the percolation theory are widely used to describe phenomana in composite materials and disordered porous media.

One of the main tasks in percolation theory is determination of the percolation threshold since its value is not a universal constant. Moreover, if we consider a percolation problem on non-orientable surface, conformal field theory cannot give us any result. In this case the problem of the percolation threshold calculation can be solved only by numerical simulations.

In this work we present the results of calculation of the percolation threshold fro the fully permeable disks on projective plane. It is shown that the calculated percolation threshold for the projective plane coincides with the thresholds on torus and Klein bottle, indicating that this threshold is topologically invariant.

2. Method of the calculation

The Monte Carlo calculations were performed on square systems of various sizes with the Newman–Ziff algorithm [1]. This algorithm is currently the fastest algorithm of the search for the percolation threshold. We use completely permeable disks with the unit diameter $d = 1$ as objects. The addition of a disk to the system is reduced to the generation of two random numbers determining the position of its center. The random numbers were generated according to the Mersenne–Twister algorithm having a period of $2^{19937} - 1$.

The topology of the surface on which calculation is performed is determined by the boundary conditions. Projective plane can be obtained from the square by "gluing" according to Fig. 1. In terms of boundary conditions, this means that points on the upper and lower sides as well as

Figure 1. Gluing of the boundaries providing the projective plane.

the left and right sides of the square are identified in the "mirror" manner. In order to verify the percolation conditions in each of the directions, we used the method proposed in [2] adapted for continuum percolation [3]. On an non-orientable surfaces this algorithm must be adapted in the following manner [4]. Instead of the calculation in the square calculation domain with the size L, we performed the calculation in a square domain with a size of $2L$ divided in four squares. When adding a new disk, two random numbers x, y were generated in the interval $0 \div L$, the disk was added to the lower left square of the calculation domain and three another "copies" of this disk was added to the upper left, bottom right and upper right squares with allowance for mirror reflection. Example of the calculation domain for the size $L = 4$ is shown on fig. 2. Digits and solid border disks in the lower square, whereas primed digits and dashed border indicate their reflections.

If each calculation ends at the number of disks n that results in the first appearance of the wrapping cluster, the probability of the existence of the wrapping cluster in the canonical ensemble $P_L(n)$ is equal to the fraction of the calculations ending at the number of disks no more than n. The wrapping probability R_L in the grand canonical ensemble is a function of the dimensionless concentration (degree of filling) $\eta = nS/L^2$ (S is the area of one disk) and is obtained by the convolution of the probability P_L with the Poisson distribution with the mean value $\lambda = \eta L^2/S$ [3]:

$$R_L(\eta) = e^{-\lambda} \sum_{n=0}^{\infty} \frac{\lambda^n}{n!} P_L(n) . \tag{1}$$

At a fixed size of the system L, each calculation was performed until percolation occurred in both axes and the number of disks resulting in the first appearance of the wrapping cluster in each of the axes was fixed. We calculated six different probabilities R_L of the appearance of the wrapping cluster:

(i) In any of the axes R_L^e.
(ii) In both axes simultaneously R_L^b.
(iii) In the horizontal axis R_L^h independently of the presence of a cluster in the vertical axis.
(iv) In the vertical axis R_L^v independently of the presence of a cluster in horizontal axis.
(v) In the vertical axis, but not in the horizontal axis R_L^{1v}.
(vi) In the horizontal axis, but not in the vertical axis R_L^{1h}.

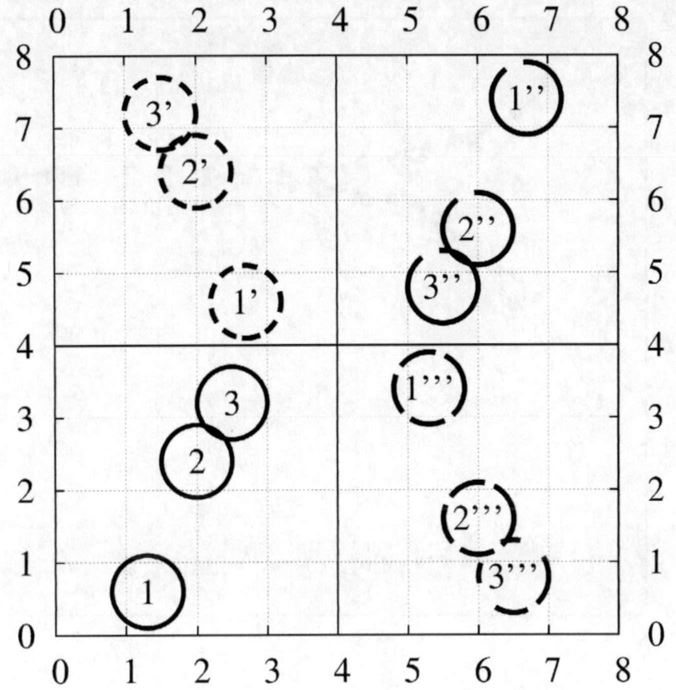

Figure 2. Example of the square calculation domain for $L = 4$. Digits and solid border disks in the lower square, whereas primed digits and dashed border indicate their reflections.

In order to calculate percolation threshold η_c we used two methods. The first one, proposed in [1] based on a nonmonotonic behavior of the functions R_L^{1v} and R_L^{1h}. The second one use the expression [5, 6]

$$R_L(\eta_L) = R_{L/2}(\eta_L) \tag{2}$$

to calculate the critical concentration at a given L value. The sequence of critical concentrations η_L thus obtained should converge to the percolation threshold in the limit $L \to \infty$.

3. Results and discussion

The calculations were performed for different sizes of systems in the range from $L = 8$ to $L = 3072$. For each size, we performed from $N \simeq 10^6$ to $N \simeq 10^9$ depending on the size of the system. After that we calculate critical concentrations η_L for each system size and percolations threshold η_c using two methods mentioned above. Calculated results are shown on fig. 3.

The percolation threshold obtained in this work is $\eta_c = 1.128086(4)$, which almost coincides with the percolation threshold in the case of continuum percolation of permeable disks on the torus [3] and klein bottle [4]. This indicates that the percolation threshold η_c is independent of the topology of the surface and is a topological invariant.

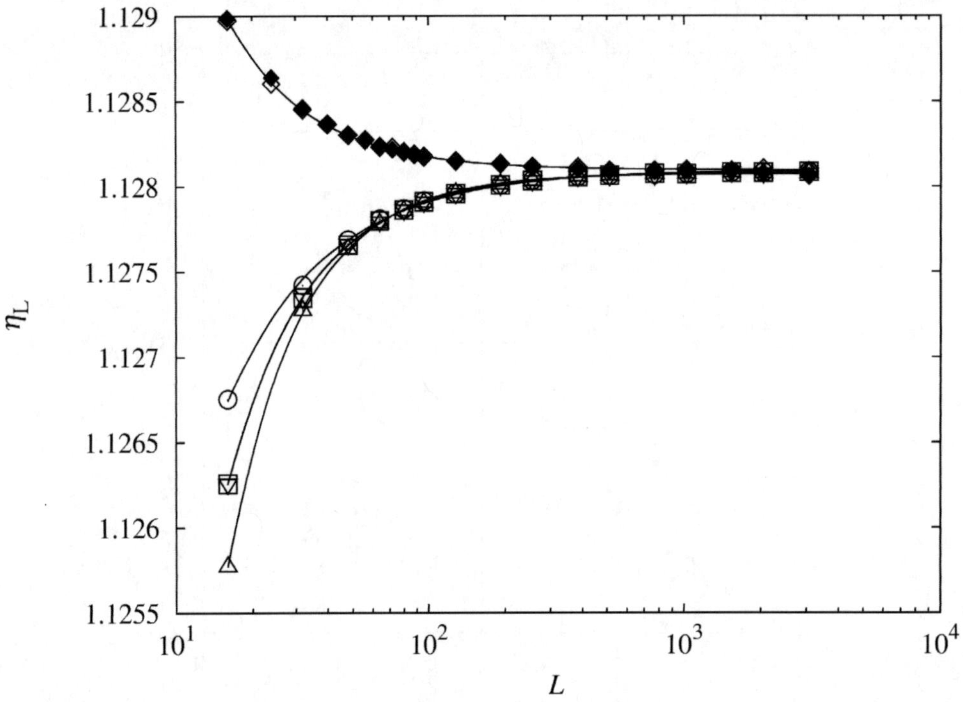

Figure 3. Estimated critical filling factors for continuum percolation on a projective plane, derived from the maxima of R_L^{h1} (open diamonds), R_L^{v1} (closed diamonds) and from (2) for R_L^e (upward pointing triangles), R_L^b (circles), R_L^h (squares), and R_L^v (downward pointing triangles).

4. Conclusions

The percolation threshold η_c for the continuum percolation of permeable disks on the projective plane have been obtained. It has been shown that the percolation threshold η_c is independent of the topology of the surface and coincide with the respective values obtained for continuum percolation on the torus and Klein bottle.

Acknowledgements

The work has been performed under the support of the Ministry of education and science of Russia, grant no. 3.726.2014/K. The computer simulations have been partially performed on the computer cluster of National research nuclear university(MEPhI), Basov farm. This work was also performed within the framework of the Center "Physics of nonequilibrium atomic systems and composites" supported by MEPhI Academic Excellence Project (contracts No. 02.a03.21.0005, 27.08.2013).

References

[1] Newman M E J, Ziff R M 2001 *Phys. Rev. E* **64** 016706
[2] Machta J, Choi Y S, Lucke A, Schweizer T and Chayes L M 1996 *Phys. Rev. E*, **54** 1332

[3] Mertens S, Moore C 2012 *Phys. Rev. E* **86** 061109
[4] Borman V D, Grekhov A M, Tronin V N and Tronin I V 2015 *J. Phys. A-Math. Theor.* **48** 475002
[5] Reynolds P, Stanley H and Klein W *J. Phys. A-Math. Gen.* **11** L199
[6] Reynolds P, Stanley H and Klein W *Phys. Rev. B* **21** 1223

Investigation of application possibility of quasi-softening for scale formation decrease

E A Ananeva and D Y Feklistov
National Research Nuclear University MEPhI (Moscow Engineering Physics Institute), Kashirskoe sh. 31, 115409, Moscow, Russia.

E-mail: fateev@isp.nsc.ru

Abstract. Methods of water treatment were developed using activators, which lead to decrease of scale formation due to the conversion of hardness salts into aragonite crystal form. The installation was created on the base of conducted research, which can be used in systems of circulating water supply and allows to provide the decrease of scale formation.

1. Introduction

Precipitate formation process of calcium carbonate precipitation is a major process of scale formation during water usage in thermal engineering. The aim of the work is to create mobile installation, using water treatment methods. Water treatment methods allow decreasing scale formation on the walls of heat exchanger. Preparatory experiments had been put in the base of installation development. Using possibility of such ways of water treatment leads to the decrease of scale formation without water softening.

Crystalline form of hardness salts, size distribution of crystals, and other properties of nascent scale depends on different factors. For example, with the help of synchrotron emission it was detected that at temperatures 7 – 25 °C fast dehydration of nanoparticles of amorphous calcium carbonate and its crystallization as vaterite takes place on the first stage; on the second stage vaterite transforms into calcite [1].

Well-known nonreagent magnetic water treatment [2] facilitates to directional crystallization of calcium carbonate in the form of niddle aragonite. These crystals have bad adhesion to the support, poor grip with each other and can be located in solution for a long time. In this case magnesium and calcium salts, contained in water, don't form solid deposit on the walls of heat exchanger pipes. However, decrease effect of scale formation by magnetic field is unstable and in some cases is lost with time. Magnetic and electrochemical water treatment unequally influence on the shape, size, and crystals precipitation rate of calcium carbonate. Transformation effect of electrochemically formed calcium carbonate crystals influenced by magnetic field into sheaf-like crystals of aragonite was detected [3].

It was established, that filtration of earth water through microporous materials influences on the shape of calcium carbonate crystals, formed at water heating.

2. Main text

Research of microfiltration influence on scale formation process and on the shape of hardness salts of crystals was carried out, using filtration unit «Millipore» with membrane, which channel size is

0,45 μm. Model hardness solutions were mixed solutions of calcium chloride and sodium bicarbonate in tap water. Hardness of water was determined on the inlet and outlet of solutions from installations using method of complexometric titration with trilon B. The presence of scale reducing effect was detected by the value of carbonate hardness in supernatant fluid, and also by scale formation on glasses, which had been contacted with treated and untreated water on the evaporation stage. If the hardness in supernatant fluid after boiling of already treated water was higher, than in case of analogue experiment with untreated water, this fact circumstantially detects the presence of the effect of scale decrease. In this case, micro particles of calcium carbonate are in suspension state.

Relative amount of scale, formed on glasses at their immersion to initial evaporated and filtered water, was evaluated by the change of optical density (D) using spectrophotometer Specol 1300. Glasses were immersed into vessels, filled with test water samples. Water was boiled and evaporated to half in order to create the supersaturation to calcium carbonate. Effect was interpreted as scale reducing, if glasses optical density, which were boiled in treated water samples was lower, than glasses optical density in analogue experiments with initial water samples. The shape of nascent calcium carbonate crystals was evaluated using microscope METAM PB-21-1.

Verification results of «quasi-softening» effect presence in model solution, using filtration unit "Millipore" with membrane 0,45 μm are presented in table 1 and fig.1

Table 1. Verification results of «quasi-softening» effect

Indicator name, unit of measurement	Initial water (hydrocarbonate water with high value of hardness)	Water after treatment	Notation
pH	7.3	7.3	«Quasi-softening» effect was detected
Hydrocarbonates, mg/l	520	500	
Water hardness before boiling, °H	7.3	7.3	
Supernatant fluid hardness after boiling, °H	4.6	4.8	
Scale formation on glasses, increment of optical density, ΔD	0.035	0.005	Effect of scale decrease was detected
Crystal shape	Calcite crystals	Aragonite crystals	Aragonite crystals prevail fig.1

Conducted researches show, that microfiltration doesn't lead to water hardness changing, but post heating of this water doesn't lead to scale formation. Detected effect of «quasi-softening», earlier was mentioned at filtration through polymer with space-globular structure. (SGS) [4].

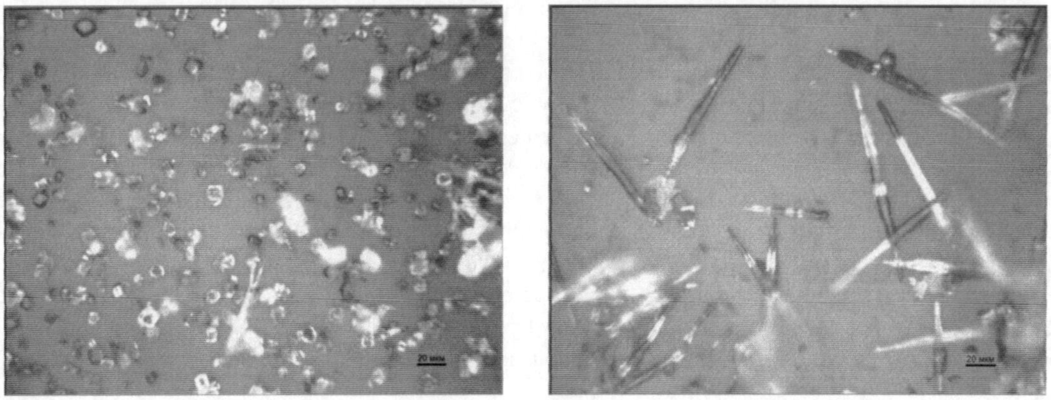

a) b)

Figure 1. Shape of calcium carbonate crystals, formed at water boiling: a) initial water – calcite; b) after filtration though nylon membrane 0,45 μm – aragonite.

For the creation of stable «quasi-softening» effect at installation designing activating additives in the form of ions of ferric iron and hydrocarbonate. Ions Fe (+3) promotes crystallization of calcium carbonate in the form of aragonite. Photos of calcium carbonate crystals, formed at initial water boiling and water with the addition of ferric chloride, was shown in fig. 2.

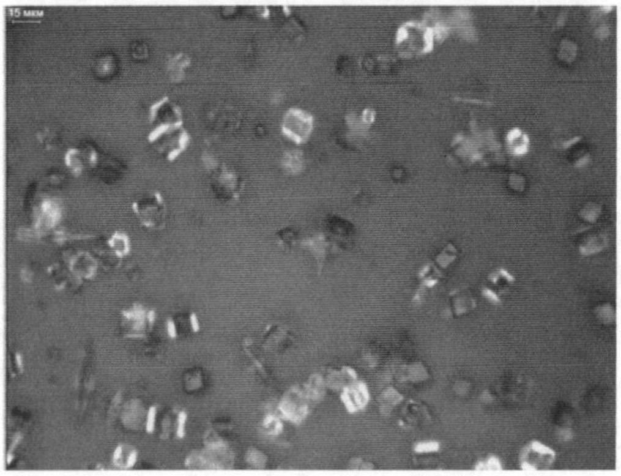

a)

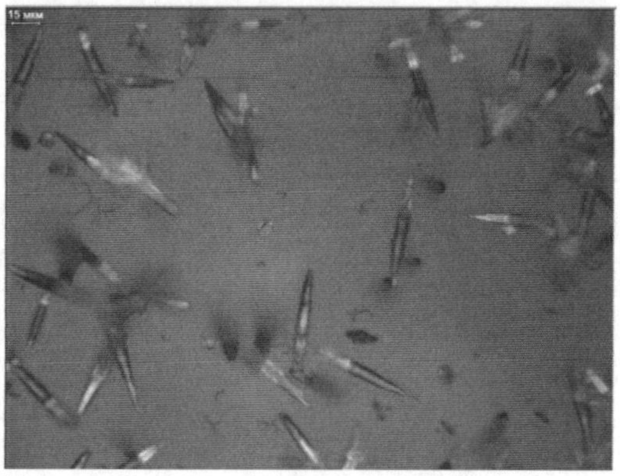

b)

Figure 2. Calcium carbonate crystals shape, formed at water boiling: a) initial water – calcite; b) initial water with activating additive of ferrous ions (+3) – aragonite.

Activating additive of sodium hydrocarbonate allows to regulate pH of initial water and accelerates the process of hydrolysis of ferric chloride. Hydrolysis takes place with the forming of ferric chloride, which coagulates and gives a precipitate with high developed surface. On the surface of the precipitate $Fe(OH)_3$ adsorption of calcium and magnesium ions occurs and partly reduces the hardness of initial water.

Suspended particles, including calcite microcrystals as the most thermodynamically stable modification of calcium carbonate are removed from the water on the stage of microfiltration. As the result, conditions for for aragonite crystals formation are created, as aragonite crystals are less resistant to supersaturation and precipitate faster. [5]. It corresponds to known "steps rule" of Ostwald (at system transformation from one state to another, firstly less stable form segregates) [6]. Available energy of phase transformation of aragonite into calcite approximately is 1,1 kJ/mol. The energy barrier for such a transformation is big enough, that's why unstable form of aragonite can be maintained indefinitely long.

Nowadays, creation of circulating and closed circuit systems of water supply in order to reduce the consumption for technological purposes of pure water and to decrease environmental impact is an actual problem. The use of circulating water supply allows reducing the consumption of natural water in ten times, with significantly lower operating and capital costs. The proportion of circulating water is continuously increasing in all industries. Basically, circulating water is used in heat-exchange facilities for excess heat removing, while it is repeatedly heated to 40 - 45 ° C and then cooled in cooling tower. Considerable part in the result of evaporation and entrainment is lost, besides, it is polluted to a certain limit while passing through pipelines of heat-exchange equipment. On the inner surfaces of heat exchanger tubes and the calcium carbonate is deposited, whose solubility decreases with increasing temperature. The deposition of calcium carbonate on the surface occurs even at relatively low heating 30 - 35 ° C. The rate of deposition of calcium carbonate and other salts according to the technical standards should not exceed 0,25 $g/(m^3 \cdot h)$. In order to prevent corrosion, fouling and scaling of the recycled water is removed from the system, replacing it with fresh water from a source or purified waste water [7].

In the result of conducted research, installation was designed. It allows to provide the decrease of scale formation in the systems of circulating water supply. Principal scheme of installation is shown in the Fig. 3.

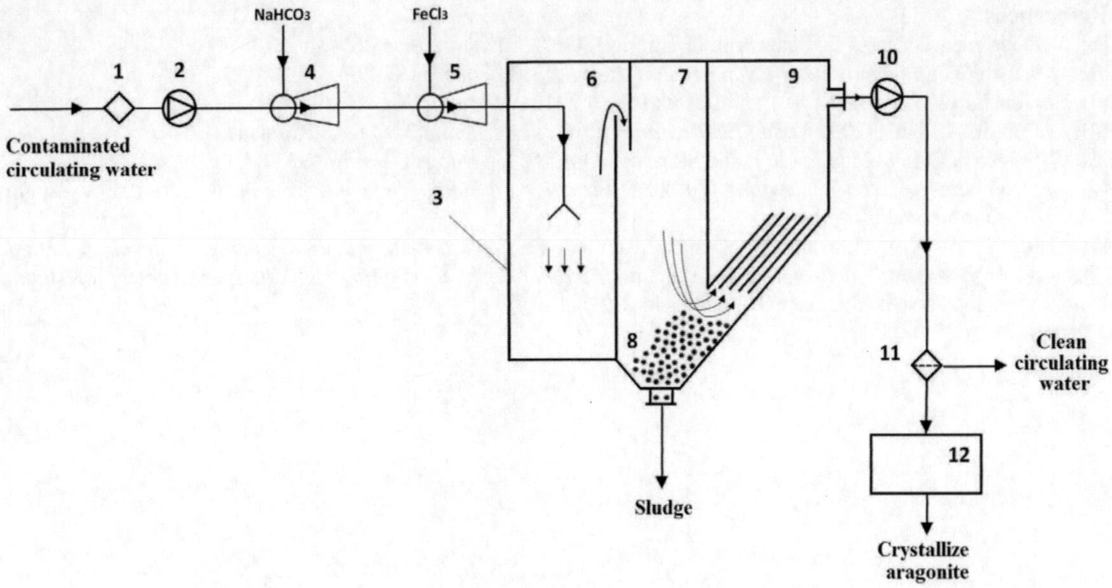

Figure 3. Principal installation scheme of scale formation decrease.

Installation works according to principal scheme. Contaminated circulating water, going through inlet strainer 1 of preliminary mechanical purification, with the help of the pump 2 this water enters to the block 3 of reagent treatment, thus it is treated successively with sodium hydrocarbonate and ferric chloride. Input of reagents is carried out using water-watered ejectors 4 and 5, operating in mixing mode. Efficiency of treatment by reagent increases due to intensive mixing of the reagents and their uniform distribution in the treated water, which leads to lower costs of reagents. Due to the fact that the input of the reagents into purified water is carried out by ejectors without the use of special metering devices (metering pump), a reduction in energy consumption and reduction in material costs of water treatment takes place. On the next step, the water flow mixed with reagents is directed to reaction chamber 6 of the reagent treatment module 3. Effective mixing of reagents with treated water and their uniform distribution in the water volume takes place in reaction chamber 6. From the reaction chamber 6 water enters the coagulation chamber 7, where the formation of coagulated sludge takes place in sludge removal zone 8, and water is directed to clarified water chamber 9. Water, already containing no flakes and slurries with the help of water pump 10, enters into the microfiltration unit 11. Microfiltration unit operates on the principle of tangential flow with a pore size of the filter element is 0,5 μm. Tangential flow of feed water provides a continuous washout of detained contaminants from the surface of the filter element, preventing the formation of a layer of filtered sludge on the surface of the filter element. Pure water is returned to the working cycle of circulating water supply system and the water flow, not passed through the filter element enters the crystallization unit 12.

Conclusion
Calcium carbonate, obtained in crystallization unit in the form of aragonite, can be used as additives in the production of various construction materials: refractory materials, cements and fillers.

Acknowledgements.
The work was financially supported by the Ministry of Education and Science, agreement № 14.575.21.0086. Unique project identifier RFMEFI57514X0086.

Journal of Physics: Conference Series **751** (2016) 012037 doi:10.1088/1742-6596/751/1/012037

References

[1] Rodriguez-Blanco J D, Shaw S and Benning L G 2011 *Nanoscale* **3** 265-71.
[2] Mosin O and Ignatov I 2014 *Nanotechnol. Research Practice* **4** 187-200.
[3] Goncharuk V V, Bagrij V A and Bashtan S Yu 2012 *J. Water Chem. Technol.* **34** 226-31.
[4] Fridkin A M et al 2004 *Voda i Ekologija: Problemy i Reshenija* **3** 22-8 (in Russian).
[5] Gas'kova O L, Solotchina Je P and Skljarova O A 2011 *Geology Geophys.* **52** 704-11.
[6] Stromberg A G and Semchenko D P 2009 *Fizicheskaja himija: uchebnik dlja vuzov* (Moskva: Vysshaja shkola) p 527 (in Russian).
[7] Rodionov A I, Klushin V N and Sister V G 2000 *Tehnologicheskie processy jekologicheskoj bezopasnosti: uchebnik dlja studentov tehnicheskih i tehnologicheskih special'nostej* (Kaluga: Izdatel'stvo N Bochkarevoj) p 800 (in Russian).

Journal of Physics: Conference Series **751** (2016) 012038 doi:10.1088/1742-6596/751/1/012038

The development of polymer membranes and modules for air separation

N E Vinogradov and G G Kagramanov

D. Mendeleyev University of Chemical Technology of Russia 125047 Moscow, Miusskaya sq., 9

E-mail: kadri@muctr.ru

Abstract. Technology of hollow fiber membrane and modules for air separation was developed. Hollow fibers from the polyphenylene oxide (PPO) having a diameter of 500 um were obtained. The permeability of the fibers by oxygen was up to 250 Ba, while the separation factor by O_2/N_2 was 4.3. The membrane module has been made by using these fibers and tested for permeability of individual gases.

1. Introduction

Membrane technology is the most rational method for producing gaseous nitrogen with a purity up to 99.5%. Plants based on this technology are produced worldwide. The performance of these plants can be up to 5000 Nm3 / h[1].

Henis and Tripodi [2] proposed to produce high-performance composite membrane covering a thin selective layer on a porous substrate. The properties of this membrane as well as other commercially available, permitted to develop the series of modules and industrial units for air separations.

In this paper the development of polymer membranes, accompanied by membrane modules, construction and their separation properties are being analyzed and discussed.

2. The analyzes and determination of the required geometric and operational characteristics

Initially, it was necessary to determine the separation surface of membrane module. For this purpose it was necessary to determine the geometrical parameters of the membrane and its packing density.

Membranes packing density ($m^2_{\text{membrane surface}}/m^3_{\text{module volume}}$) on the one hand should not be too low to present a so called "plug" mode in interfibrillar space. On the other hand a too tight packing leads to the formation of "dead" zones. In both cases, the separation performance decreases. Optimum packing density is stated by Li and al. in the range of 40-45% [3].

In order to optimize values of pressure losses the geometrical parameters of hollow fibers was chosen to be 500x350 microns.

Table 1. The input parameters and the results of calculation based on Davis Sandall method [4,5].

Initial parameters	
Separation surface	7,5 m^2
Feed air pressure	6bar
Feed air flow rate	7.5 nm^3/h
Retantat flow rate	3 nm^3/h

Residual oxygen concentration in the product nitrogen	5%
Calculation results	
Desired membrane permeability to oxygen in the module	0.32 nm^3/(m^2·h·bar)
Separation factor forO2/N2	3.92

In order to determine the required permeability of the polymer was necessary to determine the thickness of the selective layer. The thickness of the layer in the commercial polymeric (gas separation) membranes varies usually from 0.1 to 0.5 um[5–7].The dependence of the membrane permeability and the selective layer thickness has an inverse relationship. For selective layer thickness of 0.1-0.5 um required permeability values are of 11.4-57.1.

3. Selection of the polymeric material

The choice of the gas separation membrane such as the O_2 permeability coefficient accompanied be O_2/N_2 separation factor the attention was focused on permeability characteristics. Mechanical strength and operational characteristics are also the important parameters to be analyzed.

Despite the large number of polymeric materials developed for gas separation processes, the number of polymers used in industrial applications is quite limited. Results of the analysis of existing polymer materials are shown below in table 2.

Table 2. Existing membrane polymer materials

Polymer type	Internal structure	Glass temperature, °C	O_2permeability, Ba	O_2/N_2separation factor	Reference
Cellulose acetate (CA)	Glassy polymer	187	1.3	5.0	[8]
Polysulfone (PS)	Amorphous polymer	190	1.0	7.5	[9],[10]
Polyetherimide (PEI)	Glassy polymer	216	1.1	7.6	[5]
Polycarbonate (PC)	Glassy polymer	150	1.5	4.8	[11]
Polyimides (PI)	Glassy polymer	>240	1.1	7.2	[12]
Polyphenylene oxide (PPO)	Glassy polymer	218	15	4.5	[13]

Poly(phenylene oxide) became the best candidate suitable under requirements. The polymer has a permeability of about 15 Barrer, while the selectivity of O2/N2 is 4.5-4.6[13]. In this case, the thickness of the selective layer should be no more than 0.13 um.

4. Fabrication of hollow fibers

So called "dry-wet spinning" method (fig. 1)was used to obtain the hollow fiber membranes with a pronounced anisotropy (asymmetry).For this aim as the first step the state of PPO solutions was studied.

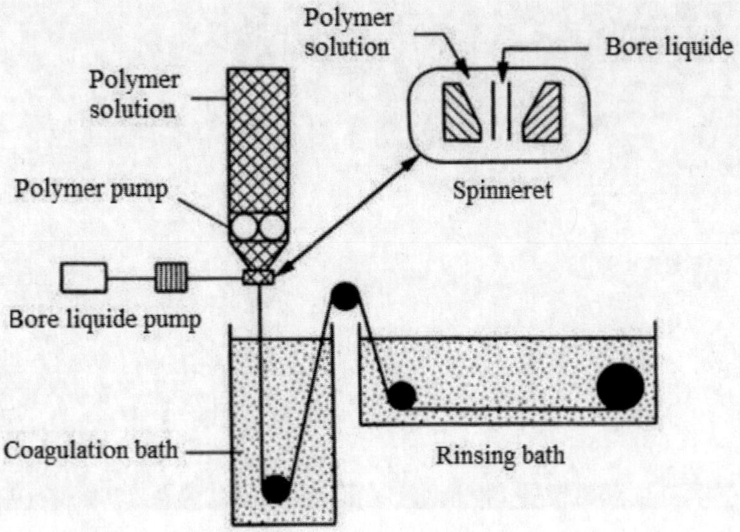

Figure 1. Apparatus for forming a hollow fiber gas separation.

Spinning process involves several stages:

1. The diffusion of internal precipitant molecules into the concentrated forming polymer solution. At the same time polymer coagulates (precipitates from solution) to form a solid phase.

2. A cross-diffusion of solvent molecules from the polymer solution into the internal nonsolvent. This speeds up the process of coagulation of the polymer and the formation of the porous structure of the fiber body.

3. Evaporation of the solvent molecules to the outer surface of the fiber into the air. The concentration of polymer on the outer surface of the fiber is increased and the formation of dense (selective) polymer layer is formed.

The polymer concentration in the spinning solution is important, since it affects the strength of the spinning solution exiting from spinneret. It should be sufficient to withstand the weight suspended in the air or immersed in the nonsolvent fiber. Furthermore, there are literature data that the selectivity of the membranes of the PPO by oxygen/nitrogen is higher, the higher the concentration of the initial spinning solution. However, a high concentration of PPO (18-20 wt.%) brought a great technological challenges.

Moreover, during forming select the optimal fiber spinning solution feed rate to a spinneret. The feed rate of bore-liquid into the fiber and the rate of filament winding onto a bobbin also controlled.

Quality control of the fibers was carried out by visual inspection of the surface and shape of fibers. With the help of micrographs of the fiber sections was evaluated their internal structure. Photomicrographs were made using a scanning electron microscope Phenom manufactured by FEI Company at maximum magnification 24000x.

As a result thin dense (selective) layer on the outer fiber surface was obtained (fig. 2).

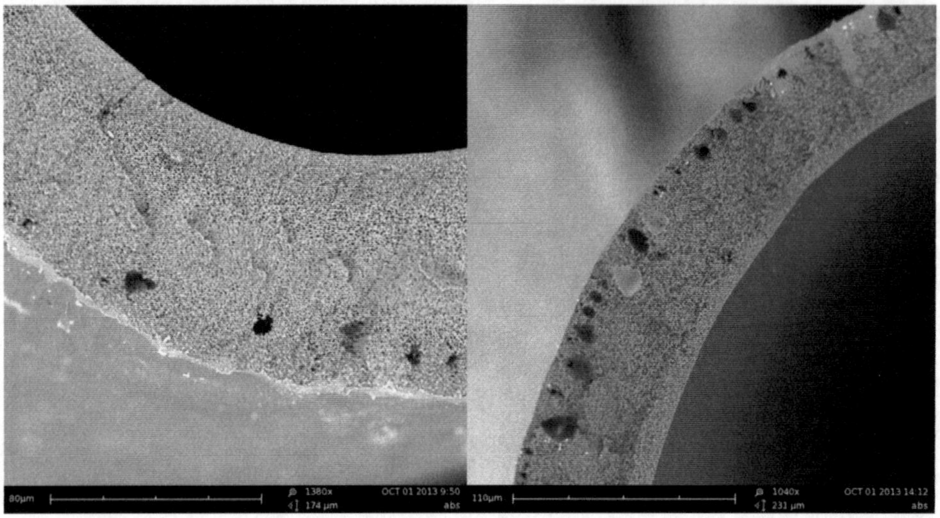

Figure 2. Micrographs of fibers sections

Fabricated fibers were selectively tested on individual gases (oxygen and nitrogen) permeability values. Average permeability of the fiber was 200-250 Ba, while the average selectivity byO_2/N_2 was 4.2. Defects in the surface of the fibers in the form of "furrows" hade no influence on the selectivity of the membranes.

The outer diameter of the resulting fiber was of 480 to 580 um and the thickness of the dense layer was of 0.1-0.2 um.

5. Fabrication of membrane module

Hollow fiber membrane module production procedure was carried out in several stages:
1) Module shell preparation;
2) Hollow fiber bundle preparation;
3) Compound tube sheets casting and ends cutting.

Aluminum was chosen as the material oamodule vessel. Fixing the end caps was accomplished by quickly removing clamps. (fig. 3)

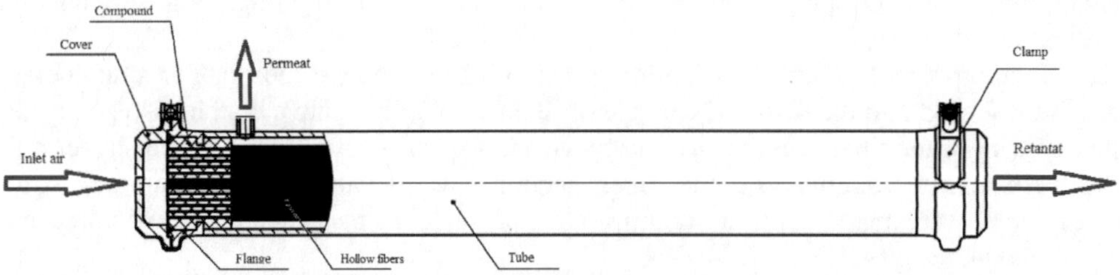

Figure 3. Hollow fiber module for air separation

For forming a common bundle of fibers used several bundles with fewer fibers. Each of these bundles was visually looked around for any visible defects. For each bundle were selected the few fibers randomly and tested for permeability to oxygen and nitrogen. Total number of fibers in the bundle was approximately 6,200pcs., and total separation surface was –4.1 m^2.

As the compound used industrially manufactured in Russia epoxydiane resin grade ED-20. After curing, the resulting epoxy matrix with fibers cut off at the ends so that the fiber lumen was open. For cutting use high cutting diamond blade. During the cut of epoxy matrix ends need be sure that there is no closing fiber lumens.

Fabricated module has been tested for both pure gases nitrogen and oxygen. The following data were obtained: oxygen permeability – 0.84 nm^3/(h·bar·module);separation factor of O2/N2 - 4.34.

Conclusion

The technology of hollow fiber membranes manufacturing for the separation of air was presented in this paper. Hollow fiber membranes with outer diameter from 480 to 580 μm were obtained. The permeability of the fibers by O_2 was 200-250 Ba, the selectivity for O_2/N_2 was 4.2-4.5. The membrane air separation module has been made and tested for permeability for individual gases. Permeability of this module was 0.84 nm^3/(h·bar), while the selectivity was 4.34.

References

[1] Dytnersky Y I, Brykov V P and Kagramanov G G 1991 *Membrane separation of gases* (Moscow: Khimia) p 344.
[2] Henis J M S and Tripodi M K 1982 *J. Memb. Sci.* **8** 75–8.
[3] Li D, Wang R and Chung T S 2004 *Sep. Purif. Technol.* **40** 15–30.
[4] Davis R A and Sandall O C 2003 *Chem. Eng. Education* 74–80.
[5] Peng N, Chung T-S, Chng M L and Aw W 2010 *J. Memb. Sci.* **360** 48–57.
[6] Peter J and Peinemann K V 2009 *J. Memb. Sci.* **340** 62–72.
[7] Baker R W 2004 *Membrane Technology and Application* (John Wiley and Sons) p 538.
[8] Vallejos E, Peresin M S and Rojas O J 2012 *J. Polym. Environ.* **20** 1075–83.
[9] Wang J and Rong R 2008 *Memb. Desalin. Technol.* **13** 391-430.
[10] Han J, Lee W, Choi J M, Patel R and Min B-R 2010 *J. Memb. Sci.* **351** 141–8.
[11] Muruganandam N and Paul D.R. 1987 *J. Memb. Sci.* **34** 185–98.
[12] Clausi D T and Koros W J 2000 *J. Memb. Sci.* **167** 79–89.
[13] Chenar M P, Soltanieh M, Matsuura T, Tabe-Mohammadi A and Feng C 2006 *Sep. Purif. Technol.* **51** 359–66.

Vapour and acid components separation from gases by membranes principles and engineering approach to membranes development

G G Kagramanov[1,2], I P Storojuk[2] and E N Farnosova[1,2]

[1] Chemical Engineering Faculty, D. Mendeleyev of Chemical Technology of Russia, 125047 Moscow, Russia
[2] "Membranica" Ltd, 125047 Moscow, Russia

Abstract. The modern commercially available polymer membranes and membrane modules for purification of gases, containing acid components, simultaneously with dehumidification of treated gas streams, were developed and commercialized in the very end of XXth century. The membranes basic properties – selectivity (separation factor) and permeation flow rates – are relatively far from satisfying the growing and modern-scale industrial need in purification technologies and corresponding equipments. The attempt to formulate the basic principles, scientific and engineering approaches to the development of prospective membranes for the purification of gases, especially such as natural and oil gases, from acid components, simultaneously with drying them, was being made. For this purpose the influence of various factors – polymer nature, membrane type, structure, geometrical and mass-transfer characteristics, etc. – were studied and analyzed in order to formulate the basic principles and demands for development of membranes, capable to withstand successfully the sever conditions of exploitation.

1. Introduction

World-wide production of natural and oil gases is constantly growing, achieving approximately 3 500 billion m^3/year. Russia, as one of world's leaders in this field, produces 640 billion m^3/year of natural gas and 65 billion m^3/year of oil gas.

All these gases, before delivering to the market, must undergo the indispensible treating procedure, namely, the purification from "acid" gas components – CO_2, H_2S, COS, CS_2, etc., as well as dehumidification [1,2]. These two processes are absolutely indispensible, as well, in production of liquefied hydrocarbons – methane up to butane etc. [3].

Purification by absorption of acid gases, well-known and spread all over the world, is being constantly replaced by membrane separations due to their evident advantages, such as lower capital and exploitation cost [4, 5]. Moreover, membrane separations allow, simultaneously with extraction of acid components from natural and oil gases, to minimize the vapor content in gas processed.

So, the advantages of membrane technology in this field are very promising and prospective. [6].

2. Membranes

The existing commercially available polymer membranes were developed mainly at the very beginning of XXI century and their properties are far from being optimal and should be ameliorated in order to correspond to nowadays demand. Their main characteristics – permeability and selectivity – are presented in Tables 1 (a) and 1 (b) [7, 8].

Table 1 (a). Commercially available membranes and their permeability

Gases	Permeability coefficient $[(mol \cdot m)/(m^2 \cdot s \cdot Pa)] \cdot 10^{15}$			
	Cellulose acetate "Gasep"	Cellulose acetate "Du Pont"	Polysulphone (bisphenol) "Ube"	Polyimide "Ube"
CO_2	5.5	13.9	2.32	4.36
H_2S	9.1	–	1.04	–
CH_4	0.18	0.58	0.084	0.134
N2	0.17	0.73	0.061	0.201
C_2H_6	0.09	–	0.074	0.027
C_3H_8	–	–	–	0.005

Table 1 (b). Separation factor (selectivity)

Material (polymer)	Separation factor				
	CO_2/CH_4	H_2S/CH_4	C_2H_6/CH_4	C_3H_8/CH_4	N_2/CH_4
Cellulose acetate "Gasep"	30.6	50.6	0.50	–	0.94
Cellulose acetate "Du Pont"	24.0	–	–	–	1.30
Polysulphone (bisphenol) "Ube"	27.6	12.4	0.88	–	0.73
Polyimide "Ube"	32.5	–	0.20	0.04	1.50

The analysis of these (Tables 1 (a) and (b)) data shows, that all materials, chosen for extraction of acid components from methane, are glassy polymers in their nature utilizing diffusion (size) selectivity mode of separation. The main disadvantage, to our mind, is that for these polymers, the permeability values for higher hydrocarbons diminish considerably in the "methane" – "ethane" – "butane" row. So, permeate is to be enriched by carbon dioxide and hydrogen sulphide; the retentate – by hydrocarbon components. Taking into account the boiling point (condensability) values, rising from methane to butane it could lead to the partial condensation of hydrocarbons on the high pressure side of membranes which is extremely undesirable for mass-transfer process, influencing (lowing) the values of permeability coefficients drastically.

The analysis of data presented in tables 1 (a) and 1 (b) shows, as well, that these characteristics are rather far from modern needs and can not satisfy the growing interest of engineering and industrial society in membrane gas separation technology.

So, the authors of presented study tried, first of all, to formulate the basic principles and demands for development of gas separation membranes – polymer nature, membrane type and structure, geometrical and mass-transfer characteristics.

The main principles for the development of polymer membranes for natural and oil gas purification and dehydratation are as follows:

- membranes are to be highly permeative for acid components and water vapor;
- polymer materials are to be highly selective towards acid components and vapor;
- membranes are to be anisotropic in structure, having extremely thin nonporous selective layer (40-100 nm), based on porous support;
- both selective layer and support should be made of the same material, i.e. to be asymmetric;
- membranes should be made in hollow fibers form;

- geometrical parameters of hollow fibers should provide the maximal packing density in membrane module – up to 20 000 m2/m3, as well as to be able to withstand the process pressure drop up to 70 – 100 bar;

Taking into account these numbers (up to 20 000 m2/m3 and ΔP=70 – 100 bar) the hollow fibers geometrical parameters should be: outer diameter of 250-500 μm and inner diameter of 100-250 μm/

- polymer material (and it's properties) should be resistant to physico-chemical interaction with gas mixture's components, i.e. resistant to so called "plastification" effect.

So, in our work we dared to analyze all physical phenomena during membrane separations, deciding to combine the main advantages of two types of polymers – glassy and rubbery – in a form of block-copolymers [9, 10].

Block-copolymers, having microheterogeneous structure and containing completely mixed amorphous and crystalline phases should provide the significant growth of permeability and selectivity values due to the combination of diffusion and solubility mechanism's advantages.

For this purpose, the block copolymers, made of polysulphone-polybutadiene, varying the molecular weights of precursors and their composition in membrane material, were synthesized and their gas permeabilities were measured and selectivities calculated (Tables 2 (a) and (b)).

Table 2 (a). Permeability coefficients for synthesized block-copolymers

Gases	Permeability coefficient $[(mol \cdot m)/(m^2 \cdot s \cdot Pa)]\ 10^{15}$		
	Block copolymer I	Block copolymer II	Block copolymer III
CO_2	21.9	18.7	18.6
H_2S	33.6	67.2	156.6
CH_4	0.22	0.21	0.21
C_2H_6	1.07	1.81	6.85
C_3H_8	26.4	40.1	74.3
N_2	0.51	0.51	0.51

Table 2 (b). Separation factor

Block-copolymers	CO_2/CH_4	H_2S/CH_4	C_2H_6/CH_4	C_3H_8/CH_4	N_2/CH_4
I	99.5	152.7	4.9	120.0	2.3
II	89.0	320.0	8.6	191.0	2.3
III	88.6	745.7	32.6	353.8	2.3

3. Conclusion

The comparison of data, obtained for "glassy" and block-copolymers (Tables 1 (a), (b) and 2 (a), (b)) shows that synthesized block-copolymers exhibit substantially and simultaneously greater values of both permeability and selectivity for acid gases extraction.

Very interesting is the controversial character of selectivity data in pairs methane – ethane (butane, nitrogen) compared with "glassy" polymer membranes. This fact, to our mind, can change considerably the engineering approach to membrane technology application in natural and oil gas treating (purification) processes – to withdraw at first separation step all "impurities" (CO_2, H_2S, C_2-C_3), obtaining in retentate high-quality methane and, at the second stage, to separate acid gases and ethane-methane mixture using "glassy" polymers.

The greater values of permeabilities, exhibiting by block-copolymers, will diminish considerable the capital cost, especially, for the first separation step.

At present, the development of hollow fiber block-copolymer membranes for these purposes is being carried out.

References

[1] *Technology of natural gas and condensate processing handbook*. 2002 (Moscow: Nedro-Bisnes)
[2] Chen G Q, Scholes C A, Qiao G G, Kentich S E 2011 *J. Memb. Sci.* **379** 479-87
[3] Scholes C A, Stevens G W, Kentich S E 2012 *Fuel* **96** 15-28
[4] Bhide B D, Stern S A 1993 *J. Memb. Sci.* **81** 209-37
[5] Bhide B D, Stern S A 1993 *J. Memb. Sci.* **81** 239-52
[6] Mohshim D F, Mukhtar H Man Z, Nasir R 2012 *J. Eng.* **2013**
[7] Dytnerskiy Yu I, Brykov V P, Kagramanov G G 1991 *Membrane gas separation* (Moscow: Chemistry) p 344
[8] Alentiev A, Shantarovich V, Merkel T C, Bondar V, Freeman B D, Yampolskii Yu P 2002 *Macromolecule*s **35** **25** 9513.
[9] Dytnerskii Yu I, Kagramanov G G , Storozhuk I P , Kovalenko N F 1989 *J. Memb. Sci.* **41** 49-54
[10] Yampolsky Yu P, Sidorenko V V, Cooper A, Storozhuk I P 1996 *High-molecular comp.* **38A** 857-67

Use of two-stage membrane countercurrent cascade for natural gas purification from carbon dioxide

I M Kurchatov[1], N I Laguntsov[1] and M D Karaseva[1,2]

[1] National Research Nuclear University MEPhI (Moscow Engineering Physics Institute), Kashirskoe highway 31, Moscow, 115409, Russia

[2] JSC Aquaservice, Kashirskoe highway 31, Moscow, 115409, Russia

E-mail: rita09karaseva@mail.ru

Abstract. Membrane technology scheme is offered and presented as a two-stage countercurrent recirculating cascade, in order to solve the problem of natural gas dehydration and purification from CO_2. The first stage is a single divider, and the second stage is a recirculating two-module divider.. This scheme allows natural gas to be cleaned from impurities, with any desired degree of methane extraction. In this paper, the optimal values of the basic parameters of the selected technological scheme are determined. An estimation of energy efficiency was carried out, taking into account the energy consumption of interstage compressor and methane losses in energy units.

1. Introduction

According to data published in the Annual Statistical Bulletin 2015, natural gas production in Russia in 2014 was amounted to 643 billion m^3/year [1]. With the aim of providing a number of export sales, using of natural gas as an alternative fuel and gas-supplying of settlements, remoted from gas pipe line, natural gas is liquefied.

Natural gas in addition to methane homologues according to their fields may include a variety of impurities, such as helium, hydrogen, argon, nitrogen, hydrogen dioxide, water vapour and other. Carbon dioxide is badly soluble in hydrocarbons. At temperatures and pressures, generated during the liquefaction of natural gas, carbon dioxide forms carbonic ice. Under the process of liquefaction, water vapour containing in the natural gas organises hydrates with hydrocarbons, which in turn affects adversely on the service life of liquefaction system. Thus, it is necessary to produce natural gas dehydration and purification of CO_2. Acceptable levels of carbon dioxide and water vapour in the natural gas are 0.01% and 0.0001%, respectively.

In order to solve the problem of drying and purification of natural gas from the carbon dioxide, a two-stage double circuit countercurrent cascade was considered and the optimal values of its basic parameters were determined.

2. Statement of the problem

On the assumption with the methane permeability through the membrane is $\pi\ (CH_4) = 1\ l/\text{atm*h*m}^2$ and carbon dioxide permeability is $\pi\ (CO_2) = 40\ l/\text{atm*h*m}^2$, the ratio of carbon dioxide permeability to the methane permeability (selectivity) corresponds to a modified polyimide membrane. Water vapour permeability of view is the same as carbon dioxide permeability.

The pressure value of the feed flow pressure of the module is 40 atm., pressure of the permeate flow is 1 atm., operating temperature is 297K. Productivity of technological scheme, using for gas drying and purification is 140 m^3/h (100 kg/h).

3. Results and discussion

In this work the results of numerical studies of the two-stage dual-stage counter-current scheme (Figure 1) are presented for the separation of a ternary mixture of methane (the main component of natural gas), carbon dioxide and water vapour. This stage can be used for natural gas purification by CO_2 [2], и He [3],[4], and its energy efficiency substantiated in [3].

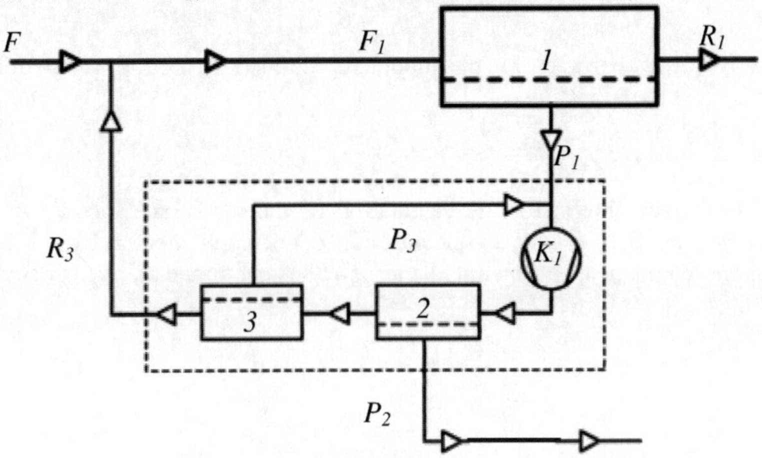

Figure 1. Two-stage double circuit countercurrent (technological scheme).

Feed flow F is supplied to the input of separation scheme - feed gas includes in it 2.50% of carbon dioxide, 0.01% of water vapour and 97.49% of methane. Module 1 of the first stage purifies the feedstock gas from carbon dioxide and water vapour in order to produce product flow - retentate R_1, which contains less than 0.01% of carbon dioxide and water vapour 0.0001%. Carbon dioxide and water vapour penetrate through the selective membrane and form a permeate flow of the first stage P_1. As the initial concentration of carbon dioxide and water vapour are small compared to the concentration of the methane, most of the permeate flow consists of methane, which is a valuable component. Use of recirculation divider as a second stage allows us to increase the concentration of carbon dioxide and water vapour in the retentate flow of the scheme, and also increases the recovery rate of methane scheme. Retentate flow of the second stage R_3, enriched with methane is combined with the feed flow of the technological scheme. It is important to note that concentrations of methane were set equal in the flow F, F1 and R3, while the numerical research had been carried out.

$$C_F(CH_4) = C_{F1}(CH_4) = C_{R3}(CH_4) \quad (1)$$

Assuming equal permeability of carbon dioxide and water vapour in the three-component approach of the equalities:

$$C_F(CO_2) = C_{F1}(CO_2) = C_{R3}(CO_2) \quad (2)$$

$$C_F(H_2O) = C_{F1}(H_2O) = C_{R3}(H_2O) \quad (3)$$

For the calculation of the first stage (. Figure 1 module 1) the following parameters were set: membrane permeability for the mixture components, R1 is equal to the product flow of 140 m^3/h, the concentration of methane (99.99% vol.), carbon dioxide (0.01 vol.) and water vapour (0.0001%) in it, the value of the high pressure p_h = 40 atm. and low p_l = 1 atm., concentration of the components in the initial mixture into feed flow of the cascade: methane - 97.49%, carbon dioxide - 2.50%, and water vapour - 0.01%

Calculation of the first stage (ordinary divider) was conducted, using «cross flow» model [5]. It was assumed that the pressure in the low pressure chamber is constant throughout the entire length of the module p_l=const, the permeability for component i is constant throughout the entire length of the module π_i=const, gas temperature is constant T=const, $\Delta p_h/p_h$<<1.

Equation of flow balance for the first stage:

$$F_1 = P_1 + R_1 \tag{4}$$

Equation of the balance of the component i:

$$C_{F1}^i = \theta_1 C_{P1}^i + (1-\theta) C_{R1}^i \tag{5}$$

where θ_1 – flow partition coefficient of the module 1, defined as the ratio of permeate flow of the module to the feed flow of the module:

$$\theta_1 = \frac{P_1}{F_1} \tag{6}$$

Selectable volume was chosen for the calculation of the membrane module in a high-pressure chamber, which includes flow q with a concentration C_q^i of component i. Partial flow $Z^i = q \cdot C_q^i$ and concentration of the component i, penetrated through the membrane C_l^i were determined from the system of equations:

$$\begin{cases} \frac{d(z^i)}{dS} = -\pi_i p_h (C_q^i - \gamma^* C_l^i) \\ C_l^i = \frac{\pi_i (C_q^i - \gamma^* C_l^i)}{\sum_j \pi_j (C_q^j - \gamma^* C_l^j)} \end{cases} \tag{7}$$

where $\gamma^* = p_l/p_h$, p_l – pressure value in low-pressure chamber, p_h – pressure value in high-pressure chamber.

Boundary conditions of given system are:

$$Z^i(s=0) = FC_F^i \tag{8}$$

$$Z^i(s=S) = RC_R^i \tag{9}$$

The system of the equations was solved by Runge-Kutta method of the fourth order. Concentrations in high-pressure chamber C_q^i and flow q were determined from equations:

$$q = \sum_j Z^j; \tag{10}$$

$$C_q^i = \frac{Z^i}{\sum_j Z^j}. \tag{11}$$

The second step is a sequential two-module recirculating divider consisting of the modules 2 and 3. The permeate flow P_1 of the first stage, combining the recirculating flow with the second step P_3, consists the feed flow of the module 2 (see Figure 1.). The flow, which did not penetrate through the membrane of the second module, inputs to the module 3. The permeate flow P_2 of the second module is equal to a retentate flow. Flow, which didn't permeate through the membrane of the third module, is combined with the feed flow F of the technological scheme.

The following parameters were set for calculating the second stage: temperature of the gas permeability of the membrane for carbon dioxide, methane and water vapour (the same as for the first-stage module), the value of the high pressure p_h = 40atm. and low pressure p_l = 1 atm., feed flow of the second stage is equal to a permeate flow of the first stage P_1. Concentrations of carbon dioxide, water vapour and methane in the feed flow of the second stage also were set, basing on the obtained concentrations in the permeate flow of the first stage. Methane concentration in the flow - $C_{R3}(CH_4)$ methane extraction degree of the second stage $\varphi(CH_4)_2$ were set. Methane extraction degree is calculated by the equation:

$$\varphi(CH_4)_2 = \frac{R_3 C_{R3}(CH_4)}{P_1 C_{P1}(CH_4)} * 100\% \tag{12}$$

Taking into account the balance equation for total flows:

$$P_1 = P_2 + R_3 \tag{13}$$

Equation (12) can be expressed this way:

$$\varphi(CH_4)_2 = \left(1 - \frac{P_2}{P_1}\right) * \frac{C_{R3}(CH_4)}{C_{P1}(CH_4)} * 100\% \tag{14}$$

Flows P_2 and R_3 are determined from equations (13), (14).

Calculation of membrane modules 2 и 3 was performed similarly to membrane module 1 (similarly to the first stage) using model «cross flow» model. The amount of substance which has entered into the second module $F_2 C_{F2}^i$, is determined by the equation:

$$F_2 C_{F2}^i = P_1 C_{P1}^i + P_3 C_{P3}^i \tag{15}$$

An initial approximation of the amount of substance $F_2 C_{F2}^i$ was set, entered the second module. Based on this value according to the equations (7), the second module is calculated with a predetermined value θ_2 approaches the split ratio. Thus, the flow of the third module and component concentrations in streams P_2 and the third module was determined. To determine the partition coefficient θ3 of the third module, the permeate flow of the third module P_3 and concentrations of the components in it, system of differential equations (7) with the boundary condition was solved:

$$Z^i(s = S) = R_3 C_{R3}^i(CH_4) \tag{16}$$

Then the system (15) was solved by Newton's method. Then recalculate was performed for the second module, wherein the flow partition coefficient θ_2 defined by the method of relaxation as long as the desired degree methane extraction was achieved. As a result of performed calculations all module parameters were determined, were carried out with the complete flow balance equation (13) and the balance equation for the components of the mixture for the second step:

$$C_{P1}^i = \frac{P_2}{P_1} C_{P2}^i + \frac{P_2}{\theta_2 P_1}(1 - \theta_3)(1 - \theta_2) C_{R3}^i \tag{17}$$

Feed flow of the cascade F is determined by the equation:

$$F = F_1 - R_3 \tag{18}$$

Methodology presented in [3] is applied for analysing of energy consumption of the scheme. According to this technique, the power consumption K_1 - W_1 [kW] of interstage compressor is calculated, the loss of methane in energy units is accounted and the total energy of given technological scheme is determined.

Calculation of energy consumption K_1 of interstage compressor was carried out in the approximation of adiabatic compression:

$$W_1 = \frac{(P_1 + P_3)}{\eta_{ac}} \frac{\kappa}{\kappa - 1} RT \left[\left(\frac{p_h}{p_l}\right)^{\frac{\kappa-1}{\kappa}\mu} - 1\right] \tag{19}$$

where P_1+P_3 – compressed flow, [mole/s]; R – universal gas constant [J/mole*K]; T=297°K – gas temperature, [K]; κ=1,32 - adiabatic index for natural gas, η_{ac}=0,85 Efficiency factor of compressor operating in an adiabatic compression mode; p_h – flow pressure at the outlet of the compressor equals to 40 atm., p_l - flow pressure at the compressor inlet to 1 atm. As a natural gas is a real gas in term

(19), gas compressibility coefficient μ = 0,87 297°K was accounted for at a temperature and a pressure of 40 atm. [7].

It is important to consider the energy that can be produced by combustion of methane in the permeate flow entered the second stage P_2 (retentate flow of the scheme) in the gas turbine. Such energy was named: methane loss in energy units W_P [kW] and was calculated by equation:

$$W_P = P_2 C_{P_2} \eta_{GT} Q \quad (20)$$

Where P_2 is a permeate flow of the second module (fig.1), mole/s; C_{P2} – methane concentration in the flow P_2; η_{GT}=0,35 – gas turbine efficiency factor; Q=800 kJ/mole – methane combustion heat.

The total energy consumption for the given technological scheme on 1kg of product gas (natural gas in permeate flow) E [kWh / m^3] is equal to:

$$E = (W_1 + W_P)/R_1 \quad (21)$$

where R1 - product stream, according to the task, it is 140 Nm^3 / h and contains 99.99% methane.

Under realizing calculations of the second stage degree of extraction of methane φ (CH_4) was varied in the range from 2 95.50% to 99.99%. At methane extraction degree of the second stage 95% recirculation circuit becomes a single divider [3].

On the Fig. 2 the dependence of total reduced energy consumption to methane extraction degree of technological scheme at compressor operation mode, working in an adiabatic compression, is shown. Recalculation of the recovery is carried out according to the formula (22) basing on equation (18):

$$\varphi(CH_4) = \frac{R_1 C_{R1}(CH_4)}{(F_1 - R_3) C_F(CH_4)} * 100\% \quad (22)$$

Varying the methane extraction degree of the second stage allows varying methane extraction degree of the whole scheme

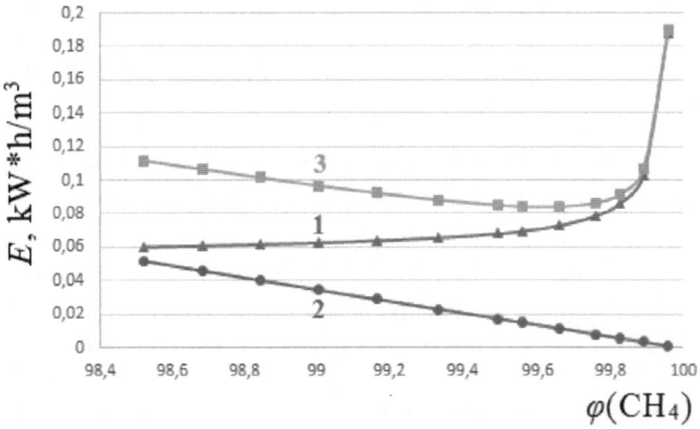

Figure 2. The dependence of the total reduced energy consumption to the degree of extraction of the technological scheme of methane, 1 - energy consumption of compressor K1; 2 - methane loss in energy units; 3 - total reduced energy consumption.

Reduced total energy consumption are low and are determined by competitive factors - the growth of energy consumption of the compressor K_1 and decrease of methane losses with the rise of extraction degree. Thus, the energy consumption at the minimum with all energy costs is 84.1 Wh/m^3. This value for given pressure depends on the ratio of permeability and is independent of their specific values.

Interestingly, the ratio of the concentrations of carbon dioxide and water vapour in the permeate stream and feeding are equal:

$$\frac{C_{CO2}^F}{C_{H2O}^F} \approx \frac{C_{CO2}^P}{C_{H2O}^P} \quad (23)$$

Journal of Physics: Conference Series **751** (2016) 012040 doi:10.1088/1742-6596/751/1/012040

Equation (23) based on equal permeability of carbon dioxide and water vapour, and low concentration of water vapour contained in initial feed gas. Importantly, in this case carbon dioxide, and water vapor may be considered as one component, and the initial concentration of the component is defined as the sum of the concentrations of carbon dioxide and water vapour.

Conclusions

In this paper we considered the problem of drying and purification of natural gas from carbon dioxide, using two-stage double circuit countercurrent membrane cascade. The technique for calculating of this stage was presented. The optimal values of the basic parameters of such a process were determined under given initial parameters.

References

[1]. OPEC *Annual Statistical Bulletin Organization of the Petroleum Exporting Countries*
[2]. Cnop T, Doong S J, Farr D and Xomeritakis J K 2012 *US Patent 2012050816*.
[3]. Laguntsov N I, Kurchatov I M, Karaseva M D and Solomahin V I 2014 *Petrol. Chem.* **8** 673.
[4]. Solomakhin V I, Laguntsov N I, Kurchatov I M and Davydov Y S 2014 *RF Patent 150520.*
[5]. Laguntsov N I, Talantseva E V and Teplyakov V V 1997 *Theor. Found. Chem. Eng.* **31** 463.
[6]. Laguntsov N I, Gruzdev E B, Kosykh E V and Kozhevnickov V Y 1992 *J. Membr. Sci.* **67** 15-28.
[7]. Stull D R, Westrum Jr E F, Sinke G C 1987 *The Chemical Thermodynamics of Organic Compounds* (Robert E. Krieger publishing company: Malabar Florida).

Journal of Physics: Conference Series **751** (2016) 012041 doi:10.1088/1742-6596/751/1/012041

Investigation of hybrid membrane-sorption technologies for air fractionating

N I Laguntsov[1], I M Kurchatov[1], M V Korolev[2] and A V Tishin[2]

[1] National Research Nuclear University MEPhI (Moscow Engineering Physics Institute), Kashirskoe highway 31, Moscow, 115409, Russia
[2]JSC Aquaservice, Kashirskoe highway 31, Moscow, 115409, Russia

E-mail: aquaserv@mail.ru

Abstract. Main aim of the work is to develop and to research single-circuit hybrid membrane-sorption system for enriching air with oxygen. The developed system allows to produce air, enriched with air up to 50%, purified from dust of any size, including nanoscale dust received in consequence of sorbent abrasion. In the course of the work the research of existing systems for air enrichment with oxygen, and the possibility of combining two methods of gas separation: membrane, and sorption, was conducted. The developed system differs from its analogues in that it has improved energy efficiency compared to methods of the membrane and sorption separation. Also work presents method of cyclogram determining of the hybrid system. In this methodic an algorithm for calculating of the cycles number, and determining of the stages duration in order to obtain the desired performance was presented.

1. Introduction

The single-circuit membrane sorption hybrid system was investigated in this paper. Numerical studies of separation features and energy efficiency of the system for enriching air with oxygen at different parameters of dividing cells were conducted. System with an increased pressure of permeate flow, designed for feeding of apparats of artificial lung ventilation, was considered.

Air enriched with oxygen is necessary to create mobile air separation systems that can be used in order to develop an artificial breathing atmosphere. Due to the use of single-circuit membrane-sorption scheme the energy efficiency of such a system can be improved.

The artificial breathing atmosphere has a wide range of applications: medicine, atmosphere generators, etc.

Hybrid membrane-sorption system (Fig. 1) consists of a compressor, two alternately working adsorbers with the sorbent, the valve system and the membrane module. The air after passing through the ejector fills one receiver and is enriched with oxygen, while the other receiver takes place desorption of the sorbent [2,3], and then, the air, enriched with oxygen, enters into the membrane unit.

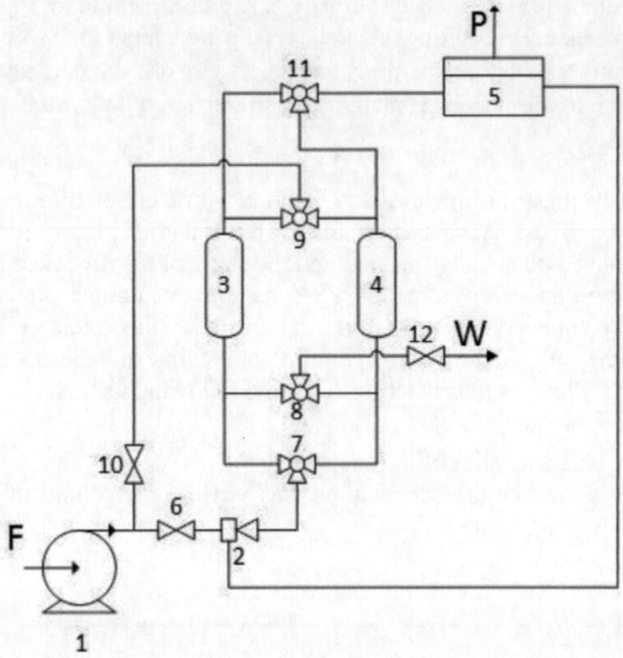

Figure 1. Schematic diagram of hybrid system: 1 - air compressor; 2 - ejector; 3.4 - absorbers; 5 - membrane module; 6 – 12 - solenoid valve;

In works [1,4,5] different variations of hybrid system schemes was investigated. The dependence of the selection on the pressure of the product flow was analyzed and it was shown that it is better to use a circuit with a feed of a power flow through the ejector shown in Figure 1 in order to obtain a flow of enriched air at high pressure. As it is shown by the diagram, first feed flow passes through the ejector, then it is enriched with the oxygen in sorption unit, then goes to a membrane module, wherein the part of the flow is enriched with the oxygen and goes in the product flow, afterwards a retentate flow of the membrane unit is returned into the ejector where it is mixed with the feed flow. Since the retentate flow is already enriched with oxygen, then as a consequence, the output of the ejector concentration of the oxygen becomes higher than the natural, which increases energy efficiency of the system.

A system of equations describing the two-component gas mixture separation is used for the calculation of the membrane module under the assumption of ideal displacement in the cavity of high pressure by perpendicular outflow in the cavity of the low pressure [1]. This model describes well the process of gas separation in hollow fiber membrane modules. Calculations of the sorption stage were carried out in assumptions about ideal model of linear sorption [1,6,7].

In the paper [6] an algorithm for calculating the hybrid membrane adsorption system and a numerical research of the energy efficiency depending on the selectivity of separating elements is presented. It is shown that the primary separation operation does sorption part, while membrane portion serves as a divider of the product flow, part of which is directed to the filling of adsorber for the increase of the concentration of oxygen in the feed flow and allows to provide a product flow completely purified from nanometer-sized particles. These particles get into the flow due to the abrasion sorbent.

2. Methods of determining the sequence diagram of the hybrid membrane-sorption system

2.1. The assumptions in the construction of the model.

The problem of determining sequence diagram of the hybrid membrane-sorption system is the multiparameter. In constructing the cyclogram, consider the following:

- Sorbent parameters, namely sorbent porosity ε, sorption constant σ, the capacity of the sorbent, the sorbent form (granules, cuttings), geometrical parameters of the sorbent;
- Membrane parameters: gas permeability and selectivity of the membrane α;
- Accommodation of displacement stage and filling stage to ensure proper operation of the system;

To determine the cyclogram of the system (Figure 2), problem of coordination of unsteady stages of the system should be solved: it is necessary that by the end of displacement stage in one adsorber, in the other the filling stage had already finished and the stage of displacement began. This is necessary to ensure continuous product flow after the PSA stage to membrane module working in a steady regime at constant parameters operating mode. To meet this requirement, it is necessary to displacement stage time equal to the amount of filling phase time and the desorption stage time.

The duration of the filling stage is determined by the following factors:

1) Adosrption pressure p_{ads};
2) The geometric characteristics of the adsorber;
3) Characteristics of the compressor used: productivity and the volume of the receiver.

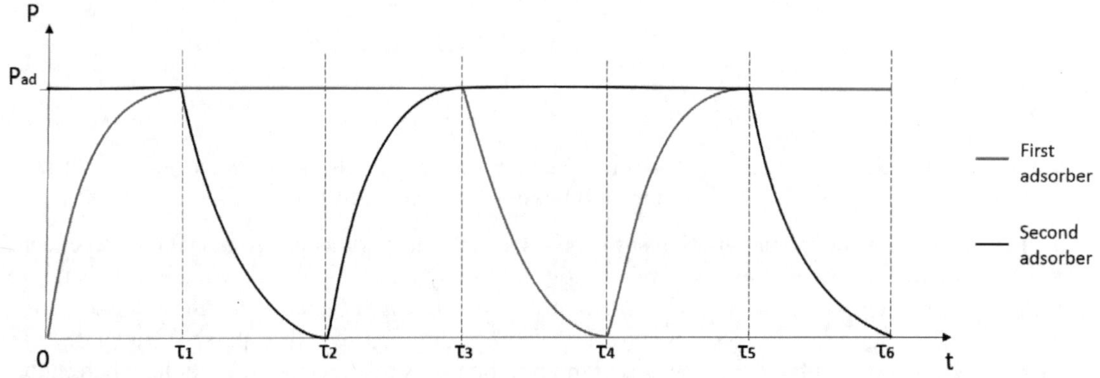

Figure 2. Cyclogram of the hybrid membrane-sorption systems. 0-τ1 - first stage of filling the canister, τ1- τ3 - displacement stage, τ3- τ4 - desorption stage from the first adsorber.

As the displacement stage is carried out at a small differential pressure Δp, the duration of the displacement stage determines by the maximum of pressure drop. The pressure differential is selected so that the adsorption rate satisfies the following criteria:

- minimum gas velocity is determined by the Peclet number cause if the speed is too low, the adsorption front will be blurred;
- maximum velocity of the gas is determined by the Reynolds number, cause if the speed is too high, then there will be a turbulence and mixture isn't enriched to the desired concentration;

The duration of the adsorption stage is determined by desorption pressure p_{des} and geometric characteristics of adsorber. The duration of filling stage is determined by the geometrical characteristics of the adsorber, the flow from the compressor, the pressure of the feed flow.

As can be seen from the graph, minimum duration of the displacement stage is determined by the sum of the durations of the stages of filling and desorption.

2.2. Algorithm for calculating the number of cycles and their duration

Calculation algorithm of the system was proposed in order to construct a cyclogram. Firstly, the product flow parameters are set: productivity P, the pressure of the product flow and the oxygen concentration in this flow. From these data volume adsorber is selected and such membrane and a sorbent that can provide the necessary enrichment is selected. After that, the calculation of useful product flows and concentrations according to the formulas described in work [6].

Under calculating flows and concentrations of useful product in flows, the number of cycles of the system can be determined. If the necessary performance P is known, the number of cycles N is determined by dividing the value of the resulting product flow to a predetermined system performance. After determining the number of cycles, it is possible to determine the duration of each cycle by dividing the time, which should be achieved for a given performance to the calculated number of cycles.

Then the duration of the filling stage and desorption stage is determined experimentally. The sum of these durations determines the minimum duration of the displacement stage. If calculated cycle time by the above described algorithm is not equal to the sum of the durations of the filling stage and the desorption stage, it is due to the fact that the selected volume adsorber V is not optimal for a preselected performance. If the calculated duration is more than experimentally determined duration, it is necessary to reduce the volume of adsorber, if less, it is necessary to increase the volume of adsorber.

2.3. Evaluation of filling time of adsorber by the gas

Adsorber filling with the gas cannot be described by a simple analytical dependence. But it is possible to describe the dynamics of the gas filling the adsorber empirically.

In the literature [8], the case of filling the tank with gas is described. If sorption is fast enough, then the adsorber filling with the sorbent volume V_0 can be viewed as the filling of the empty adsorber with volume $V = \varepsilon V_0$, ε - porosity of the sorbent. This process is described by the following equation:

$$\gamma R T_0 dW_0 = Vdp + \gamma p dV, \tag{1}$$

where γ - adiabatic index, W_0 - the mass of the gas, which is pumped into the volume V, T_0 - temperature of the gas.

To determine the filling time of the volume V, it is necessary to substitute in the equation (1) $dW = Gdt$, where G - the mass flow rate. The mass flow depends on the pressure ratio, which is defined in the following ways:

In subcritical mode (when $\sigma_i < \sigma_*$, $\sigma_i = \frac{p_i}{p_0}$, $\sigma_* = (\frac{2}{\gamma+1})^{\frac{\gamma}{\gamma-1}}$):

$$G = K\mu f \frac{p_0}{\sqrt{RT_0}} [\sqrt{\sigma^{\frac{2}{\gamma}} - \sigma^{\frac{\gamma+1}{\gamma}}}] \tag{2}$$

where

$$K = \sqrt{\frac{2g\gamma}{\gamma-1}},$$

in the supercritical mode (when $\sigma_i > \sigma_*$):

$$G_* = K_* \mu f \frac{p_0}{\sqrt{RT_0}}, \tag{3}$$

where

$$K_* = K\sqrt{\sigma_*^{\frac{2}{\gamma}} - \sigma_*^{\frac{\gamma+1}{\gamma}}}.$$

After integration of the equation (1) with corresponding definitions of the G mass flowrate, according to the equations (2) and (3), the following equation for the estimated fill time [Hz] is obtained:

for the subcritical mode

$$t = \frac{2V}{(\gamma-1)\mu f K\sqrt{RT_0}}\left(\sqrt{1-\sigma_1^{\frac{\gamma-1}{\gamma}}} - \sqrt{1-\sigma_2^{\frac{\gamma-1}{\gamma}}}\right) \tag{4}$$

and for the supercritical mode

$$t = \frac{V}{\gamma\mu f K_*\sqrt{RT_0}}(\sigma_2 - \sigma_1) \tag{5}$$

2.4. Evaluation of gas expiration time from the adsorber

Using the ideal gas equation, next equation can be obtained:

$$dp = \frac{RT}{\mu V}dm, \qquad dm = -\rho S w dt,$$

where μ - molar mass, dm - mass of the gas flowing in time dt through the opening area S.

The formula for the instantaneous rate of gas outflow from the reservoir [9]:

$$w = \sqrt{2\frac{\gamma}{\gamma-1}pv\left[1-\left(\frac{p_{out}}{p}\right)^{(\gamma-1)/\gamma}\right]},$$

where γ - adiabatic index, p - gas pressure in the tank, v - specific volume of gas in the tank, p_{out} - external pressure. The dimension w - *m/s*. The specific volume of gas is calculated using the formula:

$$v = \frac{1}{\rho},$$

where ρ - density of gas.

Using a linear relationship between pressure and density:

$$\rho = \rho_0\frac{p}{p_0}$$

where indicators index 0 correspond to the initial:

$$dp = \alpha p\sqrt{1-\left(\frac{p_{out}}{p}\right)^{(\gamma-1)/\gamma}}\,dt, \tag{6}$$

where

$$\alpha = -\frac{SRT}{\mu V}\sqrt{2\frac{\gamma}{\gamma-1}\frac{p_0}{\rho_0}}.$$

Equation (6) shows that it was possible to establish a connection between the change in pressure in the adsorber and time. According to the solution of this equation, conclusion about the dynamics of the adsorber desorption can be done, as well as the duration of the desorption process can be determined.

Conclusion

The method of cyclogram determining of the hybrid membrane-sorption system was presented. Then, an algorithm was proposed for calculating the number of system cycles and duration of each cycle required to obtain the desired performance of the system. The methods for evaluating of the duration of the filling stage and the duration of the desorption stage of the gas from the adsorber were proposed.

Acknowledgements

The work was financially supported by the Ministry of Education and Science, agreement # 14.579.21.0046. Unique project identifier RFMEFI57914X0046.

Journal of Physics: Conference Series **751** (2016) 012041 doi:10.1088/1742-6596/751/1/012041

References

[1] Onyestyfik G, Valyon J and Rees L V C 2005 *Stud. Surf. Sci. Catal.* **158A** 1019

[2] Sircar S 1988 *Sep. Sci. Tech.* **23** 2379-96.

[3] Kudinov A N, Kurchatov I M, Laguntsov N I and Neshchimenko Yu P 2014 *Theor. Found. Chem. Eng.* **48** 832–6.

[4] Esteves I A A C and Mota J P B 2007 *Ind. Eng. Chem. Res.* **46** 5723-33

[5] Suk D E and Matsuura T 2006 *Sep. Sci. Tech.* **41** 595-626.

[6] Kudinov A N, Kurchatov I M and Laguntsov N I 2014 *Theor. Found. Chem. Eng.* **48** 352–9.

[7] Makeev E A, Zelenko V L and Kheifets L I 2007 *Moscow Univ. Chem. Bull.* **62** 136.

[8] Hertz E V, Kreynin G V 1975 *Calculation of pneumatic actuators* (Moscow "Mechanical engineering") p 272.

[9] Abramovich G N 1991 *Applied gas dynamics* (Moscow "Nauka") p 601.

Journal of Physics: Conference Series **751** (2016) 012042 doi:10.1088/1742-6596/751/1/012042

Regularities of asymmetric gas transfer in PVTMS-membranes

I E Skuridin[1], I M Kurchatov[2] and N I Laguntsov[2]

[1] JSC Aquaservice, Kashirskoe highway 31, Moscow, 115409, Russia
[2] National Research Nuclear University MEPhI (Moscow Engineering Physics Institute), Kashirskoe highway 31, Moscow, 115409, Russia

E-mail: aquaserv@mail.ru

Abstract. In this work the dependence of permeability of the polymer asymmetric gas-separation membrane on its orientation relative to the areas of high and low pressure was investigated. It is shown that under different directions of gas supply, permeability values are not the same due to the characteristics of the asymmetry of the object.

1. Introduction

Membrane permeability measured for various gas flow directions may be the same only in the case of identical left and right borders, that is, for symmetric membranes. For asymmetric membranes is commonly observed phenomenon of direct asymmetry, when the flow from the diffusion layer exceeds the flow from the porous support. Typically, the direct effect of permeability asymmetry was observed in ceramic SHS and asymmetric gas separation nonporous polymeric membranes [1]. However, there are objects, where the phenomenon of reverse asymmetry is observed, one of which we discuss in detail in this article. For example, the asymmetry of the gas transfer through PVTMS-membrane has a reverse direction. The phenomenon of reverse anisotropy was firstly observed in [2].

In general, it follows that the measured values of permeability are higher in the case of the gas supply from the side, characterized by a high value of the kinetic factor, i.e. constants ratio of sorption and desorption speeds. The increase of kinetic factor will mean the increase of stationary surface concentration of transported molecules through asymmetric membrane. In turn, the increase of surface concentration due to the sorption differences of the porous support and the film.

2. Main text

Investigated object, on which reverse asymmetry is observed, is an asymmetric gas separation membrane PVTMS. PVTMS is conventionally expressed by three layers: a non-porous diffusion (selective) with a thickness of 70 ÷ 100 nm, fine-pored layer with the thickness 10 ÷ 15 micron and a pore size of 300 nm and a layer with a large (up to 4 mm) transport pores [3]. Thus PVTMS membrane is a gradient porous structure, i.e. medium consisting of a porous layer – support with a thickness change of pore radius and porosity, and non-porous diffusion layer. Diffusion layer and porous support made of the same material. (See. Figure 1).

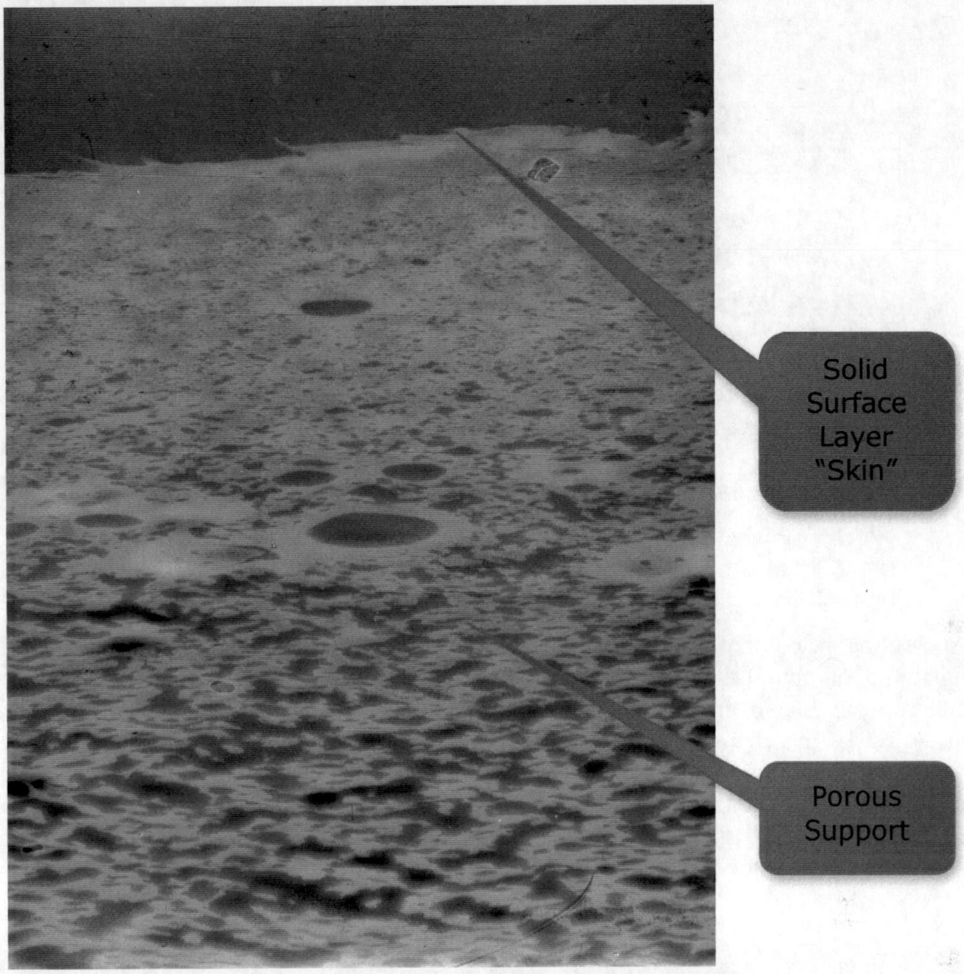

Figure 1. Cross-section view that illustrates the structure of an asymmetric gas separation membrane, made of PVTMS.

This article is devoted to the description of the statistical data of the experimental results using the phenomenological parameters characterizing the process of the transmembrane transport of gases through PVTMS structure.

Test bench in order to determine the permeability of the flat membrane was designed for the experiment. The cell with the flat asymmetric membrane is attached through the gas transportation system of channels with a gas balloons. Cell construction allows performing a series of straight and reverse permeability measurements at different pressures without orientation change of the membrane, that is, the direction of the gas flow, by the use of the electromagnetic valve. For determining the output flow of the membrane system the mass flow sensor with automatic switching depending on the flow is used. Membrane permeability at its different orientation and for different gases can be calculated from the flow values. This method for measuring of permeability is called a differential.

Reverse asymmetry gas transport effect in PVTMS membrane was detected and verified for all gases: N2, O2, CO2, and He, in the pressure range [0.03 - 0.96] MPa, namely. Figure (2) shows the measured values in differential mode for various gas permeability in their forward and reverse feed.

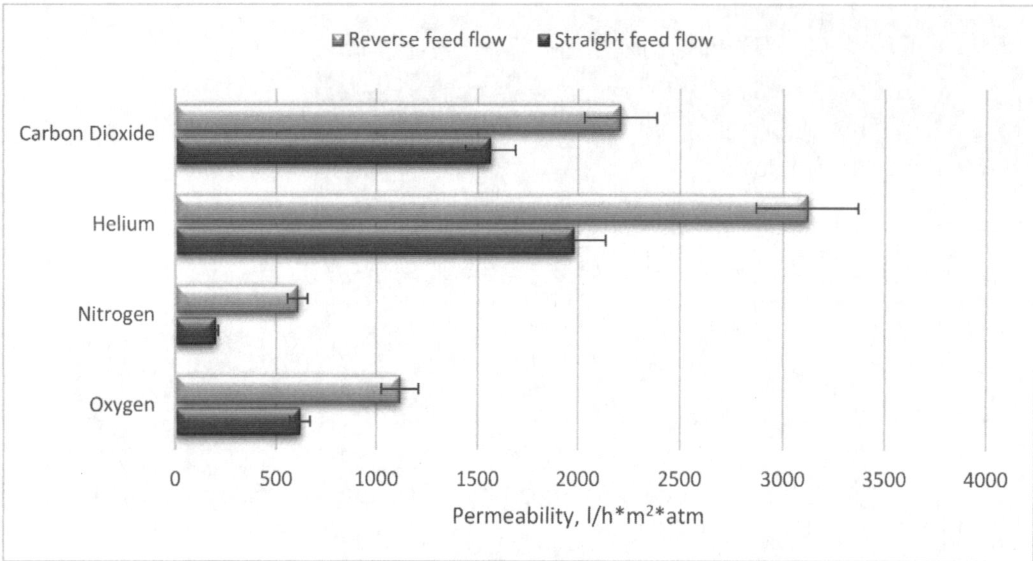

Figure2. Gas permeability values depending on the orientation of the membrane

It should be noted that most clearly reverse asymmetry transport effect appears during the investigation of N_2 transport through the membrane. Gas permeability is essentially independent of the pressure drop generated over a wide range, but for small values of $P < 2$ kPa permeability increases up to a certain value. This change can be attributed to the change of permeability mode, since a transition from a quasi-equilibrium state to disequilibrium with increasing of the pressure drop. The concentration ceases to be determined by its corresponding Henry constant and begins to depend on the state of the polymer surface of the structure [4]. The following Table. (1) shows the calculated values of asymmetry for the transport of gases, ASV, (asymmetry value):

Table. 1 The asymmetry values of selected gases

Gas	O_2	N_2	He	CO_2
Asymmetry value	1,80	3,05	1,57	1,41

Values of selectivity under straight and reverse gas supply to the membrane were also calculated. Results are shown in Table. (2, 3)

Table. 2. Selectivity values under the straight direction of the gas flow (gas is supplied on the diffusion layer).

a (O_2/N_2)	a(O_2/He)	a(O_2/CO_2)	a(N_2/He)	a(N_2/CO_2)	a(He/CO_2)
3,11	0,312	0,394	0,1	0,12	1,266

Table. 3. Selectivity values under the reverse direction of the gas flow (gas is supplied on the porous support).

a (O_2/N_2)	a(O_2/He)	a(O_2/CO_2)	a(N_2/He)	a(N_2/CO_2)	a(He/CO_2)
1,84	0,358	0,504	0,193	0,274	1,412

Using specialized software environment MGS, the initial parameters of the process were set, such as the feed stream F = 1000 l(stp)/h, the optimal division ratio = 0.5 which is corresponded to the extraction degree of oxygen in the flow of permeate FrP = 65%. Concentrations and the extraction degree of air components for a single element of a membrane module used in oxygen concentrators were calculated. The calculation results are shown in Table. 4, from which it can be seen that with a slight degradation of the separation parameters, membrane area decreases by half due to the greater permeability, hence the use of the reverse supply with the same effective membrane area can improve the performance twice of the single element in the membrane module.

Table. 4 Separation parameters association with effective membrane area, depending on its orientation. (calculation data)

	Straight feed, P, l(stp)/h – 395.750 FrP(O_2), % - 65.07; S, m^2 – 1.682E-01		Reverse feed, P, l(stp)/h – 500.500 FrP(O2), % - 65.05; S, m2 – 7.987E-02	
Gas	Pi, l(stp)/h	Cp, %	Pi, l(stp)/h	Cp, %
CO_2	1563.00	2.19	2208.00	1.67
O_2	617.00	32.88	1115.00	26.00
N_2	198.00	64.92	605.00	72.33

One way to describe the transfer of gas through the polymer membrane is dissolution-diffusion model [5]. Apparently, in identical conditions, the amount of adsorbed molecules in the surface layer from the side of porous support is greater than in the outer selective layer. Molecules of gas dissolve in a whole volume of the porous support, wherein the inner side of the selective layer has a well-developed surface area that requires a larger number of sorption centres than at the surface. Obviously, if we compare the sorption surface area at the interface between diffusion layer and porous layer to the surface of the film, they will be different.

Conclusion

Based on this effect, in particular, on the data of permeability, selectivity, and asymmetry value of the various components of the gas mixture it becomes possible to create an apparatus for producing of artificial atmosphere with the possibility of changing CO_2 level. In this work permeability of gases through the polymer membrane depending on its orientation to the high-pressure area was investigated experimentally. The phenomenon of asymmetric gas transfer is detected, and it is shown that the reverse effect of asymmetry is observed, when the flow value from the side of the porous support is greater than flow value from the selective layer: the change of the flow value, depending on selected gas can vary up to three times. Permeability, selectivity and asymmetry values were also represented for these gases.

Acknowledgements

The work was financially supported by the Ministry of Education and Science, agreement № 14.579.21.0046. Unique project identifier RFMEFI57914X0046.

References

[1] Kurchatov I M, Krjukov A V, Laguncov N I 2012 *Membr. Membr. Technol.* **2** 124-229
[2] Kurchatov I M and Laguntsov N I 2010 *Proc. of the XXV International symposium on physico-chemical methods of separation* ARS Separatoria (Torun, Poland) p 286
[3] Dytnerskij J I, Brykov V P and Kagramanov G G 1991 *Membrannoe razdelenie gazov* (Moscow: Chemistry) p 307
[4] Timashev S V *Fiziko-himija membrannyh processov* (Moscow: Chemistry)
[5] Mulder M 1991 *Basic Principles of Membrane Technology* (Kluwer Dordrecht) p 372.

Journal of Physics: Conference Series **751** (2016) 012043 doi:10.1088/1742-6596/751/1/012043

Use of ejector for recirculating flow organization in hybrid membrane-sorption system

I M Kurchatov[1], N I Laguntsov[1], V N Gurkin[1] and I E Skuridin[2]

[1] National Research Nuclear University MEPhI (Moscow Engineering Physics Institute), Kashirskoe highway 31, Moscow, 115409, Russia

[2] JSC Aquaservice, Kashirskoe highway 31, Moscow, 115409, Russia

E-mail: aquaserv@mail.ru

Abstract. In this paper the possibility of ejector using for recycle organization in the hybrid membrane adsorption system is considered. An algorithm of system solution that connects its dimensions and output parameters of the flow was described. The dependence of output pressure on the ejection coefficient under the selected geometry of the ejector was obtained.

1. Introduction

Oxygen plays an important role in human life. Besides the fact that it is necessary for respiration, oxygen is used in many spheres of human activity: energetics, metallurgy and chemical industry. Therefore obtaining of oxygen is an important and necessary objective.

Currently, an obtaining of oxygen is performed advantageously from air. One of the most efficient systems, according to the average air enrichment with oxygen, is a hybrid membrane-sorption system. This system allows to obtain air enriched with oxygen up to 50-60%. Such air may be used in the medical field and also, in order to create an artificial environment.

Hybrid membrane sorption system includes two separation steps - sorption and membrane. The most perspective scheme of the hybrid system is a recycling system without compression [1], wherein the recycle flow is mixed into the feed flow, which exits from compressor, using the ejector mixer.

Principal scheme of the system is shown in Fig. 1.

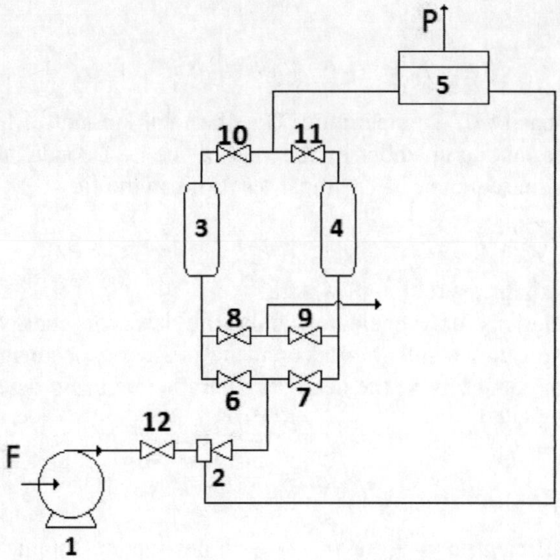

Figure 1. Schematic diagram of the recirculation of Single hybrid membrane-sorption installation. 1 – air compressor. 2 – ejector. 3,4 – absorbers. 5 – membrane module. 6 - 12 – adjustable valves.

In this system ejector mixing of flows is used. Ejector design is shown in Fig. 2.

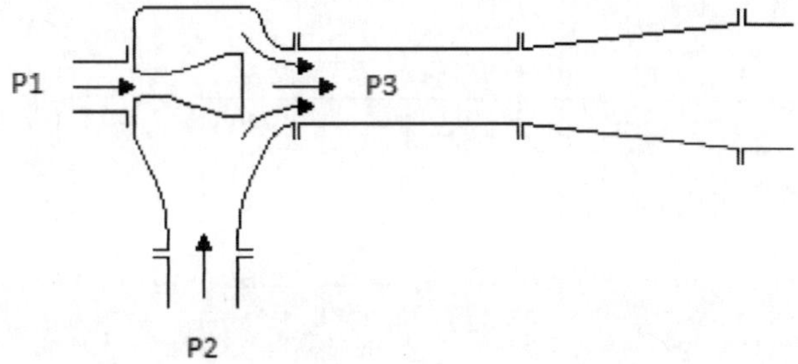

Figure 2. Schematic diagram of the ejector.

Gas ejector consists of inlet nozzle of high-pressure flow, low-pressure flow nozzle, flow part or mixing chamber and outlet diffuser. One of the main advantages of this device is the lack of moving parts.

As applied to the installation shown in Fig. 2, high-pressure jet is the flow, that will be fed to the compressor. Low-pressure jet is retentate flow from the membrane module. Obviously, the output flow has a pressure value which lies between the values of the high-pressure and low-pressure flows, and its value depends on the flow values and the geometric parameters of the ejector. Therefore, under known values of the high-pressure and low-pressure flow the main task is to calculate the geometric parameters of the ejector: high-pressure nozzle cross-sectional area, cross-sectional area of low-pressure nozzle and the area of the mixing chamber section.

Under calculations we will use a cylindrical mixing chamber profile as cylindrical mixing chamber provides a greater degree of pressure recovery compared with a chamber of other profile. [2] Furthermore cylindrical chamber is easy in manufacturing.

For a description of the output parameters of the flow using determined input parameters is sufficient to use the laws of energy, mass and momentum conservation [3].

$$G_3 = G_1 + G_2,$$

where G –is the mass flow;

$$G_3 c_{p3} T_3^* = G_1 c_{p1} T_1^* + G_2 c_{p2} T_2^* + Q,$$

where c – specific heat capacity, T^*- temperature, Q – the total amount of heat, which is supplied to the gas in 1 second into the mixing chamber by the way of the heat conductance through the walls of the chamber or allotting in consequence of chemical reactions in the flow;

$$G_3 w_3 + p_3 F_3 = G_1 w_1 + p_1 F_1 + G_2 w_2 + p_2 F_2 ,$$

where w – enthalpy, p – static pressure, F – flow area.

While providing calculations of the ejector, only the laws of energy, mass and momentum conservation were used, we can not take into account the processes occurring in the flow part of the ejector. From the system of equations all the necessary parameters using determined input parameters according to [3] can be calculated:

$$\frac{T_3^*}{T_1^*} = \frac{1+n\theta+\vartheta}{n+1}, \quad (1)$$

где $n = \frac{G_2}{G_1}$ – ejection coefficient, $\theta = \frac{T_2^*}{T_1^*}$, $\vartheta = \frac{Q}{c_p T_1^* G_1}$ – per-second supplied amount of heat to heat content of per-second ejecting gas discharge ratio;

$$\sqrt{(n+1)(1+n\theta+\vartheta)}\, z(\lambda_3) = z(\lambda_1) + n\sqrt{\theta} z(\lambda_2), \quad (2)$$

where λ – ratio of the real flow velocity in a section under consideration to the value of the critical velocity, therein: $z(\lambda) = \lambda + \frac{1}{\lambda}$;

$$\frac{p_3^*}{p_1^*} = \frac{\sqrt{(n+1)(1+n\theta+\vartheta)}}{1+\frac{1}{\alpha}} \frac{q(\lambda_1)}{q(\lambda_3)}, \quad (3)$$

where q – gas current density; $\alpha = \frac{F_1}{F_2}$;

$$n = \frac{1}{\Pi_0 \alpha \sqrt{\theta}} \frac{q(\lambda_2)}{q(\lambda_1)}, \quad (4)$$

where $\Pi_0 = \frac{p_1^*}{p_2^*}$.

Thus, obtained system consists of the ejecting flow, a mass flowrate G1 and the static pressure P1, stream ejected with mass flow G2 and static pressure P2. The temperature in the whole the system is assumed to be constant. Adiabatic constant is equal to 1,4.

Values G1, G2, T, P1, P2 are considered to be predefined. Quantities α, λ_1, λ_2 remain unknown.

According to [2] for any preselected geometric parameters of the ejector, it is possible to realize a work mode in which ejection coefficient was the highest under given values of the pressure at the outlet of the ejector. This mode is called critical. It occurs when the ejected flow is moving in the narrowing channel bounded by the walls of the mixing chamber and expanding cause of the ejecting flow and accelerates to the acoustical velocity, that this flow can not exceed. This fact determines the maximum of the ejection coefficient.

Critical mode has great practical interest. This ejector operation mode in majority cases, is the most advantageous. In this mode, an ejector under given ejection coefficient value provides the greatest total pressure of the gas mixture and has the greatest ejection coefficient under predetermined value of the total pressure. Also, ejector, designed to work at a critical mode, has the lowest relative size of the mixing chamber under given value n.

Calculation of the critical operation modes of the ejector is conducted as follows, using the continuity condition and the fact that ejected gas, driving on the narrowing channel, doesn't mix with the ejecting flow, an expression [1] can be obtained:

$$q(\lambda_2) = q(\lambda_2')\left\{1 - \alpha\left[\frac{q(\lambda_1)}{q(\lambda_1')} - 1\right]\right\} \tag{5}$$

In critical mode $\lambda_2' = 1$ and $q(\lambda_2') = 1$.

Equation (3) shows that at a constant geometry the increase of reduced velocity $\lambda_1' > 1$ leads to decrease of the speed of ejected flow at the inlet of mixing chamber. if λ_1' increases on, that

$$q(\lambda_1') = \frac{\alpha}{\alpha+1}q(\lambda_1) = \frac{F_1}{F_3}q(\lambda_1),$$

than $q(\lambda_2) = 0$ and $\lambda_2 = 0$.. Consequently, there is no longer ejection process.

In this case $F_1 = F_3$. This phenomenon is called locking the ejector.

Expression (3) can be rewritten as:

$$q(\lambda_2) = q(\lambda_2')\left\{1 - \frac{1}{\Pi_0 n\sqrt{\Theta}}\frac{q(\lambda_2)}{q(\lambda_1)}\left[\frac{q(\lambda_1)}{q(\lambda_1')} - 1\right]\right\}$$

or

$$n\sqrt{\Theta} = \frac{1}{\Pi_0}\frac{\frac{1}{q(\lambda_1')} - \frac{1}{q(\lambda_1)}}{\frac{1}{q(\lambda_2)} - \frac{1}{q(\lambda_2')}} \tag{6}$$

under calculations values $\lambda_1 = 1$, and $\lambda_2' \leq 1$ limit the range of possible values λ_2 and n.

As the interaction with the walls surface does not give forces directed along the flow direction, than, gas motion is determined by the gas pressure difference at the boundaries of the motion area.

The equation of kinetic momentum can be written as:

$$G_1(w_1 - w_1') + G_2(w_2 - w_2') = p_1'F_1' + p_2'F_2' - p_1F_1 - p_2F_2$$

Using equation $Gw + pF = \frac{k+1}{2k}Ga_{\kappa p}z(\lambda)$, next expression is obtained

$$G_1a_{\kappa p1}[z(\lambda_1') - z(\lambda_1)] = G_2a_{\kappa p2}[z(\lambda_2) - z(\lambda_2')]$$

or

$$n\sqrt{\Theta} = \frac{z(\lambda_1') - z(\lambda_1)}{z(\lambda_2) - z(\lambda_2')} \tag{3}$$

Fig. 3 shows the calculated dependence of the output pressure of the ejector according to the ejection coefficient on different values of high-head gas inlet pressure. According to Fig. 3 the pressure at the outlet of the ejector for preselected ejection coefficient n can be determined, on condition that a critical mode is reached in the ejector.

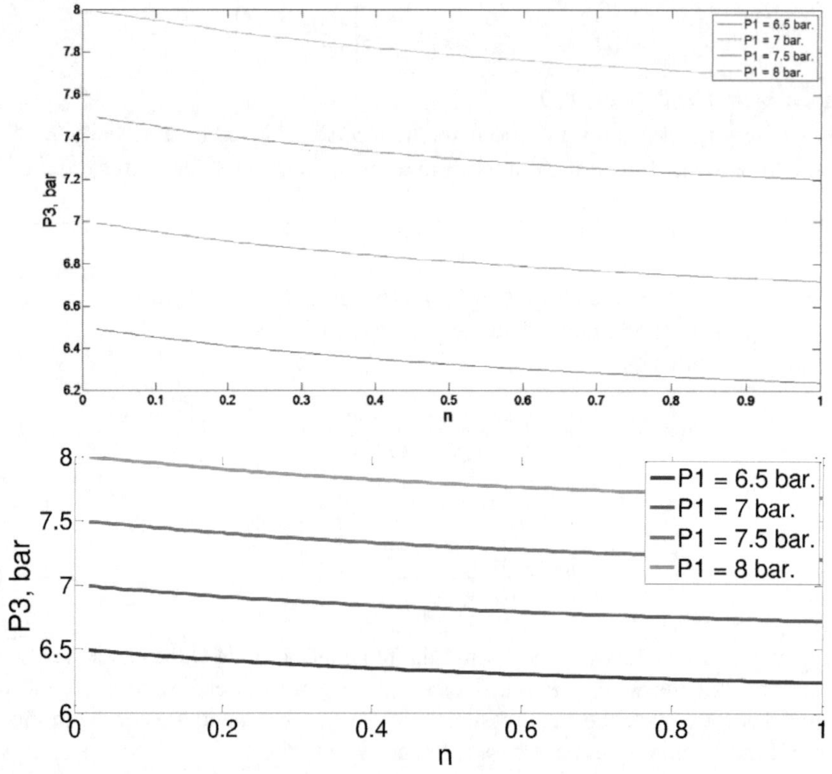

Figure 3. The dependence of outlet pressure P3 on ejection coefficient *n*, operating in the critical mode, at different values of the supplied flow pressure P1.

On the Fig. 3 is shown, that, as expected, then the ejection coefficient *n* tends to zero; i.e., in the absence of the recycle flow entering from a membrane module, the pressure at the outlet of the ejector tends tends to pressure P1 of the feed flow. When the ejecting coefficient *n* increases, the pressure at the outlet of the ejector will decrease. It can be seen that the outlet pressure is reduced slightly, which proves the appliance possibility of ejector in recirculating hybrid membrane-sorption gas separation system.

Conclusions

In this paper one of the most perspective systems of air enrichment with oxygen was considered. It is a hybrid membrane system with recycle. Recycle flow is returned to the system via an ejection mixer. This method of flow returning improves the characteristics of the system without requiring additional energy.

In this work system of equations for calculating the parameters of an ejection mixer operating in a critical mode was described. It was shown that pressure in a such mixer decreases slightly versus input pressure. That fact proves the appliance possibility of ejector in recirculating in the hybrid membrane-sorption gas separation system.

Acknowledgements

The work was financially supported by the Ministry of Education and Science, agreement № 14.579.21.0046. Unique project identifier RFMEFI57914X0046.

References

[1] Kudinov A N, Kurchatov I M, Laguntsov N I, Neshchimenko Y P 2014 *Theor. Found. Chem. Eng.* **48** 832–6.
[2] Abramovich G N 1976 *Applied gas dynamics* (Moscow: Nauka) p 888.
[3] Sokolov E Ya, Zinger N M 1989 *Struinye apparaty* (Moscow Jenergoatomizdat) p 352

Journal of Physics: Conference Series **751** (2016) 012044 doi:10.1088/1742-6596/751/1/012044

Nanofiltration properties of PTMSP in binary organic solvents mixtures

A A Yushkin[1], A A Kossov[1] and V V Volkov[1,2]

[1] A.V.Topchiev Institute of Petrochemical Synthesis RAS, Moscow, Russia
[2] National Research Nuclear University MEPhI (Moscow Engineering Physics Institute), Kashirskoe highway 31, Moscow, 115409, Russia

E-mail: polymem@yandex.ru

Abstract. In this study, the stability and nanofiltration performance of poly[1-(trimethylsilyl)-1-propyne] (PTMSP) in ethanol solutions of butylaldehyde, 1-decanal, 1-hexene, 1-decene was evaluated. It was found that PTMSP was insoluble in all aldehyde solutions, but it was soluble at olefin concentration of 80% or higher. Nanofiltration experiments demonstrate that binary mixtures of 1-decanal and ethanol viscosity are not the parameter affecting on membrane permeability and rejection of solute as well as swelling degree. In the case of decanol/ethanol solutions both solution viscosity and molar volume demonstrate the best fit of experimental data. It was shown that with the decrease of ethanol content in the feed, the rejection of anionic solute Remazol Brilliant Blue R (MW 626) increases from 94 up to 97%.

1. Introduction

Effective recovery of homogeneous catalysts from reaction mixture for its further recycling is a challenging task in homogeneous catalysis. For example, catalysts used in organic synthesis are based on expensive and toxic complexes of transition metals (Pt, Pd, Ru, Ph, etc.). Conventional separation approaches are based on energy-intensive distillation or extraction processes, which includes the phase transitions. The separation of the catalyst by such methods is quite often accompanied by its partial or complete deactivation. Organic solvent nanofiltration (OSN) is already well-known and still rapid growing area of membrane technology due to its great potential and advantages over the traditional separation methods like distillation or extraction [1-3]. In some cases, OSN can be considered as a unique separation process that could provide effective and almost quantitative recovering of target compounds in different areas, including chemical, petrochemical, and food industries. However, the major drawback towards rapid implementation of this technology is rather a limited number of membranes that possess sufficient stability in organic media, high permeability for selected solvents as well as high retention of target compounds. For successful and efficient implementation, the membranes for OSN process should have sufficient mechanical stability in organic solvents, demonstrate high retention values and transport of the organic solvent. One of the perspective membrane materials for OSN is the glassy polymers with intrinsic microporosity that naturally formed during the membrane casting as a result of rigid nature of polymer chains. For example, poly[1-(trimethylsilyl)-1-propyne] (PTMSP) has been already applied to fabricate high-performance OSN membranes due to its good mechanical properties [2-4]. This polymer is stable in the solvents like alcohols, ketones, and aldehydes. However, the recycling of homogeneous catalysts requires the stability of the membrane material during the contact with not only the products but also the raw materials that are usually used for flushing of the retentate. By this means, if the membrane is supposed to be used in hydroformylation process, it should be stable in reaction mixtures contained aldehydes and olefins, respectively. Since the presence of co-solvents such as ethanol or acetone could improve the reaction rate of conversion of olefin to aldehyde [5], the goal of this work was the study of stability and nanofiltration performance of PTMSP membranes in decene-1/ethanol and decanal/ethanol mixtures.

2. Experimental part

2.1. Membrane preparation

PTMSP used in the work was synthesized on $TaCl_5$ catalyst with the cocatalyst $Al(i\text{-}Bu)_3$, which provides a preponderance of the trans-structures of the polymer macromolecules [6]. To prepare the casting solution, polymer samples was dissolved in chloroform. Due to the high intrinsic viscosity for solutions of the PTMSP polymer concentration was 0.5 wt.%. Dense PTMSP films of a thickness of 30 ± 5 microns were obtained by casting PTMSP solution in chloroform with a polymer content 0.5 wt.% on cellophane, followed by slow drying at room temperature for 3-4 days.

To remove residual solvent obtained membrane was soaked for 2 days in 1-butanol, after which the sample was immersed in ethanol for at least 24 hours. After that, the membrane was washed sequentially in water-ethanol solutions with a gradual decrease of alcohol concentration and subsequent drying to constant weight. Immediately before filtration membrane was immersed in the corresponding solution for 1 hour.

2.2. Nanofiltration experiments

Nanofiltration experiments were performed with dead-end filtration cell. The cell was equipped with a magnetic stirring system to avoid concentration polarization. The active membrane area (S) was 10.7 cm^2. Helium was chosen to create a pressure because of its low solubility in organic solvents in a wide range of pressures. Filtration experiments were carried out at transmembrane pressure 30 bar. The volume of the liquid was 50 cm^3. During the experiment, 25 cm^3 of the solution was allowed to permeate through the membrane. The permeate flux (J) was determined by gravimetric method and calculated as follows:

$$J = \frac{m}{S \cdot \Delta t}, \tag{1}$$

where m – mass of permeate, Δt – time. Permeability coefficient (P) was calculated as follows:

$$P = \frac{J \cdot l}{\Delta p}, \tag{2}$$

where l – membrane thickness, Δp – applied pressure.

The composition of 1-Decene and 1-Decanal solutions in ethanol was measured via refractometer. Remazol brilliant blue R dye was chosen as a "model" of the catalyst because it has a molecular weight (MW 626) and it is easy to measure its concentration in solution. The concentration of dye was measured via a spectrophotometer.

2.3. Swelling degree and solubility tests

Polymer swelling degree (S_D) in the corresponding mixture was obtained by measuring a sample of polymer film thickness (h), and two perpendicular dimensions (d_1 and d_2) were measured. Sample with measured dimensions was soaked in corresponding liquid for 7 days. In swelling measures samples were 5*5 cm with thickness 50 microns. Swelling degree was calculated as follows (index 0 means value for dry sample):

$$S_D = \frac{h \cdot d_1 \cdot d_2 - h_0 \cdot d_{10} \cdot d_{20}}{h_0 \cdot d_{10} \cdot d_{20}} \quad (3)$$

Solubility tests were carried out in the same way. Small polymer film sample (1 cm2) was placed in corresponding liquid for 7 days. The sample was considered as insoluble if after 7 days it was possible to remove it from solution without damaging.

3. Results and discussion

3.1. PTMSP solubility
An important factor that must be considered in the filtration process is the membrane resistance to the filtered mixture. PTMSP-membranes stable in aldehydes and alcohols but soluble in olefins (see Table 1).

Table 1. PTMSP stability in binary ethanol solutions ("+" – stable, "-" – dissolved).

Solvent	Solvent content in ethanol, wt.%										
	0	10	20	30	40	50	60	70	80	90	100
Decanol	+	+	+	+	+	+	+	+	+	+	+
Butylaldehyde	+	+	+	+	+	+	+	+	+	+	+
1-Decanal	+	+	+	+	+	+	+	+	+	+	+
1-Hexene	+	+	+	+	+	+	+	+	+	-	-
1-Decene	+	+	+	+	+	+	+	+	+	-	-

In the hydroformylation process, reaction mixture to be separated might contain substrate (olefin), product (aldehyde) and solvent (ethanol). Therefore, in the beginning, the stability of PTMSP membranes was evaluated in the binary solutions of butyraldehyde, 1-decanal, 1-hexene or 1-decene in ethanol. For this purpose, PTMSP film samples were soaked in corresponding solutions for 7 days and then the membrane stability was visually analyzed. As can be seen from Table 1, PTMSP was insoluble in all aldehyde solutions, but PTMSP was soluble for olefin concentration of 80% or higher. Taking into account the fact that the reaction mixture in the large-scale hydroformylation processes can contain up to 85-90% of inert solvent, it can be concluded that PTMSP-based membranes show required solvent stability for this particular application.

3.2. Nanofiltration of binary solutions through PTMSP
Since catalyst should be separated from solution product filtration of 1-decanal solutions in Ethanol was carried out with PTMSP membranes (Figure 1).

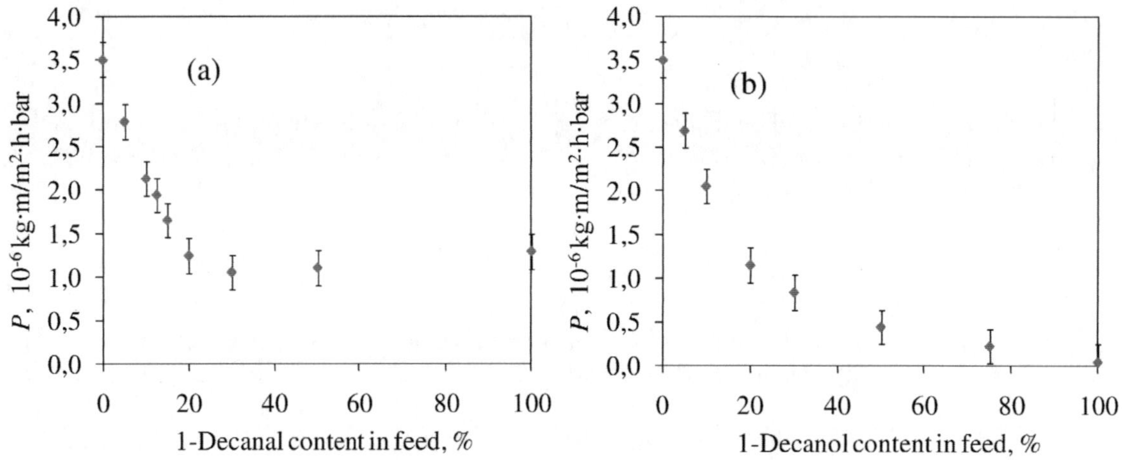

Figure 1. Permeability coefficient of 1-decanal (a) and decanol (b) solutions in ethanol through PTMSP membranes

As can be seen from Figure 1, PTMSP shows the highest permeability in ethanol and addition of 1-decanal up to 3- wt.% leads to about 3-folds reduction in liquid transport across the membrane. No significant changes in membrane permeability were observed with further increasing of 1-decanal content in the feed. In both cases of 1-decanal and decanol ethanolic solutions, no change of mixture composition occurred during the filtration. Therefore, it can be expected that the reaction mixture composition would remain the same as was reported previously for water-ethanol mixtures [7]. Such finding might be important in the case of solubility of homogeneous catalyst in the limited range of substrate-solvent or product-solvent mixture. In the case of decanol solutions, concentration monotonically decreased with increasing of decanol concentration in the whole range of concentrations. The permeability of decanol was 30 lower than for decanal. These results can be explained by the changes of solution viscosity (η) which increases in the following order: ethanol (1.19 mPa·s)<1-decanal (17.9 mPa·s) <decanol (138 mPa·s).

Table 2. Correlation of PTMSP permeability

Parameter	Correlation (R^2)	
	Decanal/Ethanol	Decanol/Ethanol
1/η	0.0631	0.8899
S_D	-0.347	-0.539
S_D/ η	-0.286	0.793
V_m	-1.753	-0.801
1/V_m	0.6274	0.6868
1/ η V_m	0.4483	0.9913

However, the viscosity, as well as swelling degree (SD), was not the particular case for binary mixtures of 1-decanal and ethanol (see Table 2). In fact, maximum swelling degree (75%) takes place in 50/50 1-decanal/ethanol mixture but very low permeability and lower swelling degree for ethanol (61%) and 1-decanal (55%). The best fit was obtained in the case of reverse molar volume (V_m). A Larger molecule of 1-decanal moves slowly through PTMSP membrane and block transport channel for ethanol since no composition change take place.

In the case of decanol/ethanol solutions both solution viscosity and molar volume demonstrate the best fit of experimental data. It should be noted that viscosity itself correlates with molar volume, but the only combination of two this parameters takes good correlation with experimental data. As well as for 1-decanal in the case of decanol solutions in ethanol swelling degree is not the parameter affecting on permeability since it monotonically increases with decanol concentration in contrast with permeability. More detailed explanation of permeability results obtained in this work is the object of further investigation.

3.3. Nanofiltration of ternary solutions through PTMSP

Ternary solutions containing binary solution mixture and the solute Remazol Brilliant Blue R (MW 626), 1-decanal/decanol and ethanol was used to determine membrane rejections. Remazol Brilliant Blue R concentration in solution was 10 mg/l. It was obtained that increasing of decanal content lead to a decrease of Remazol Brilliant Blue R rejection by PTMSP membrane (Figure 2).

The same result was obtained for decanol. In both cases, a minimum of rejection takes place at 5% concentration. As well as in case permeability swelling degree increases in investigated range of concentrations and does not affect by rejection. This result means that steric effect is not the main parameter for rejection of this dye by PTMSP. This result can be explained by specific interactions for the solvent(s)-solute-membrane system. As was shown before for rejection of different solutes from its ethanolic solution by PTMSP [8], there is a competition sorption of solvent and solute molecules into membrane material. Since the ethanol has lower affinity to PTMSP than more hydrophobic molecules of 1-decanal or 1-decanol, it can be expected that the sorption selectivity for the solvent over the anionic solute would increase with decrease ethanol content in the feed. Thus, an increase in sorption selectivity would also lead to an increase of overall selectivity of nanofiltration process.

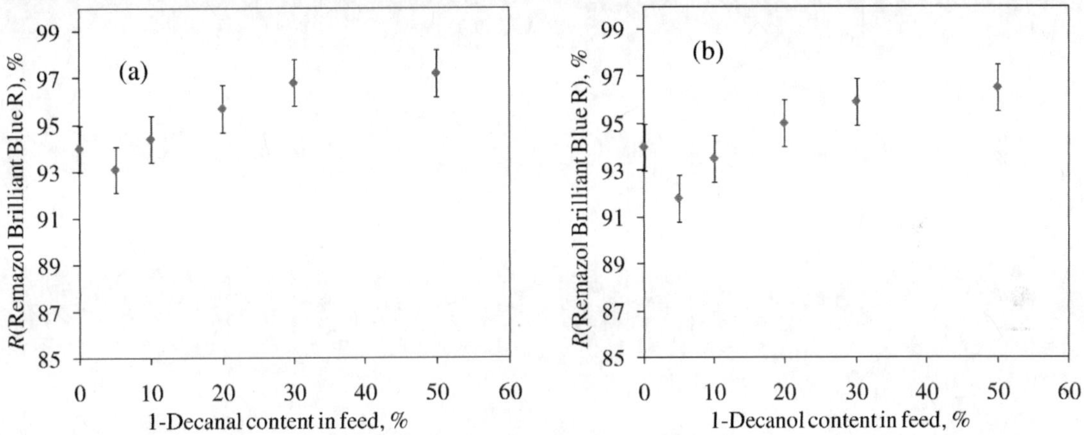

Figure 2. Rejection of Remazol Brilliant Blue R from 1-decanal (a) and 1-decanol (b) solution with ethanol by PTMSP membrane

4. Conclusions

In this study, the stability and nanofiltration performance of poly[1-(trimethylsilyl)-1-propyne] (PTMSP) in ethanol solutions of butyraldehyde, 1-decanal, 1-hexene, 1-decene was evaluated. It was found that PTMSP was insoluble in all aldehyde solutions, but it was soluble at olefin concentration of 80% or higher. Nanofiltration experiments demonstrate that binary mixtures of 1-decanal and ethanol viscosity are not the parameter affecting on membrane permeability and rejection of solute as well as swelling degree. In the case of decanol/ethanol solutions both solution viscosity and molar volume demonstrate the best fit of experimental data. It was shown that with the decrease of ethanol content in the feed, the rejection of anionic solute Remazol Brilliant Blue R (MW 626) increases from 94 up to 97%.

Journal of Physics: Conference Series **751** (2016) 012044 doi:10.1088/1742-6596/751/1/012044

Acknowledgments
The work was supported by the Ministry of Education and Science of the Russian Federation, contract no. 14.607.21.0083 (ID: RFMEFI60714X0083).

References
[1] Vandezande P, Gevers L E M and Vankelecom I F J 2008 *Chem. Soc. Rev.* **37** 365–405.
[2] Volkov A V, Korneeva G A and Tereshchenko G F 2008 *Russ. Chem. Rev.* **77** 983–93.
[3] Marchetti P, Jimenez Solomon M F, Szekely G and Livingston A G 2014 *Chem. Rev.* **114** 10735
[4] Volkov A V, Parashchuk V V, Stamatialis D F, Khotimsky V S, Volkov V V and Wessling M 2009 *J. Membr. Sci.* **333** 88–93.
[5] Deshpande M, Raghunath M and Chaudhari V 1988 *Ind. Eng. Chem. Res.* **27** 1996-2002.
[6] Khotimsky V S, Tchirkova M V, Litvinova E G, Rebrov A I and Bondarenko G N 2003 *J. Polym. Sci. Part A*. **341** 2133-55.
[7] Volkov A, Yushkin A, Grekhov A, Shutova A, Bazhenov S, Tsarkov S, Khotimsky V, Vlugt T J H and Volkov V 2013 *J. Membr. Sci.* **440** 98-107.
[8] Tsar'kov S E, Malakhov A O, Litvinova E G and Volkov A V 2013 *Petrol. Chem.* **53** 537–45.

Journal of Physics: Conference Series **751** (2016) 012045 doi:10.1088/1742-6596/751/1/012045

Optical microinterferometry method for evaluation of phase state and diffusion in ternary systems: phase separation in cellulose/N-Methylmorpholine-N-oxide/non-solvent mixtures

V V Makarova[1], S V Antonov[1], T A Anokhina[1] and V V Volkov[1,2]

[1] A.V.Topchiev Institute of Petrochemical Synthesis RAS, Moscow, Russian Federation

[2] National Research Nuclear University MEPhI (Moscow Engineering Physics Institute), Kashirskoe highway 31, Moscow, 115409, Russia

E-mail: polymem@yandex.ru

Abstract. Microinterferometry technique was used to evaluate the phase state and diffusion of cellulose solutions N-Methylmorpholine-N-oxide upon contact with non-solvents (water and aqueous solutions of isopropyl alcohol). The method was helpful in visualization of the structure of the forming cellulose film in connection with the diffusivity of the components of the systems. The interdiffusion coefficient were determined. Isopropyl alcohol addition to water slows down the diffusion of the coagulant into the cellulose solution thus delaying cellulose precipitation. Increase of temperature leads to formation of less dense cellulose film morphology with large vacuoles.

1. Introduction

The membranes used in different separation processes are mainly made of a wide range of rubbery and glassy polymers due to their lower price in contrast to inorganic membranes. A typical membrane consists of the thin selective layer on top of a thick porous support layer that provides the required mechanical properties. In the case of glassy polymers with a low or moderate price like polysulfone or polyamides, the selective and porous support layers can be fabricated from the same polymeric material by phase inversion technique [1]. The polymeric solution film with a required thickness is formed on the glass plate or non-woven support by casting knife followed by immersion in the non-solvent bath. The contrariwise diffusion of non-solvent and solvent molecules leads to a change in the composition of the polymeric solution resulting in the phase separation and formation of a polymer film. At the same time, polymer precipitation and change of the viscosity of the solution affect on the further diffusion of non-solvent and solvent molecules in the bulk of polymeric solution and lead to the formation of an asymmetric porous structure. Nowadays, the greater attention is focused on the development of membrane materials and membranes from sustainable sources, and cellulose can be considered as a promising candidate due to its renewable origin and biodegradability. For example, cellulose-based membranes can be used for gas separation [2,3], organic solvent ultrafiltration [4,5] and nanofiltration processes [6,7].

The optimal casting conditions that allow obtaining the membrane with target properties such as flux, selectivity, and mechanical resistance are usually determined experimentally when the polymer solution composition, as well as solvent and non-solvent nature, is varied. It should be noticed that availability of a ternary (polymer/ solvent/non-solvent) phase diagram could significantly assist in this approach, meanwhile, the determination of 3-component diagram for an unknown system might be time and effort costly routine. Optical methods are usually used to determine the starting moment of polymer precipitation from its solution (so-called cloud point) or to study the polymer structure (presence of crystalline regions etc.). At the same time, the optical interferometry method allows analyzing polymer-polymer and polymer-solvent interactions [8]. The principle of this method is based on the multipath interference from two surfaces of the film of the polymeric solution or the polymer placed between two glass plates having the inner surface covered with a semitransparent metallic layer to reinforce its reflection capability. By applying of monochromatic beam through the film, it is possible to obtain the interference pattern. The main requirements of the method are follows: transparency of at least one of the contacting phases and difference in refraction indices of contacting phases not less than 10^{-3}. An important advantage of microinterferometry is an opportunity to visualize the process of phase separation and the morphology of the forming phases in the course of an experiment, which seems very useful for membranes manufacturing.

Microinterferometry is generally applied for binary systems [8, 9], and the developed method of mathematical interpretation of interferograms has been regarded to be truly valid only for this case. Quite recently, however, the applicability of the microinterferometry technique was successfully extended to some cases of pseudobinary systems where one of the contacting phases was gas oil - a mixture of completely miscible components [10]

In this study optical interferometry method was used for evaluation of phase state and diffusivity in a ternary polymer/solvent/coagulant system – cellulose solution in N-methylmorpholine-N-oxide with water and isopropanol aqueous solutions selected as coagulating agents (non-solvents).

2. Experimental part

Sulfate viscose cellulose with a polymerization degree of 600 produced by the Baikal pulp-and-paper plant (Russian Federation) was used in this work and have the following properties in a accordance to manufacturer specification: the α-cellulose content of 94..96 wt%, the equilibrium moisture content 8 wt% at RH 60%. Cellulose was used as a powder with 200 μm average particles size. N-methylmorpholine-N-oxide (NMMO) monohydrate with 10-13 wt% water was supplied by Demochem (China). Solid-phase cellulose solutions in NMMO were prepared by the grinding in a mortar of cellulose and NMMO monohydrate powders for 7-10 min for activation of the mixture as described elsewhere [11]. The cellulose concentration in NMMO solutions was 12 wt% as a typical composition used for films, fibers, and membrane formation. The films of 12wt% cellulose solution in NMMO were prepared using a HLCL-1000 coater-laminator (Cheminstruments, USA). The completion of the dissolution of cellulose in NMMO was controlled by the using the optical polarizing microscope Boetius equipped with the heating table. Distilled water and 20 and 50 wt% isopropyl alcohol (IPA) aqueous solutions were selected as coagulating agents (non-solvent for cellulose).

Cellulose coagulation from the solution in NMMO at contact with coagulating agents was studied at 25 and 65°C. Before the experiments, the diffusion cell with the solid-phase solutions was thermostated for 15 min at 90°C until complete cellulose dissolution in NMMO. Then the cell was cooled down to the required temperature, and then the coagulation agent was brought in contact with the system. The interference patterns (interferograms) were recorded by the digital camera. Mutual diffusion of the components leads to the drift of the refraction index ion the diffusion zone and consequently, to bending of interference bands. Using interferograms and refractions indices of the contacting phases, it was possible to plot the concentration profiles of components in the diffusion zone. The resulting concentration profiles were then used to calculate the diffusion coefficients following the Matano-Boltzmann method [12, 13] The schematic representation of the principle of the optical microinterferometry method and obtained data are shown in Fig. 1 and 2, respectively.

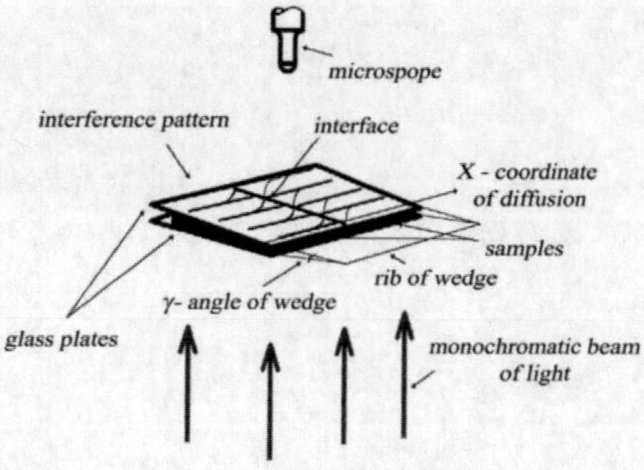

Figure 1. Principle of the optical microinterferometry method (interferometer).

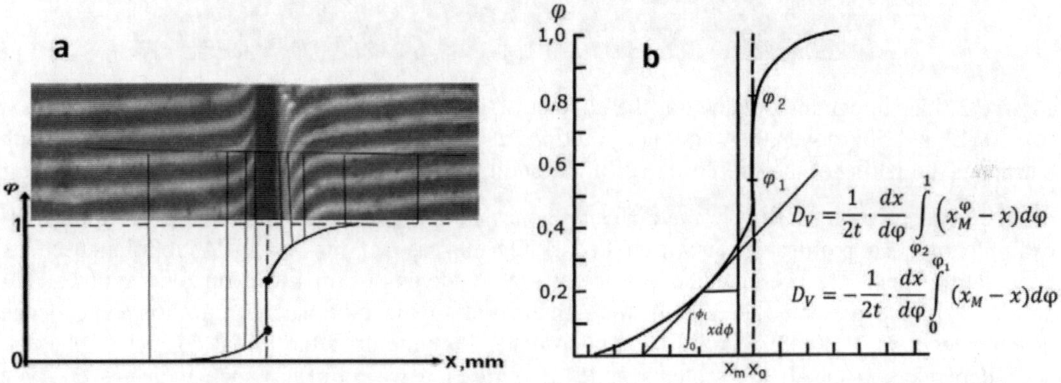

Figure 2. An example of plotting the concentration profile from the interferogram (a) and calculation of the interdiffusion coefficients (b).

3. Results and discussion

3.1. Cellulose coagulation at 25°C

Interferograms obtained at the contact of 12 wt% cellulose in NMMO solution with water, 20 and 50 wt% of IPA in water at 25°C are presented in Fig. 3. It could be noticed that all of them are quite alike with continuous interference bands which indicate on the complete miscibility of all components except for cellulose which forms a transparent and almost invisible film in the diffusion zone. However, the boundary of the film (depicted in Fig. 3) is quite distinct, so it can be speculated that a dense surface sublayer is formed at the initial contact of the cellulose solution with a non-solvent. The increase of IPA in the coagulating bath seems to slow down the non-solvent diffusion and cellulose film formation and can explain less pronounced film boundary in the case of 50 wt% IPA solution.

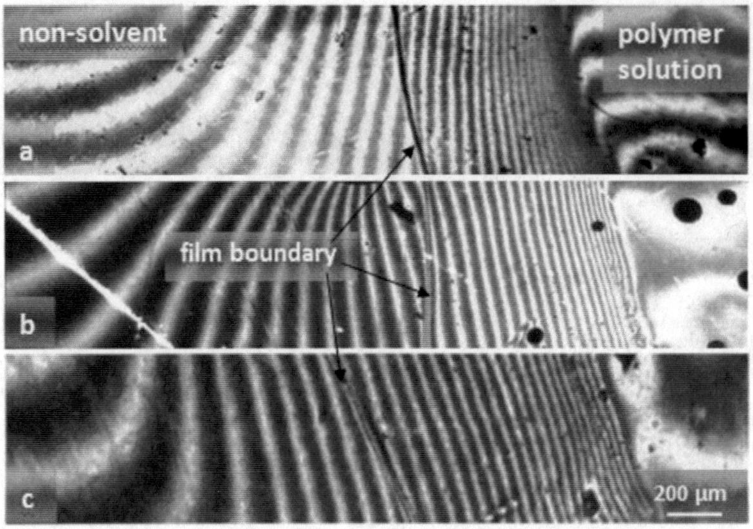

Figure 3. Interferograms obtained at the contact of 12 wt% cellulose in NMMO solution with water (a), 20 (b) and 50 (c) wt% IPA aqueous solutions at 25°C. Dark circles on the photos are air bubbles entrapped at melting of the cellulose in NMMO solid solution

The concentration profiles in the diffusion zones can be used to estimate the process kinetics. An example of such profiles is shown in Fig. 4 for the case of 50 wt% IPA solution used as the coagulating agent. The presented curves demonstrate widening of the diffusion zone in time. The data similar to those presented in Fig. 4 were than used to the plot isoconcentration dependences in coordinates x vs. $t^{1/2}$ (here x is diffusion coordinate, t is time after the initial contact of phases). All such dependences for all the systems under investigation were linear which indicates on the truly diffusive mixing of the components without chemical interaction.

It should be noted that Fig. 4, in fact, presumes system to be quasi-binary and uses "polymer solution" and "non-solvent" as independent basic components of the system. Therefore, the concentration and of interdiffusion coefficients profiles presented below should be regarded as some effective ones describing the overall kinetics of the process. The calculated interdiffusion coefficients for the systems are shown in Fig. 5.

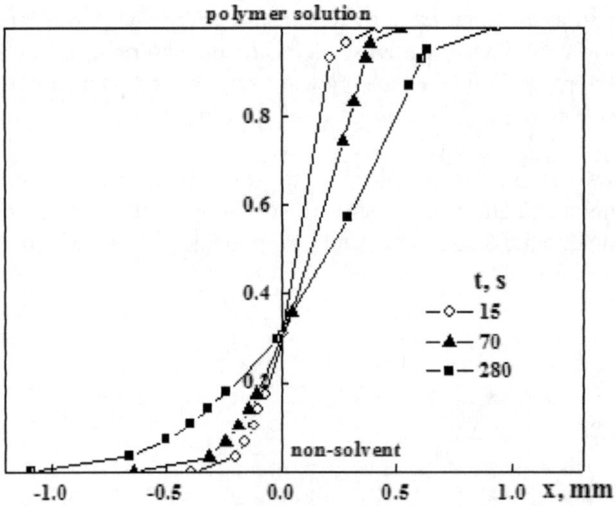

Figure 4. Concentration profiles of components for 50 wt% IPA solution at 25°C and 15, 70, 280 s after contact

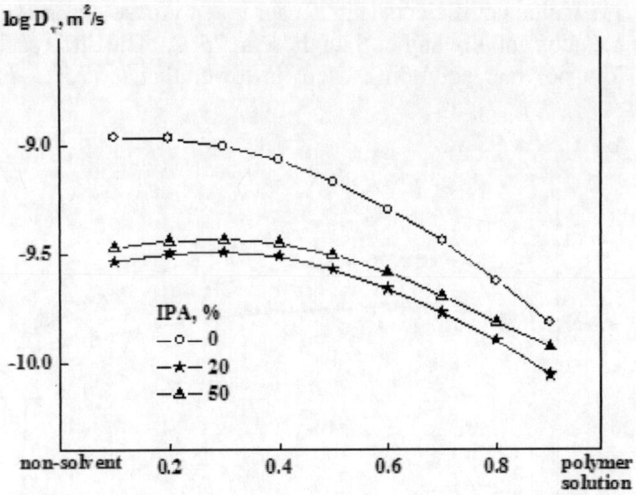

Figure 5. Interdiffusion coefficients for cellulose solution – non-solvent systems at 25°C

The cellulose film precipitated from the solution at contact with a non-solvent works as a membrane between the solution and non-solvent, so its morphology would influence on the diffusion process kinetics. Dv values in the region enriched with a non-solvent (left part of Fig. 5) are half an order of magnitude higher than those in the region of the polymer solution (right part of the plot). This is true for all the systems studied and may be explained either by higher viscosity of the polymer solution or by the dense morphology of the forming cellulose film. IPA addition leads to a noticeable decrease in the interdiffusion coefficients which may be connected with some increase in the non-solvent viscosity [14].

3.2. Cellulose coagulation at 65°C

The interferograms of the same systems obtained at 65°C (Fig. 6) look different as compared with those at 25°C (Fig. 3). One may notice large vacuoles in the forming cellulose film beyond the dense surface layer followed by a network-like structure. The vacuoles are oriented in the diffusion direction. The addition of IPA changes the pattern; vacuoles become less distinct and quite rare.

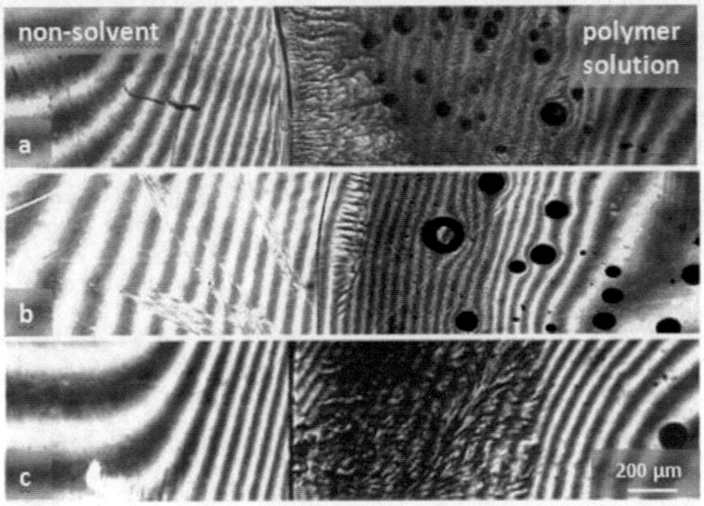

Figure 6. Interferograms obtained at the contact of 12 wt% cellulose in NMMO solution with water (a), 20 (b) and 50 (c) wt% IPA aqueous solutions at 65°C. Dark circles on the photos are air bubbles entrapped at melting of the cellulose in NMMO solid solution.

The calculated interdiffusion coefficients (Fig. 7) agree with these observations. First, the values of diffusion coefficients are substantially higher than those at 25°C. The difference between Dv values in areas with high and low polymer solution content is lower than at 25°C and decreases with IPA content.

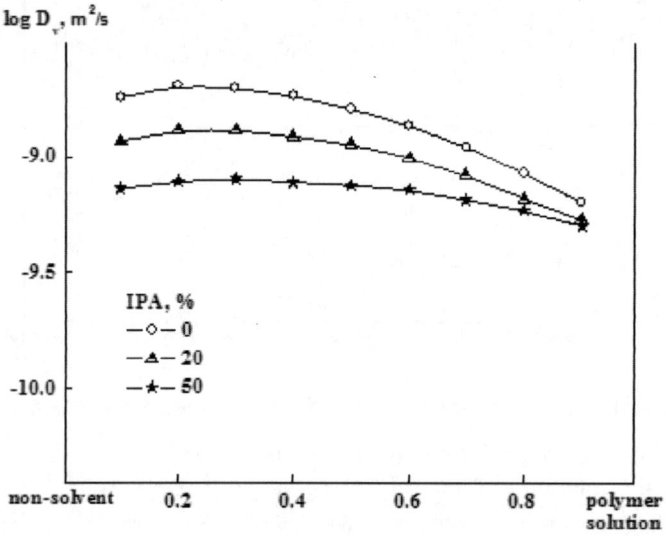

Figure 7. Interdiffusion coefficients for cellulose solution – non-solvent systems at 65°C

4. Conclusions

Application of microinterferometrry for studying polymer/solvent/non-solvent systems may be very helpful as a means of visualization of the structure of the precipitating polymer as well as assessing the kinetics of the process in spite of intrinsic limitations of the method.

Acknowledgments

This work was done at TIPS RAS with financial support from Russian Science Foundation (project no. 14-19-01775).

References

[1] Mulder J 2012 *Basic principles of membrane technology* (Kluwer Academic Publishers Dordrecht, Boston, London) p 75.
[2] Wu J and Yuan Q 2002 *J. Membr. Sci.* **204** 185.
[3] Jie X, Cao Y, Lin B and Yuan Q 2004 *J. Membr. Sci.* **91** 1873.
[4] Zhang Y, Shao H, Wu Ch and Hu X 2001 *Macromol. Biosci.* **1** 141.
[5] Li J H, Cao M T, Qin J J, Jie M X, Wang H T, Liu H J, Yuan Q, Zhang Y 2006 *J. Membr. Sci.* **279** 328.
[6] Anokhina T S , Yushkin A A, Volkov V V, Antonov S V, Volkov A V 2015 *Phys. Proc.* **72** 171.
[7] Yushkin A A, Anokhina T S and Volkov A V 2015 *Pet. Chem.* **55** 746.
[8] Makarova V and Kulichikhin V 2012 *Application of Interferometry to Analysis of Polymer-Polymer and Polymer-Solvent Interactions* (InTech) ed I Padron p 395.
[9] Malkin A Ya, Chalykh A E 1979 *Diffusion and viscosity of polymers. Measurement techniques* (Moscow: Khimia) in Russian.
[10] Pakhmanova O A 2012 *Catalytic cracking of polymer-containing oil feedstock* (Moscow: PhD thesis, A.V. Topchiev Institute of Petrochemical Synthesis, RAS) in Russian.
[11] Golova L K, Romanov V V, Lunina O B, Platonov V A, Parkov S P, Khorozova O D, Yashkin V V, Belasheva T P, Sokira A N 1992 Patent SU 1645308.
[12] Boltzmann L 1894 *Ann. Phys* **289** 959.
[13] Matano C 1993 *Jap. J. Phys* **8**. 109.
[14] Lide D R 2007 *CRC Handbook of Chemistry and Physics* (Boca Raton: Taylor and Francis).

Journal of Physics: Conference Series **751** (2016) 012046 doi:10.1088/1742-6596/751/1/012046

Investigation of alumino-silicic reagent interaction with iron in water by Mössbauer spectroscopy

D Y Feklistov[1], V P Filippov, I M Kurchatov, N I Laguntsov and V A Salomasov[2]

[1] National Research Nuclear University MEPhI (Moscow Engineering Physics Institute), Kashirskoe highway 31, Moscow, 115409, Russia
[2] JSC Aquaservice, Kashirskoe highway 31, Moscow, 115409, Russia

E-mail: aquaserv@mail.ru

Abstract. Mössbauer spectroscopy is a required analytical technique in the investigations of water treatment process mechanism for purification from iron. In this paper the new-made alumosilicic reagent for purification of water was described. This reagent was used for water purification from iron dissolved complexes. The precipitated sediment is obtained and investigated. The iron valence state and relative concentrations of these states were detected in sediment. The presence of superparamagnetic particles is established.

1. Introduction

Environmental human safety depends on the quality of the used water. Currently, the present problem is the treatment of man-made and waste waters everywhere. The sources of water for household needs are: surface waters and artesian waters. Artesian waters contain a complex of minerals which are various forms of ion-molecules, colloidal states, organic substances and iron compounds. The surface waters are still more vulnerable to anthropogenic pollution. In general, water under treatment contains inorganic salts, organic substances, dissolved and hydrated gases, aquatic organisms such as plankton, bacteria and viruses.

Today, there are no technological scheme of natural and wastewater treatment which works without using coagulants and flocculants [1]. This is proved due to the fact that the majority of natural and waste water in addition to suspended allocated solids released by devices of mechanical treatment (grit chamber and others) containing fine-grained pollutions, which removal is effectively possible after consolidation with coagulants and flocculants only.

It is widely known that artesian waters of Moscow and Moscow region are contaminated with iron. Sometimes the contamination index is higher than several times of the maximum permissible concentration which is 0.3 mg per liter. This scientific and applied research work is aimed to study the properties of a new reagent in alumosilicic reagent (flocculant-coagulant) [3] containing aluminium sulfate and active silicic acid on example of removal of iron compounds. Create the physico-chemical model of the sorption artesian water from high iron content is the main purpose. Also it is needed to say that the excess iron ions are toxic for the organism, they react with the hemoglobin in the blood, cause of which decreases in blood oxygen levels also form insoluble compounds contaminating blood.

In this paper exactly there is a special interest of is the analysis of sediment water after reaction with reagent and the study of the valence state of iron of this sediment which contain last and its mechanism of sorption in the process of interaction with the reagent. Iron interact with the natural waters which contains there as mineral and organic substances present in the water dissolved, colloidal and suspended which form a complex set of compounds. The main form of iron in artesian waters is complexes of ferrous iron in divalent state with dissolved inorganic and organic compounds.

The most effective and direct method to determine the chemical state of iron and its correlation between formation of nanostructure matrix and follow adsorption iron is the method of nuclear gamma-resonance Mössbauer spectrometer which supplies information about the isomer shift, quadrupole splitting. Research precipitated reagent (pellet) carried out by Mössbauer spectroscopy in transmission mode.

Nuclear gamma resonance method is able to detect the magnetic state of the valence iron in the reactants before and after reaction with water and to determine the quantity of iron (less than 0.1%) in the test sample also. Therefore, this method is suitable for analysis of substances with a low content of iron atoms. The aim of this work is to reveal the results of the interaction of the alumosilicate coagulant-flocculant with the iron contained in the water. Mossbauer spectroscopy is used to study the process of interaction of the contaminated water with reagent.

2. Experimental technique

2.1. Method of preparation of sorption-coagulation-flocculation reagents and reaction process during cleaning

The alumosilicic reagent used in this paper, refers to a group of binary composites, acting simultaneously as a coagulant-flocculant. This reagent is produced from cheap alumosilicate raw materials is of mining waste containing nepheline, syenite minerals [3]. The interaction nepheline with dilute sulfuric acid thus formed a mixture with potassium alum $KAl(SO_4)_2 \cdot 12H_2O$ and $NaAl(SO_4)_2 \cdot 12H_2O$ and hydrated silicic acid (nanoparticles)

$(Na, K)_2SO_4Al_2 (SO_4) 3 \bullet mH_2O + 2SiO_2 \bullet nH_2O$.

This active reagent is applied as an aqueous solution with different concentrations of the main components which are nepheline and syenite. The addition of this made reagent in a weakly acidic, neutral or alkaline (pH range of the reagent of 5 or higher) causes hydrolysis processes occur, resulting in the formation of aluminum hydro having coagulation properties and hydrated silicic acid as is an inorganic flocculant. As a result of their chemical interaction is an association and formation of silico-supramolecular structures with high rate adsorption. Then self-organization of these structures is performed, possibly forming macrocomplexes such zeolite matrix (flocculation). Dynamics of synergetic effect in the alumosilicic reagent is poorly studied. The studies of reagent particle sizes carry out using the laser particle size analyzer Nanotrac Ultra (ISO 13321). In our case the particle size range is 40-200 nm.

The last resulting macrocomplexes adsorb pollutants as impurities contaminated water and consequently increase in size and in weight as result. In this process, there were an intense flocculation and a formation of flocs which are dropped to the bottom of the container. Then flocs are taken and dried and thus the obtained sediment takes the form of the little crystals which are used for investigations.

3. Mössbauer spectroscopy sample preparation.

These crystals of the sediment may have the size up to several mm. The crystal density is 2.0 g cm^{-3}. These crystals of the sediment are pulverized with using of a porcelain mortar for obtain the same-dimension crystals in form of a powder. The resulting powder filled in the molten paraffin and then mixed until a homogeneous mass. This mixture contains a known amount of paraffin and a known amount of the test material. The mixture warmed to the viscous state and placed in a cuvette with known parameters such as the area of a circle. After the cooling the obtained mixture can be used as a Mössbauer absorber. Paraffin is used for a Mössbauer absorber preparation as it is neutral substance and it does not entail the reaction with the powder. The Mössbauer absorbers have prepared, their thickness is equal to 98 ± 2 mg cm^{-2}.

The Mössbauer spectrometer (named YAGRS-6 "Persey") operating in the mode of constant velocities is used ^{57}Co(Rh) was used as a source of resonant radiation. Isomer shifts are given relating to metallic iron. . Mossbauer radiation detector is a scintillation detector with a thin NaJ (Tl) crystal (250 microns). The spectrometer is calibrated using the standard thin absorber of alpha-iron. The fitting of spectra is carried out using programme Univem MS.

The total iron content in the powder was determined by x-ray fluorescence spectrometry (RF-spectrometry). X-ray fluorescence spectrometer of system INNOV-X Series ALPHA SERIES ™ was used.

RF spectrometry is used to identify elements in the substance, an element is identified by his characteristic wavelength (λ) or energy (E) of X-rays. The concentration of the identified element is determined by measuring the intensity of the line of its characteristics. RF-spectrometry ultimately determine the elemental composition of the material.

The selected powder pellet with weight of 100 mg has been used for analysis. The analysis had showed that the powder contains elements such as Fe, Zn, Zr, Cu and other heavy elements. Heavy elements are in very small quantities. The iron content in the powder is approximately 2% of a total weight. The content of the Mössbauer isotope ^{57}Fe will be 0.02x0.0214 = 0.000428, where 0.0214 is the percentage of ^{57}Fe in natural iron. The ^{57}Fe iron density in the absorber is 0.000428x98 (mg cm^{-2}) = 0.04 mg cm^{-2}. This low content of the ^{57}Fe requires much time to obtain a Mössbauer spectrum.

4. Results and discussion.

The results of Mössbauer investigations are shown in Figures 1, 2 and tables 1, 2. Figure 1 shows the Mössbauer spectrum of ^{57}Fe in the sediment produced from reaction of the alumosilicic with water. The spectrum is recoded at 293 K. The shape of the spectrum is the superposition of lines of hyperfine magnetic splitting of the magnetic phases and lines of quadrupole splitting of the paramagnetic phases. The Mössbauer spectra parameters obtained in paper [4] were used for fitting. The Mössbauer spectra parameters such as values of the hyperfine magnetic field Heff, quadrupole splitting QS, isomers shift IS and areas of spectra lines are used for phases and compounds identifications. One could see that the lines of magnetic phases have low intensities.

The Mössbauer spectrum (Fig.1) is fitted using the model of two paramagnetic phases. The obtained Mössbauer spectrum parameters are presented in Table 1. The main share of the iron (~86%) is in the trivalent paramagnetic state (Is=0.34±0.02 mm s^{-1}, Qs=0.78±0.02 mm s^{-1}) and another part (~14%) is in the paramagnetic phase in divalent state (Is=1.14±0.02 mm s^{-1}, Qs=2.38±0.02 mm s^{-1}).

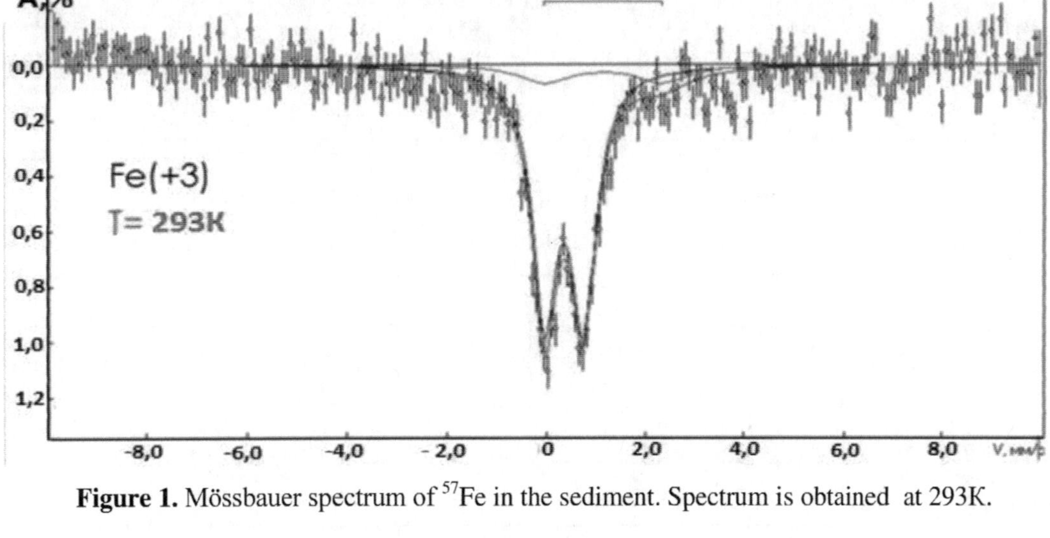

Figure 1. Mössbauer spectrum of ^{57}Fe in the sediment. Spectrum is obtained at 293K.

It is supposed that in initial water the iron is in divalent state than one can concludes that during reaction with reagent (alumosilicate coagulant-flocculant) the main share of the iron transforms in to the trivalent state. The rest iron reacts with reagent too and it is found in sediment in the divalent state. For identification of the obtained phases the spectrum is registries at liquid nitrogen temperature.

Table 1. Mössbauer spectrum parameters at 293 K
of the sediment produced from alumosilicic reagent.

Phase	IS, mm/s	QS, mm/s	S,rel.,%
Fe^{3+} doublet 1	0.35±0.02	0.78±0.02	88±1
Fe^{2+} doublet 2	1.14±0.02	2.38±0.02	12±1

Figure 2 shows the Mössbauer spectrum of ^{57}Fe in the sediment. The spectrum is recoded at temperature of liquid nitrogen. This Mössbauer parameters are presented in Table 2. Mössbauer spectrum obtained at 77 K shows the superposition of lines of hyperfine magnetic splitting of the magnetic phases and lines of quadrupole splitting of the paramagnetic phases. The intensities of magnetic phase lines are increased. This indicates the existence of superparamagnetic particles in the sediment. It follows from knowledge that with decreasing of temperature the part of superparamagnetic particles shows hyperfine magnetic splitting.

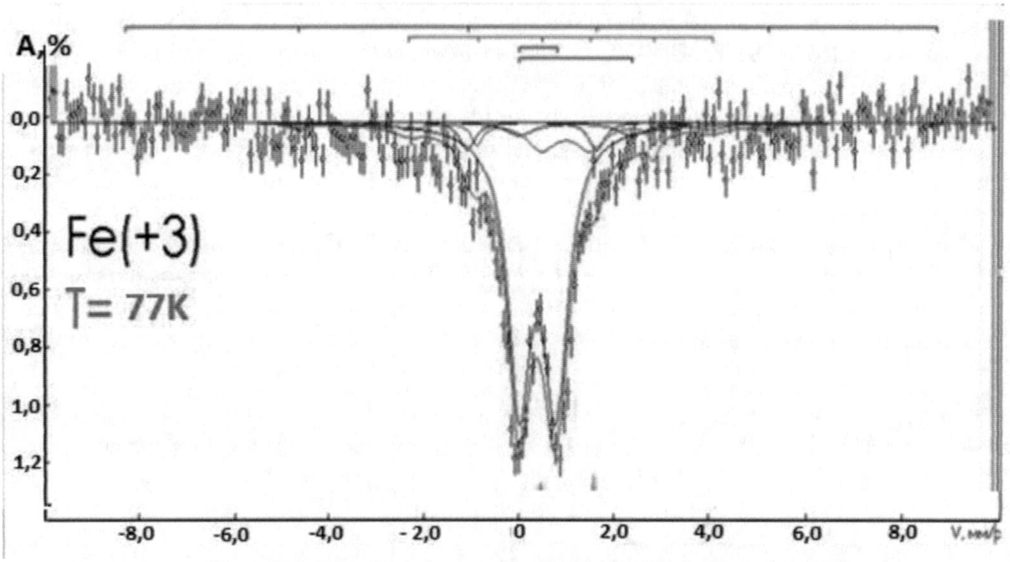

Figure 2. Mössbauer spectra of 57Fe in the sediment at 77K.

Table 2. Mössbauer parameters if the sediment produced from alumosilicic reagent at 77 K.

Phase	IS, mm/s	QS, mm/s	H_{eff}, T	S,rel.,%
Fe^{3+} doublet 1	0.45±0.05	0.89±0.05	----	77 ±1
Fe^{2+} doublet 2	1.21±0.05	2.43±0.05	----	4±1
Fe^{3+} sextet 1	0.34±0.05	0.02±0.05	51.8±0.3	14±2
Fe^{2+} sextet 2	0.93±0.05	0.02±0.05	19.2±0.6	5±2

The Mössbauer spectrum parameters of the paramagnetic phase with trivalent iron state (IS=0.34±0.02 mm s^{-1}, QS=0.78±0.02 mm s^{-1} S=88±2) at room temperature and magnetic phase with trivalent iron state (IS=0.34±0.02 mm s^{-1}, QS=0.02±0.05 mm s^{-1} , H_{eff}=51.8 T, S=14±2) at liquid nitrogen temperature present the same iron composition in superparamagnetic state. The iron composition as α-Fe_2O_3 at the same temperatures has approximately the same Mössbauer [4] spectra parameters.

5. Conclusion

1. The alumosilicic reagent is prepared and its reactions with water is carried out, and products of the reaction such as sediment are investigated.
2. The product contains elements such as Fe, Zn, Zr, Cu and other heavy elements in small concentrations.
3. It has been shown that main part of iron (88%) is in trivalent state, another part (~12%) is in divalent state.
4 The existence of superparamagnetic iron containing particle in the sediment are established.

Acknowledgements

The work was financially supported by the Ministry of education, agreement № 14.575.21.0086 dated 20 October 2014. Unique project identifier RFMEFI57514X0086

References

[1] Laguntsov N I, Neschimenko Yu P, Feklistov D Y 2008 *Proc. of Nanotech. Int. Forum "Rusnanotech 08"* 609-11.
[2] Kim V E, Laguntsov N I, Karpukhin V F, Lisujk B S 1998 *The method of water purification* Patent RU 2114787. B.I. # 19.
[3] Kurchatov I M, Laguntsov N I, Neschimenko Yu P, Feklistov D Y 2015 *Phys. Proc.* **72** 89-92.
[4] Fultz B 2011 Characterization of Materials *"Mössbauer Spectrometry"* ed E Kaufmann (John Wiley, New York).

Journal of Physics: Conference Series **751** (2016) 012047 doi:10.1088/1742-6596/751/1/012047

Use of alumosilicic reagent for water purification

S N Tikhonov[1], I M Kurchatov[1], V A Byrkin[1], D Y Feklistov[2] and N I Laguntsov[1]

[1] National Research Nuclear University MEPhI (Moscow Engineering Physics Institute), Kashirskoe highway 31, Moscow, 115409, Russia
[2] JSC Aquaservice,
Kashirskoe sh. 31, Moscow, 115409 Russian

E-mail: tisenik@mail.ru

Abstract. Workability of the hybrid reagent based on aluminium salts and the use of active silicic acid for the purposes of water treatment was investigated in this paper. The research of the residual aluminium concentration in the water was conducted after the introduction of the reagent into the model solution. The optimum concentration ASFC and the pH value was determined at which the coagulation process is intensified. The approaches of the interaction of the dispersed particles, specified method for calculating the interaction potential of the dispersed particles in the circumstance were described.

1.Intoduction

Currently, complex reagents are widely applied in water purification systems and perform the function of coagulant, flocculant, precipitant and adsorbent. This allows using the benefits and advantages of the individual components. In this work a reactant hybrid alumosilicic reagent based on materials containing aluminium sulphate and an active silicic acid was used. The main task of the reagent is a dispersion system destabilization [1].

During the hydrolysis the aluminium compound coagulates with particles formation, sedimented with pollutants. The coagulating effect of active silicic acid is based on the mutual coagulation of oppositely charged particles, wherein the silicic acid molecules are the centres of the coagulation system. Under certain conditions, these centres form a cell as a result of their continued engagement. Next, the structure becomes larger due to the coalescence of these links. As a result, the self-organization process produces complex spatial three-dimensional structure in which the flocculation ability is increased cause of the ability to absorb pollutants in the cavity between the units. High efficiency of alumosilicic reagents is conditioned, on the one hand, as an additive reagent coagulant properties of components and on the other hand, the formation of mesoporous alumosilicic structures, leading to flocculation.

2. Chemical processes

ASFC are prepared from raw materials containing nepheline by treating it with sulphuric acid, what is not a great technical problem in manufacturing. Such rocks have a greater concentration of the necessary components SiO_2 and Al_2O_3.

Table 1. The chemical composition of nepheline raw materials

Element	SiO_2	Al_2O_3	$Na_2O + K_2O$	Fe_2O_3
Concentration, %	44	28	19	2.3

The main stage of the destabilization of the colloidal system is the hydrolysis of aluminium salts and precipitation $Al(OH)_3$. Chemical reaction scheme:

$Al_2(SO_4)_3 + 2H_2O \leftrightarrow 2AlOHSO_4 + H_2SO_4$ $Al^{3+} + H_2O \leftrightarrow AlOH^{2+} + H^+$ + $2AlOHSO_4 + 2H_2O \leftrightarrow [Al(OH)_2]_2SO_4 + H_2SO_4$ $AlOH^{2+} + H_2O \leftrightarrow [Al(OH)_2]^+ + H^+$ + $[Al(OH)_2]_2SO_4 + 2H_2O \leftrightarrow$

$2Al(OH)_3\downarrow + H_2SO_4$ $[Al(OH)_2]^+ + H_2O$

Formed protons H^+ connects with HCO_3^{2-} of natural water. This reaction can be represented by the stoichiometric equation:

$$Al_2(SO_4)_3 + 3Ca(HCO_3)_2 \leftrightarrow 2Al(OH)_3\downarrow + 3CaSO_4 + 6CO_2\uparrow$$

Under addition of alumosilicic flocculant with pH = 6 - 7 takes place an allocation of aluminium ions, which are included in the exchange reaction with the cation impurity diffusion layer hydrophobic particles. This diffusion layer is compressed reducing the stability of the system, resulted in coagulation of the medium particles. Silicic acid is a flocculant, forming a bridge structure "particle – micro ion - multiply charged counter ion – micro ion - particle", passed through the coagulation stage.

The presence of free aluminium is an important indicator of water quality treatment, as aluminium is toxic and carcinogenic, is not oxidized and is not excreted from the body. In accordance with WHO recommendations, the EU and the standards of developed countries [2] content of residual aluminium in water should not exceed 0.2 mg / l. residual aluminium problem exists for any aluminium-containing reagents (aluminium sulphate, ASFC oxychloride aluminium $Al_2(OH)_5Cl{\cdot}6H_2O$ etc.)

Table 2. Doses ASFC

ASFC dose (per litre of treated water), ml/l	**pH** model solution	**Residual aluminium, mg/l**
0	7,3	<0,04
0,5 ml/l	6,4	<0,04
1 ml/l	6,2	<0,04
2 ml/l	5,5	0,13 ± 0,05
3 ml/l	4,5	0,31 ± 0,05
4 ml/l	4,1	> 0,4

The results show that the aluminium content of the water is primarily dependent on the pH of the aqueous circumstance, which in turn is related to the concentration of hydrogen ions and the acidity of the initial water solution introduced ASFC.

The optimum pH value for coagulation is between 6.0 - 8.0. The coagulation process finishes in the formation of poorly soluble hydrocomplexes of aluminium. This process is accompanied by the formation of virtually no residual aluminium as goes through hydrolysis, i.e. all aluminium ions are involved in the formation of hydroxyl complexes. At lower pH values (acidic medium) is decelerating aluminium hydrolysis process. The reduced alkalinity causes hydrolysis is stopped at an intermediate stage in the appearance of the clarified water unhydrolyzed free aluminium ions with a charge of +3.

With a decrease in the pH of the contribution of this process increases, respectively, increasing the quantity of residual aluminium.

It was found that with increasing doses ASFC from 0.5 to 2 ml / l speed purification processes increases. Velocity decreases with further increase in the concentration, and at a dose of 4 ml / l brightening process is slowed dramatically. This slowdown is due to a decrease in pH and as as consequence, finishing model solution hydrolysis process.

3. Simulation method

ASFC is a colloidal particle system. The interaction of particles responds to van der Waals forces of attraction and repulsion of Coulomb forces. Coagulation centres are electrolyte ions having a negative charge. Counter ion concentration causing coagulation system is called threshold coagulation. This value is calculated according to the rule Schultz-Hardy [3], based on the structure and properties of the boundaries of the electric double layer (EDL). Driving EDL structure shown in Fig. 1.

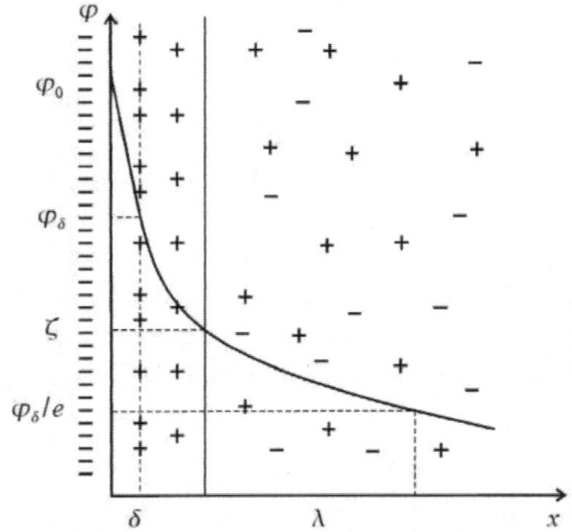

Figure. 1. The scheme of distribution of charges and potentials
δ- Helmholtz layer (adsorption layer)
λ- layer by Gouy (diffusion layer)
The adsorption layer counter ions are held in the charged surface by electrostatic and, to some extent, the adsorption forces. In diffusion layer counter ions are held only cause of opposite charges.

In order to simulate the coagulation process, it is necessary to present the interaction of particles as the Coulomb force of the same or different charges. Thus the charges of the particles vary depending on the dispersed state of the system as a whole. Since pH is easily measurable parameter of the system, it will select for implicit determination of the potential on the surface of the electrical double layer $\zeta = f\ (pH))$. Figure 2 shows the results of the study the average zeta potential agglomerates ASFC at different pH of the medium. Measurements of ζ –potential are carried out for hybrid reagent layer 200 microns thick. The NaOH solution was added a solution of 10 l / 10 sec, to pH = 10 with external conditions: T= 20-25^0 C, p = 1 atm.

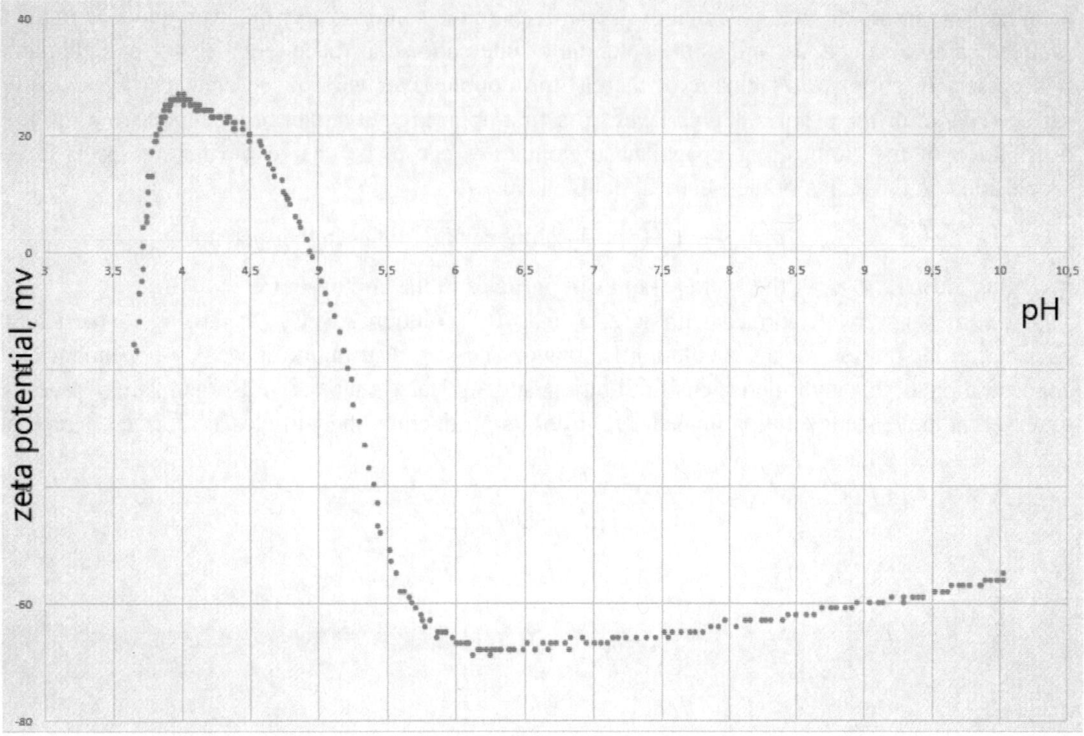

Figure 2. Dependence of ζ potential of 1-% ASFC solution from pH.

In this case, besides the fact that the zeta potential ζ determines the surface charge of the particles, it is also a criterion for the stability of the dispersed system, determines the nature and extent of interaction between the particles. The colloidal theory emits the characteristic value of the zeta potential of 30 mV. [4] Above this limit the system has a stable state, and below coagulation and flocculation is possible.

The interaction of particles is determined by the potential U (r), which will consist of the van der Waals attraction potential (W^-) [5] and the Coulomb potential repulsion or attraction (W^+) [6].

$$W^+ = k\frac{q_i' \cdot q_j}{r_{ij}}, i \neq j \tag{1}$$

q_i' - micelle charge ($q_i' = \frac{\zeta \cdot N}{\sum_i^N r_i}$, N - the number of particles in the system)

q_i- charge of the system of ions surrounding the particles of ASFC

r_{ij} – the distance between the particles with indexes i and j

$$W^-(z, R_i, R_j) = -\frac{A}{6}\left[\frac{2R_iR_j}{z^2-(R_i+R_j)^2} + \frac{2R_iR_j}{z^2-(R_i-R_j)^2} + Ln\left(\frac{z^2-(R_i+R_j)^2}{z^2-(R_i-R_j)^2}\right)\right], \tag{2}$$

where A- Hamaker constant, R_i, R_j- the radius of the spherical bodies, $z = R_i + R_j + r$, r- the distance between the centres of the spherical bodies.

Potential interactions in which the Coulomb force is repulsion forces are described for like-charged particles. For heterogeneous systems with particles of opposite charges of the electrostatic forces would be attractive forces at all distances, caused by irreversible coagulation [7]. For pairs of particles of like charges is a potential barrier, depending on the capacity of less than the charged surface.

When measuring the zeta potential of the particles in the solution, only the average value in the test sample can be obtained, as well as the potential of interactions has been recorded for particle pairs, it is necessary to correlate the charge of the electric double layer with the reactant particle sizes. The surface charge of the electric double layer, to a first approximation, can be obtained from the initial distribution of the alumosilicic coagulant agglomerates cause the charge on the surface is directly proportional to the radius of the electrical double layer:

$$q_i' = \frac{\zeta \cdot N}{\sum_i^N r_i} \cdot \left(\frac{r_i}{<r>}\right)^3, \text{ } N\text{- number of particles in the system,} \quad (3)$$

r_i- radius of EDL, $< r >$ - the average radius of particles in the circumstance

Experiments were performed with 1, 2, 5 and 10% solution ASFC. On the Fig. 3 particle size distribution in solution for a 1% solution is shown. The size distribution is weakly dependent on the concentration of the hybrid reagent in solution and indicates that the self-organization process is observed in the reagent solution immediately after its production, the particles are formed, which may reach sizes of 180 nm.

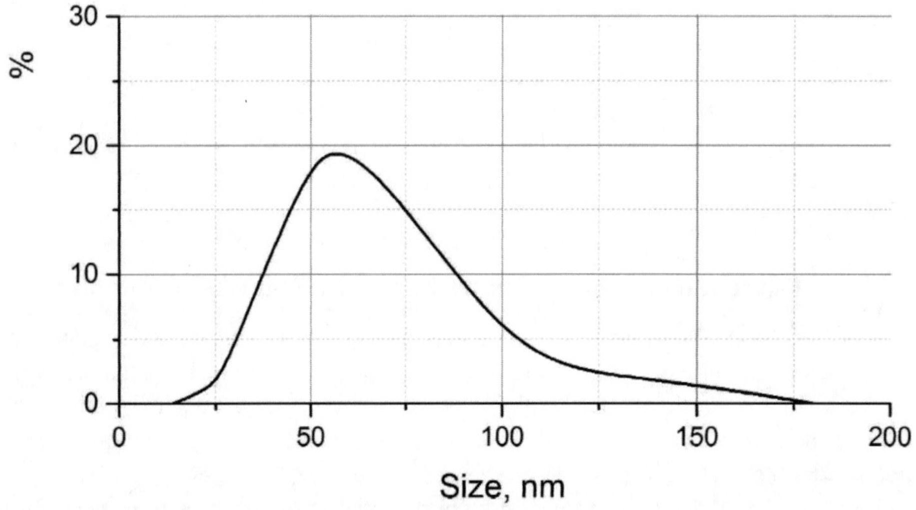

Figure 3. Particle size distribution of the reagent solution.

3. Conclusion

During the investigations it was established that pH parameter is implicitly affects the concentration of free aluminium in the sample. The results have the following qualitative explanation: While introducing ASFC into the water, hydrolysis of aluminium cation takes place, the hydrolysis products are polynuclear hydroxocomplexes, having coagulating properties and also, these products are the centers of the formation of primary flocs, from which floccules grow subsequently.

With the view of identifying optimal ASFC concentration, it is necessary to take into account an economic viability, and also the fact that the increase of aluminium ions concentration leads to inversion of the particles charge on the surface of the electric double layer as a result of adsorption of hydrated aluminum complexes. The result of this process is stabilizing of dispersed system and next shutdown of coagulation and flocculation [8].

Acknowledgements

The work was financially supported by the Ministry of Education and Science, agreement № 14.575.21.0086. Unique project identifier RFMEFI157514X0086.

References

[1] Hubbard A T 2004 *Encyclopedia of Surface and Colloid Science* (CRC Press) p 4230.
[2] World Health Organization *2004 Guidelines for water quality*.(Geneva: SNP Best-set Typesetter Ltd. Hong Kong).
[3] Overbeek JT G 1980 *Pure Appl. Chem.* **52** 1151—61
[4] Gorshkov V S, Saveliev V G, Fedorov N F 1988 *Physical chemistry of silicates and other refractory compounds* (Moscow: Vysshaya shkola).
[5] Barash S 1988 *Van der Waals force* (Moscow: Nauka) p 344.
[6] Landau L D, Lifshitz E M 1975 *Course of Theoretical Physics Volume 2 The Classical Theory of Fields* (Butterworth-Heinemann) p 402.
[7] Deriagin B V 1979 *Rus. Chem.* **48** 675-721.
[8] O'Melia Ch R, Stumm W 1967 *J. Coll. Int. Sci.* **23** 437-47.

Effect of agglomeration of carbon nanotubes on gas permeability of PVTMS/CNT mixed matrix membranes

A M Grekhov[1,2], Yu S Eremin[1], D Bakhtin[2] and V V Volkov[1,2]

[1]National Research Nuclear University MEPhI (Moscow Engineering Physics Institute), Kashirskoe highway 31, Moscow, 115409, Russia
[2]A.V.Topchiev Institute of Petrochemical Synthesis, Russian Academy of Sciences, Russia

E-mail: yseremin@ya.ru

Abstract. Mixed matrix membranes (MMMs) with unique transport characteristics can be prepared by the addition of the minor amounts of carbon nanotubes. Qualitative (critical, effective, marked) changes in the membrane performance are shown to be provided by the formation of a percolation cluster composed of nanotubes. For MMMs based on poly(trimethylvinylsilane) (PVTMS) containing carbon nanotubes (CNT), due to the formation of the CNT percolation cluster, gas permeability increases by a factor of 5-15. When the CNT content in the MMMs is higher than the percolation threshold, gas permeability remains on the same level or even decreases. Numerical simulation proves that the above negative changes are provided by the agglomeration of nanotubes and subsequent deterioration of the percolation structure in the membranes.

1. Introduction

Polymer-based membranes are widely used in various industrial-scale membrane technologies to separate and purify gases and liquids [1, 2]. Despite the fact that non-organic membranes based on either zeolites or carbon materials have higher permeability and selective values, polymer-based membranes have better mechanical, strength and technological characteristics. Besides, recently, new polymer materials have been synthesized, of which transport characteristics are close to non-organic materials [4]. However, the synthesis of new polymers is quite a long process, therefore alternative methods for improving the transport properties of polymer membranes still remain urgent, for instance the development of "mixed matrix membranes" (MMM) that are polymers containing non-organic nanoparticles [22-28].

Particles embedded into MMM can be split into several types: non-porous particles (SiO_2, TiO_2, etc.), nanoporopus particles (zeolites, porous earth silica, MOF) and nanoparticles with a high aspect number (carbon fiber and nanotubes). Carbon nanotubes (CNTs) have some specific features as compared to other particles that makes CNT-based membranes the most promising MMM samples. Firstly, high strength characteristics of CNTs increase greatly the membranes strength [29]. Secondly, a significant change in the CNT-based MMM permeability is observed at a concentration of nanotubes of less than 2% that is several times less than the target concentration of other particles. However, it appeared that at high CNT concentration, no further improvement on the MMM transport properties is observed but some worsening in the membranes parameters. For instance, the permeability of gases (He, H_2, CO_2, O_2, N_2, CH_4) through polydymethylsiloxane-based membranes increases as 2% of CNTs is embedded and changes insignificantly as the CNT concentration increases up to 10%. [2]. The same phenomenon is observed as the CNT concentration increases from 2 to 10%wt in PES-based membranes; the permeability of N_2 and CO_2 increases insignificantly, while the selectivity decreases [9]. The pervaporation permeability of the mixture water/ethanol (10/90) through hydrophilic polymer PVA increases monotonously as 1-5%wt of CNTs is embedded [10], however, the membrane selectivity decreases significantly when the CNT concentration is more than 3%. The minimal permeability and the maximal selectivity are observed at a CNT concentration of 2%wt in PVA during the water/isopropanol (20/80) mixture pervaporation as the CNT concentration increases from 2 to 4%wt [30]. The maximal change in the permeability was observed for some polymers as the CNT concentration raised. Such changes were observed for MMM with CNTs based on poly 2,6-dimethyl-1,4phenylene oxide [31.], polysulphone [32], polyimidesiloxane [2]. This non-linear change in the permeability with the CNT concentration increase can not be explained by the traditional models, where the change of membrane properties is in proportion to the concentration of embedded nanoparticles. Some non-linear effects, for instance the permeability threshold change, can be explained by the percolation theory [2], polysulphone [32].

At a low CNT concentration, isolated areas of modified polymer of high permeability appear in the membrane and molecules successively go through the areas of high and low permeability. That is why the change in flow through the membrane is in proportion to the CNT concentration and, at low CNT concentration, it increases insignificantly. At a certain concentration, the embedded nanotubes generate a cluster that bounds two membrane surfaces. There appear an additional "parallel" flow through this cluster and the membrane permeability increases significantly that determines the threshold change in the permeability of MMM with embedded CNTs. However, to explain the non-linear changes in the permeability with further CNT concentration, one need to analyze the conditions of the percolation cluster formation and its structure. To explain the ultimate changes in the permeability, some papers give suppositions about the relation of the membrane permeability drop to the nanotubes agglomeration as the CNT concentration rises. The interaction between molecules and agglomerates causes an increase in the diffusion path tortuosity and decrease in the membrane permeability. However, according to the calculations, the nanotubes agglomeration also causes changes in the percolation cluster parameters that intensifies worsening in the membrane transport characteristics.

This paper presents the research in the permeability of membranes based on glassy polymer polyvinyltrimethylsilane PVTMS with embedded MWCNT and shows a percolation mechanism of permeability change at a CNT concentration of 0.45wt. However, at a concentration of more than 1%wt, the permeability remained the same; this phenomenon was explained by the numerical modeling of the transport characteristics of the CNT-based MMM with the account of nanotubes agglomeration and changes in the percolation cluster structure.

2. Experimental

2.1 Materials

In this work, we used the multiwalled carbon nanotubes (CNT) (TAUNIT™, NanoTekhCenter LLC, Russian Federation) prepared by chemical precipitation from a gaseous phase. According to technical specifications, TAUNIT™ carbon nanomaterial is composed of fibrils based on polycrystalline graphite with a length below 2 microns; their outer diameter varied from 20 to 70 nm, specific surface was 120-130 m^2/g, and the bulk density was 0.4-0.6 g/cm^3. After the treatment with nitric acid according to the standard procedure [12], density of the CNT samples was 2.26±0.03 g/cm^3, specific volume was $V = 0.20 \pm 0.02$ cm^3/g, and the surface area was $S = 150 \pm 5$ m^2/g. The inner diameter of the nanotubes was calculated by the BJH method from the low-temperature nitrogen sorption data, and this value was 3-6 nm. For the model of cylindrical channels, the outer CNT diameter was 20-60 nm.

We used the commercial PVTMS sample with a density of 0.86 g/cm^3: the permeability coefficients were $P(H_2) = 220$, $P(O_2) = 44$, $P(CH_4) = 13$ barrer [13].

2.2 Membranes
Mixed matrix membranes were prepared mixing PVTMS and CNT solutions in chloroform. Carbon nanotubes were dispersed in chloroform by the ultrasound treatment for 15 min. The dimensions of agglomerates and individual CNT units were estimated by the dynamic laser scattering measurements [14]. The dimensions of agglomerates in the solution fit the dimensions of individual nanotubes with a diameter varying from 20 to 70 nm and with a length from 1 to 2 μm. The chloroform solution of PVTMS (3 wt %) was prepared under normal conditions then and filtered. Concentrations of PVTMS and CNTs in the chloroform solutions were controlled by their optical density. The membranes were prepared by casting the PVTMS and CNT solutions on a cellophane support. Chloroform was evaporated under normal conditions (atmospheric pressure, T = 20-24°C) until the constant weight was attained. Thickness of the prepared membranes was 25±2 μm. We prepared the PVTMS/CNT membranes containing 0.2, 0.4, 0.8, 1.2, 1.5, 2 and 3 wt % of CNTs; the error was below 0.1%. The content of CNTs in PVTMS was estimated by the helium pycnometry method.

2.3 Methods and Equipment
The density of CNTs, PVTMS, and the related mixed matrix membranes was measured using a Micro-Ultrapyc 1200e Quantachrome density helium pycnometer; their specific surface and volume of pores were estimated by the low-temperature nitrogen sorption (Nova 1200e, Quantachrome). The inner diameter of the CNT channels d_{in} was calculated according to the BJH method [15]; the specific surface was estimated according to the BET method [16].
Concentrations of PVTMS and CNT solutions in chloroform was controlled by their optical density using a Hach 500 DR Lange spectrophotometer; the accuracy was 0.01 abs. The integral absorption coefficient of PVTMS was measured in the 300-500 nm wavelength interval, thus making possible to control the PVTMS concentration with an accuracy of 0.2 wt %. The absorption coefficient of CNT was measured at a wavelength of 600 nm; the accuracy was not less than 0.05 wt %. The dimensions of CNTs in chloroform were studied by the dynamic laser scattering method (DLS) at a wavelength of 780 nm (Nanotrac 252, Microtrac Inc.).
Permeability of ethanol through the PVTMS/CNT membranes was measured by the dynamic pressure decay method [17], the initial excessive pressure gradient was 200 atm, and the temperature was 30°C. The membrane permeability was calculated from the pressure decay rate over a membrane:

$$\mathrm{P}_l = \frac{\chi \cdot V \cdot \Delta \mathrm{p} \cdot \rho}{\Delta \mathrm{t}} \cdot \frac{d}{\mathrm{S} \cdot \mathrm{p}} \qquad (1)$$

where p is the current pressure in the liquid [atm], Δp is the pressure decay within time Δt [s], V is the volume of the liquid over the membrane [m^3], χ is the fluid compressibility [atm^{-1}], ρ is the liquid density [kg/m^3], d and S stand for the membrane thickness [m] and membrane area [m^2], respectively.

Gas permeability of nitrogen, oxygen, methane, and propane through the PVTMS/CNT membranes were measured by the Дайнеса-Баррера Barrer method at a pressure gradient of 4 atm; the temperature was 19 – 24°C. The permeability coefficients were calculated as:

$$P_{gas} = \frac{V \cdot d}{S \cdot \Delta p \cdot \Delta t} \qquad (2)$$

where Δt is the time period within which the pressure changes by Δp [atm], V is the reference volume [m^3], d is the membrane thickness [m], S is the membrane area [m^2].

3. Results and Discussion

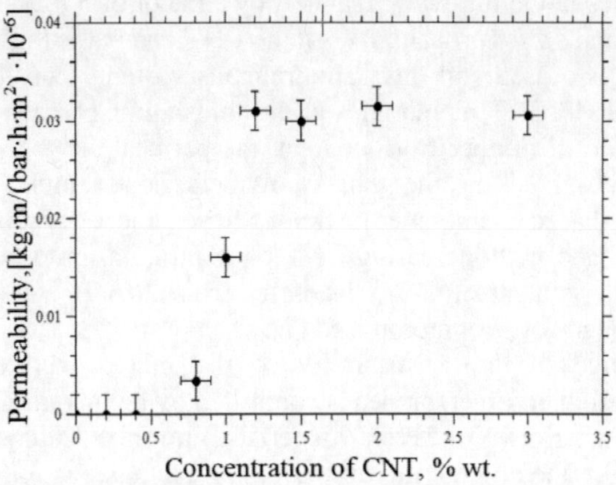

Figure 1. Permeability coefficient of ethanol for the PVTMS/CNT membranes.

Figure 1 presents the permeability coefficient of ethanol plotted against the CNT content in the PVTMS/CNT membranes and the plot shows three well-pronounced regions. When the CNT content in the PVTMS/CNT membranes is below 0.4 wt %, ethanol does not penetrate the mixed matrix membranes. The fact that the experimentally measured permeability of ethanol through the PVTMS-based membranes containing less than 0.4 wt % of CNT is absent means that the membrane does not have any regions which are permeable to ethanol. As the concentration of CNTs in the membrane is increased from 0.8 to 1.5 wt %, the permeability markedly increases by nearly a factor of 10. Hence, one can conclude that this increase in the ethanol permeability of the membrane takes place once a certain percolation CNT concentration is achieved and this concentration lies between 0.4 and 0.8 wt %. In this case, interconnected PVTMS/CNT regions are formed and these regions are permeable for ethanol. The number of these channels nonlinearly increases with increasing CNT concentration from 0.8 to 1.5 wt %. When the CNT content in the membrane is higher than 1.5 wt %, the growth in the membrane permeability is ceased. Noteworthy is that the gas permeability of the PVTMS/CNT membranes follows the same scenario.

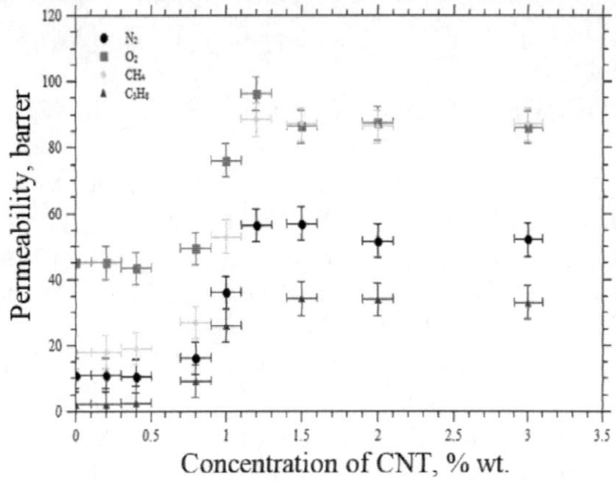

Figure 2. Gas permeability through the PVTMS/CNT membranes

Figure 2 presents the gas permeability of the PVTMS/CNT membranes for nitrogen, oxygen, methane, and propane plotted against the CNT concentration in the membrane, and this plot also shows three well-pronounced concentration intervals. When the CNT content in the membrane is lower than 0.4 wt %, gas permeability of the matrix mixed membranes for all gases under study is equal to the permeability of the neat PVTMS. When the CNT content is changed from 0.8 to 1.5 wt %, gas permeability increases in the following manner: by a factor of 5 for nitrogen, by a factor of two for oxygen, 4 for methane, and 15 for propane. When the CNT content is higher than 1.5 wt %, the growth in the gas permeability is ceased and this value remains virtually constant. This increase in the gas permeability of the PVTMS/CNT membranes at the threshold CNT concentration has been reported earlier but the fact that, at the higher CNT content, the permeability becomes invariable is new [11]. This behavior was also observed for other glassy polymers. For example, gas permeability through the poly(dimethyl siloxane)-based membranes increases when the CNT content is equal to 2% but it remains virtually unchanged with increasing CNT concentration up to 10% [6]. For the polysulfone-based membranes, as the content of CNTs increases from 2 to 10 wt %, N_2 and CO_2 permeability slightly increases but selectivity is compromised [7].

This nonlinear change in the permeability of the mixed matrix membranes suggests that permeability of the mixed matrix membranes is controlled by the parameters of the formed percolation cluster. Traditionally, when two different materials are mixed, any additive changes in their characteristics can be described by the mixture rule. For example, as follows from Fig. 3, the density of the PVTMS/CNT membranes changes linearly and can be described by the following equation $\rho_{MMM} = c_{CNT} \cdot \rho_{CNT} + (1 - c_{CNT1}) \cdot \rho_{PVTMS}$. Here, ρ_{MMM} is the density of the mixed matrix membranes, ρ_{CNT} is the density of CNTs (2.267 g/cm^3), c_{CNT} is the weight content of CNTs in the PVTMS/CNT membranes, ρ_{PVTMS} is the density of PVTMS (0.9135 g/cm^3).

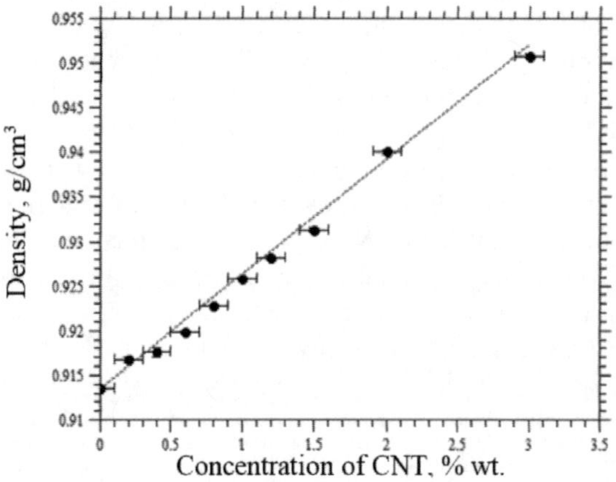

Figure 3. Density of the hybrid PVTMS/CNT membranes

By analogy, if permeability of the mixed matrix membranes depends on the content of nanotubes, this parameter can be described by the following relationship: $P_{MMM} = P_{CNT} \cdot c_{CNT} + P_{PVTMS} \cdot (1 - c_{CNT})$, where P_{CNT} is the permeability of the membrane regions containing CNTs and P_{PVTMS} is the permeability of the CNT-free PVTMS. In this case, permeability of the ideal mixed matrix membranes should change linearly with increasing CNT concentration. However, this approach fails to describe the experimentally observed changes in the permeability of the mixed matrix PVTMS/CNT membranes. If changes in the net membrane permeability are provided by the development of highly permeable open channels in the membrane, permeability starts to increase only when CNTs are organized into the percolation cluster; in this case, permeability should be proportional to the volume of the percolation cluster:

$$P_{MMM} = P_\infty \cdot x_\infty + P_{PVTMS} \cdot (1 - x_\infty) \quad (3)$$

where x_∞ is the volume fraction of the membrane occupied by the percolation cluster, P_∞ is the permeability coefficient of the percolation cluster, P_{PVTMS} is the permeability coefficient of PVTMS. Changes in the membrane permeability and x_∞ are nonlinear when concentration of nanotubes in the membrane is above the percolation threshold. At low CNT concentrations, the percolation cluster does exist or, in other words, x_∞ is equal to zero and the presence of nanotubes has no effect on the membrane permeability.

The effect of the percolation cluster on the permeability of the mixed matrix memranes was studied according to the Monte Carlo scheme which allows one to assess the conditions providing the formation of the percolation structures composed of nanotubes in the 3D space and their characteristics. We used the standard algorithms such as the Mersenne twister for the estimation of coordinates and orientation of CNTs [18], the Hashen-Koppelmann algorithm for the calculation of the CNT distribution over the clusters [19], and the method of the percolation cluster detection [20, 21].

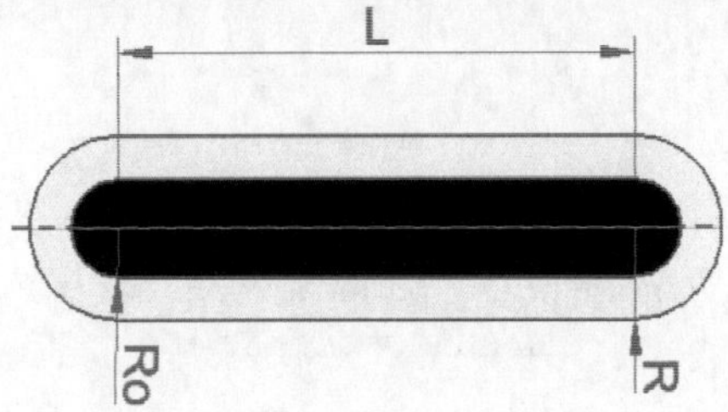

Figure 4. The model of a carbon nanotube as a capsule with an impermeable core and a permeable shell.

In the simulation experiments, nanotubes (which are modeled as capsules with a diameter varying from 0.05 to 0.08 units and with a length of 2 units) are randomly distributed on a 25 × 25 × 25 matrix. Both diameter (0.05) and length (2 units) are proportional to the dimensions of CNTs used in our experiments (the CNT diameter is 50 nm and the CNT length is 2 μm); the bigger capsules correspond to the agglomerates of nanotubes. Volume concentration of nanotubes or, in other words, the volume occupied by the capsules reads as: $x = (\frac{4 \cdot \pi}{3} \cdot R_0^3 + \pi \cdot R_0^2 \cdot L) \cdot N/V$, where R_0 is the radius of the impermeable core, L is the length of the impermeable core, N is the number of capsules, and V is the matrix volume. Each capsule is surrounded by a shell with thickness h (Fig. 4) which models the polymer regions whose permeability increases due to the interaction with the CNT surface. When the shells are overlapped, the capsules are organized into a cluster. When the cluster dimensions exceed the matrix volume, *a percolation cluster* is formed. For each fixed concentration of capsules, 1000 simulations were conducted, and the average volume occupied by highly permeable polymer regions is calculated as:

$$x_\infty = \left(\frac{4 \cdot \pi}{3} \cdot (R^3 - R_0^3) + \pi \cdot (R^2 - R_0^2) \cdot L\right) \cdot \frac{N_\infty}{V} \quad (4)$$

Here, N_∞ is the number of capsules in the percolation cluster, x_∞ is proportional to the volume of highly permeable polymer regions which control the membrane permeability. Figure 5 shows x_∞ plotted against the volume concentration of the capsules with different diameters.

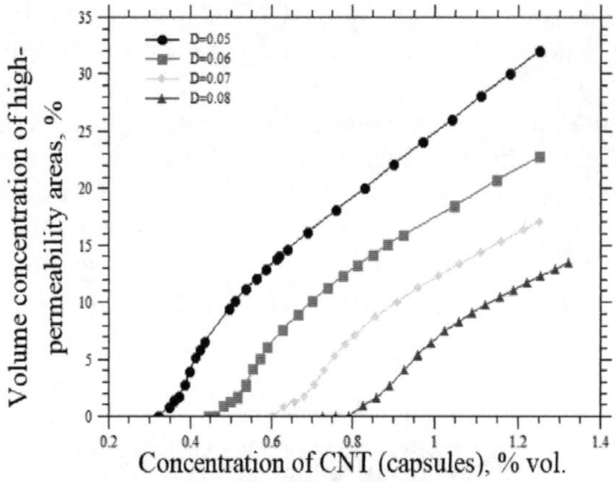

Figure 5. Volume concentration of the highly permeable regions in the matrix containing different amount of capsules

As follows from Fig. 5, for the capsules with the diameter of 0.05 which corresponds to the individual nanotubes, a percolation cluster is formed when the volume concentration of capsules increases from 0.3 to 0.35%; however, in this case, the content of the highly permeable regions is nearly zero. As the volume concentration of capsules increases from 0.35 to 0.5%, the concentration of highly permeable regions increases and x_∞ becomes as high as 15% (with respect to the volume of the matrix). When the concentration of capsules is further increased, the concentration of the highly permeable regions linearly increases. Similar changes in the dimensions of the percolation regions are also observed for the bigger capsules but, as the capsules become bigger, the concentration of capsules at which the percolation cluster is formed increases.

Figure 6 shows the permeability coefficients of the PVTMS/CNT membranes calculated after the substitution of x_∞ to Eq. (4). In these calculations, we assume that there is no gas transfer through inner CNT channels, and the increase in gas permeability is solely provided by gas transport through the highly permeable layer of the modified polymer near the CNT surface. This assumption is valid only for the permeability coefficient P_∞ but has no effect on the threshold concentrations corresponding to the formation of the percolation cluster and related changes in permeability.

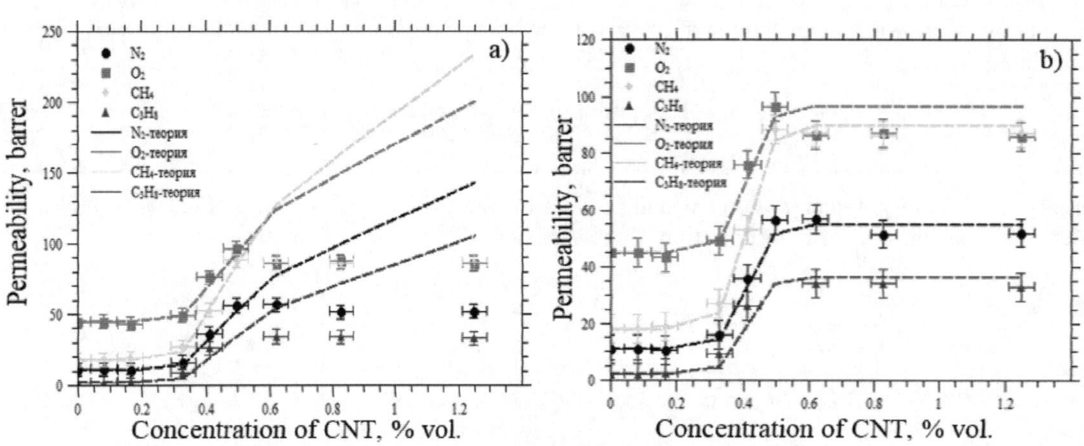

Figure 6. Gas permeability for reference (model) gases through the PVTMS/CNT membranes for (a) individual nanotubes and (b) with the account for agglomeration

Figure 6 shows the permeability coefficients of the PVTMS/CNT membranes calculated after the substitution of x_∞ to Eq. (4). In these calculations, we assume that there is no gas transfer through inner CNT channels, and the increase in gas permeability is solely provided by gas transport through the highly permeable layer of the modified polymer near the CNT surface. This assumption is valid only for the permeability coefficient P_∞ but has no effect on the threshold concentrations corresponding to the formation of the percolation cluster and related changes in permeability. The calculated P_∞ values (which are constant for each gas over the CNT concentration interval) are listed in Table 1.

Table 1. Permeability of PVTMS and percolation cluster

Gas	Permeability of PVTMS, barrer	Permeability of the percolation cluster, barrer
N_2	11	429
O_2	44	507
CH_4	18	702
C_3H_8	2	335

As follows from Fig. 6a, the calculated permeability coefficients well agree with the experimental values when the concentration of nanotubes in the matrix is below 0.5%. The development of the percolation structure and threshold changes in permeability are observed when the volume concentration of nanotubes varies from 0.3 to 0.4%. When the volume concentration of CNTs is above 0.5% (1.2wt %), the calculated permeability coefficients appear to be appreciably higher as compared with the experimental values. This disagreement can be explained by different parameters of the percolation cluster due to the agglomeration of nanotubes with increasing their content in the membrane. As follows from Fig. 6, when the average diameter of capsules increases due to the coalescence and agglomeration of nanotubes, the volume fraction of highly permeable regions at the same concentration of the capsules in the matrix decreases. For example, if the concentration of the capsules in the matrix is 0.008, as the diameter of the capsules increases from 0.05 to 0.06 (by 20%), the volume fraction of the percolation cluster decreases from 0.077 to 0.067 (by 13%); when the diameter is equal to 0.8, x_∞ approaches zero! Figure 7b shows the permeability coefficients calculated according to Eq. (4) with the account for agglomeration plotted against volume fraction of CNTs. These calculations show that, when the CNT volume concentration varies from 0.5% to 1.2%, the permeability coefficient is constant only if the diameter of the capsules linearly increases from 0.05 to 0.09 units, which corresponds to the formation of the CNT agglomerates with an average size of about 100 nm.

Nonlinear changes in permeability lead to marked changes in the ideal selectivity of the membranes with increasing concentration of CNTs (Fig. 8). For example, separation factor $\alpha(O_2/N_2)$ decreases from 4 down to 1.7 as the CNT concentration increases from 0.5 to 1.2 wt %; at higher concentrations, $\alpha(O_2/N_2)$ remains unchanged. On the other hand, $\alpha(CH_4/N_2)$ = 1.7 for all CNT concentrations within the margin of error; in the percolation interval, $\alpha(C_3H_8/N_2)$ increases from 0.2 to 0.6 and this value remains unchanged when the CNT concentration is higher than 1.2 wt %. The above changes in the separation factors prove the molecular mechanism of gas transport in the percolation membrane structure.

Conclusions

This work addresses the experimental study of gas permeability through the mixed matrix PVTMS/CNT membranes when the CNT concentration is higher than the percolation threshold. According to the experimental results, as the CNT concentration varies from 0.4% to 1.2 wt %, permeability of the CNT-containing membranes markedly increases: 5 times for nitrogen, 2 times for oxygen, 4 times for methane, and 15 for propane. However, further increase in the CNT content does not lead to any changes in permeability. This behavior (effect, phenomenon) was explained by the numerical simulation of a percolation cluster composed of capsules in the 3D matrix with the dimensions equivalent to nanotubes and membranes. Changes in the volume of the percolation cluster are shown to control the threshold changes in the permeability of the mixed matrix membranes upon the percolation transition. Experimentally observed changes in the membrane permeability at high CNT concentrations can be explained with the account for the agglomeration of nanotubes (capsules) and related changes in the dimensions of the percolation cluster.

This work shows that, for the mixed matrix CNT-containing membranes, there exists the concentration interval where gas permeability dramatically increases. At high (excessive) concentrations of CNTs, due to agglomeration of nanotubes and deterioration of the percolation structure, further increase in the CNT concentration does not provide any improvement in the membrane performance.

References

[1] Noble R 2011 *J. Membr. Sci.* **378** 393
[2] Kim S, Pechar T and Marand E 2006 *Desalination* **192** 330
[3] Andrady A, Merkel T and Toy L 2004 *Macromolecules* **37** 4329
[4] Gomes D, Nunes S and Peinemann K 2005 *J. Membr. Sci.* **246** 13
[5] Ma H and Gao X 2008 *Polymer* **49** 4230
[6] Ahmad A, Jawad Z, Low S, Zein S 2012 *J. Membr. Sci.* **451** 55
[7] Ge L, Zhu H and Rudolph V 2011 *Sep. Pur. Technol.* **78** 76
[8] Khan M, Filiz V, Bengtson G, Shishatskiy S, Rahman M, Lillepaerg J and Abetz V 2013 *J. Membr. Sci.* **436** 109
[9] Ismail A, Rahim N, Mustafa A, Matsuura T, Ng B, Abdullah S and Hashemifard S 2011 *Sep. Purif. Technol.* **80** 20
[10] Jae-Hyun C, Jonggeon J, Woo-Nyon K and Ho-Sang C 2009 *J. Appl. Polym. Sci.* **111** 2186
[11] Grekhov A, Eremin Yu, Dibrov G and Volkov V 2013 *Petrol. Chem.* **53** 549
[12] Hung N, Anoshkin I, Dementjev A, Katorov D and Rakov E 2008 *Inorg. Mater.* **44** 270
[13] Yampolskii Yu and Volkov V 1991 *J. Membr. Sci.* **64** 191
[14] International Standard ISO13321
[15] Barrett E, Joyner L and Halenda P 1951 *J. Am. Chem. Soc.* **73** 373
[16] Brunauer S, Emmett P and Teller E 1938 *J. Am. Chem. Soc.* **60** 309
[17] Grekhov A, Belogorlov A, Yushkin A and Volkov A 2012 *J. Membr. Sci.* **390-391** 160
[18] Matsumoto M. M 1998 *ACM Trans. Mod. Comp. Simul.* **8** 3
[19] Hoshen, J and Kopelman R 1976 *Phys. Rev. B* **14** 3438
[20] Rubin, F. 1974 *IEEE Trans. Comp.* **23** 907
[21] Lee M. 2007 *Phys. Rev. E* **76** 027702
[22] Yampolsky Yu, Starannikova L and Belov N 2014 *Membr. Membr technol.* **4** 231
[23] Winberg P, Sitter K, Dotremont C, Mullens S, Vankelecom I and Maurer F 2005 *Macromol.* **38** 3776
[24] Merkel T, Freeman B, Spontak R, He Z, Pinnau I, Meakin P and Hill A 2002 *Science* **296** 519
[25] Zhong J, Wen W and Jones A 2003 *Macromol.* **36** 6430
[26] Matteucci S, Kusuma V, Sanders D, Swinnea S and Freeman B 2008 *J. Membr. Sci.* **307** 196
[27] Kong Y, Du H, Yang J, Shi D, Wang Y, Zhang Y and Xin W 2002 *Desalination* **146** 49
[28] Ahmad J and Hagg M 2013 *J. Membr. Sci.* **445** 200
[29] Moniruzzaman M and Winey K 2006 *Macromolecules* **39** 5194
[30] Shirazi Y, Tofighy M and Mohammadi T 2001 *J. Membr. Sci.* **378** 551
[31] Cong H, Zhang J, Radosz M and Shen Y 2007 *J. Membr. Sci.* **294** 178
[32] Kim S, Chen L, Johnson J and Marand E 2007 *J. Membr. Sci.* **294** 147

Effect of temperature on the transport of solvents through PTMSP under ultra-high pressures

A M Grekhov[12], A A Belogorlov[12], Yu S Eremin[1], E V Pastukhova[1], A A Yushkin[2] and A V Volkov[2]

[1] National Research Nuclear University MEPhI (Moscow Engineering Physics Institute), Kashirskoe highway 31, Moscow, 115409, Russia
[2] A.V.Topchiev Institute of Petrochemical Synthesis, Russian Academy of Sciences, Russia

E-mail: yseremin@ya.ru

Abstract. Despite a large number of studies, by now there is no any definitive explanation of the solvent transport mechanism in nanostructured polymer materials. Both convective and diffusive transport of solvents can be observed in these materials. The study of the solvents permeability at different temperatures and pressures allow the variation of the physical parameters and structure of the solvent–membrane interaction thus becoming the key factor in the understanding of the fundamental aspects of the selective transport process in nanostructured polymer membranes.
The paper presents the study of ethanol, propanol and water transport through poly [1-(trimethylsilyl)-1-propine] (PTMSP) at pressures 50-150 atm and temperature up to 90°C. The study was done by the method of pressure dynamic decay. As the temperature rises, the permeability of ethanol and propanol through PTMSP is shown to increase in proportion to decreasing viscosity that denotes a convective type of transport. As for water, the permeability change is thermo-activated that is typical for a diffusive type of transport. This difference in the transport characteristics can be related to a change in the membrane structure and energetic characteristics of the solvent–polymer interaction.

1. Introduction

Industrial-scale technologies of organic solvents nanofiltration are successfully used in food, petrochemical and chemical industry as well as for water purification systems, etc. [1,2]. Further development of nanofiltration technologies is in many aspects related to the development of new polymer membrane materials, which have to possess chemical and mechanical stability in organic solvents as well as a combination of selectivity and permeability close to those of inorganic membranes. Nowadays, a particular attention is paid to vitreous polymers with a high fraction of non-relaxed free volume. High permeability, mechanical and film-forming properties of such polymers allow a formation of high-productive membranes with thin separation layers [3, 4].

Studies of nanofiltration processes in such polymers have shown that both diffusive and convective types of solvents selective transport [5-9] are possible. However, despite a large number of studies, nowadays, there is no any uniform approach to the description of mechanisms of interaction and transport of solvents in the nanostructured materials. On the one hand, in case of solvent condensation in the micro-heterogeneous structure of polymers, areas filled with liquid appear in a membrane, through which the convective transport is possible. On the other hand, is we speak about nanometer elements of free volume, the evolved free surface of polymer defines a strong interaction between the polymer and solvent at the molecular level and in this case the diffusive transfer is more probable. Besides, the mobility of polymeric chains may cause essential changes in the membrane structure under the effect of pressure and temperature or in case of interaction with solvent.

At present, when modeling nanofiltration modes, several Darcy's equation-based models are used [10]. The micro- and ultra-filtrations modes, the description of permeability in the nanofiltration mode considers the parameters defining the interaction of liquid and membrane material: viscosity factor η, superficial tension γ, porosity parameters, sorption value, swelling factor *SD*, etc. Nonlinear changes in transport properties are related to changes in structural characteristics of membranes under the pressure effect [11,12]. However, the temperature change has more significant effect on membrane structural parameters and solvents properties [18-22]. However, the analysis of temperature effects on membrane transport properties in the nanofiltration mode is rather fragmentary and does not allow this process to be described in a united way.

This paper studies a hydrophobic nanostructured vitreous polymer poly [1-(trimethylsilyl)-1-propine] (PTMSP). This polymer is extensively studied for the purpose of organic solvents nanofiltration [14] and membrane contactors [13]. The earlier published papers study in detail the convective transport of solvents in such PTMSP membranes. At the same time, the diffusive transport in PTMSP membranes is not considered because of different scale of flows. However, in a number of applications, for example in membrane contactors, the study of solvent diffusion mechanisms through a membrane have become urgent [ ]. This paper uses the method of pressure dynamic decay [24], of which unique sensitivity allows measuring the permeability of membranes in various ranges. This method was used to measure in a wide range of pressures and temperatures the permeability of solvents with different affinity to PTMSP; conditions of convective and diffusive transport in membranes have been determined as well.

2. Experiment

2.1 Polymers and preparation of membranes

The paper studies polymer poly [1-(trimethylsilyl)-1-propine] (PTMSP) synthetized in toluene solution using catalyst $TaCl_5$ with co-catalyst triisobutylaluminum (TIBA) [15[. The polymerization parameters [Monomer]/[Catalyst]=50, [Co-catalyst]/[Catalyst]=0.3, $[\text{Monomer}]_0$=1 mol/l, T = 25°C (M_w=1600000, M_w/M_n=2.9, $[\eta]^{25}_{toluene}$=6.9 dl/g). The membranes were obtained by casting 0.5-1.0%wt of the PTMSP solution in chloroform onto cellulose. The solvent was evaporated under normal conditions (atmospheric pressure, T= 20-24°C). The obtained membranes were 25±2μm thick. The PTMSP membranes characteristics defined by the helium pycnometry (Micro-Ultrapyc 1200e Quantachrome) and low-temperature absorption of nitrogen (Nova 1200e, Quantachrome) are given in Table 1. The specific area of the interior surface was calculated by the multi-pint BET and Langmuir's methods [16].

Table 1. Properties of PTMSP

Properties of PTMSP	Experiment data
Density, g/cm^3	0.9634±0,0004
Surface area, m^2/g	855±26 (BET)
Free volume, cm^3/g	0.65±0.02

2.2 Solvents

Solvents with different affinity to PTMSP have been used: chemically pure ethanol, propanol and de-ionized water (resistance not less than 17 mega ohms). Figure 1 shows superficial tension and viscosity of the used liquids at various temperatures. Viscosity of these solvents changes non-linearly as the temperature increases [18-22]. At temperatures ~ 20°C, viscosity of water is almost twice higher than the one of alcohols; at temperatures ~ 90°C, the viscosity values of all liquids decrease significantly and differ by less than 10%. The water superficial tension is also essentially higher than the one of alcohols and at a temperature of 20°C the superficial tension differs by more than 3 times. Table 2 shows PTMSP swelling and sorption coefficients in these liquids [23]. It can be seen that water is practically not sorbed into PTMSP and the water affinity is much worse than the one of alcohols.

Table 2. PTMSP swelling and sorption

Solvent	Swelling SD=$\Delta V/V^{B}$	Sorption, mol/mol
Water	0	<0.1
Ethanol	0.72	2.5
1-Propanol	1.00	2.5

2.3 Membranes permeability measurement

Permeability of solvents through PTMSP membranes were measured by the method of pressure dynamic decay (DPD) based on measuring elastic deformations of liquid in case of pressure decay [24].

Insertion of a rod into the chamber decreases the volume over the membrane and, as a result, causes compression and pressure increase in liquid. As the required pressure was provided in liquid, the rod position was fixed. Further, if liquid goes through the membrane, the amount of liquid over the membrane decreases that leads to the pressure decrease. Respectively, to define the fluid flow through the membrane, a decrease of pressure Δp for a time period Δt was measured. The membrane permeability factor was calculated by the formula:

$$P_d = \frac{\chi \cdot V \cdot \Delta p \cdot \rho}{\Delta t} \cdot \frac{d}{S \cdot p} \tag{1}$$

where V – liquid volume over the membrane [$м^3$], χ – liquid compressibility [atm^{-1}], ρ – liquid density [kg/m^3], d and S – thickness [m] and membrane area equal to $2.54cm^2$. The permeability measurement accuracy depends on the liquid compressibility. Since the compressibility factors for the majority of liquids are in the range $10^{-8} – 10^{-9}$ Pa^{-1}, this method allows recording a change in the liquid volume over the membrane with an accuracy of 0.01% as the pressure changes by 1atm. The measurements were done at an initial pressure of 180 atm for 5 hours at the same temperature with an accuracy of ±0.5°C (temperature regulator Termeh M01).

2.3 Permeability of solvents through PTMSP

Earlier, the DPD method was used to measure the permeability of ethanol and propanol through PTMSP at pressure from 10 to 200atm and temperature of 30°C [24]. It was stated that the permeability of these solvents through PTMSP at pressure from 50 to 150atm did not practically change. Therefore, to study the temperature effect on the change in PTMSP transport characteristics, the permeability of ethanol and propanol was measured at pressure from 50 to 150 atm and at temperatures less than boiling point: from 30°C to 70°C for ethanol and from 30°C to 90°C for propanol. Diagrams of ethanol and propanol flows through PTMSP at various different are given in Fig. 1. It is seen that at a constant temperature, the flow increases linearly as the pressure rises.

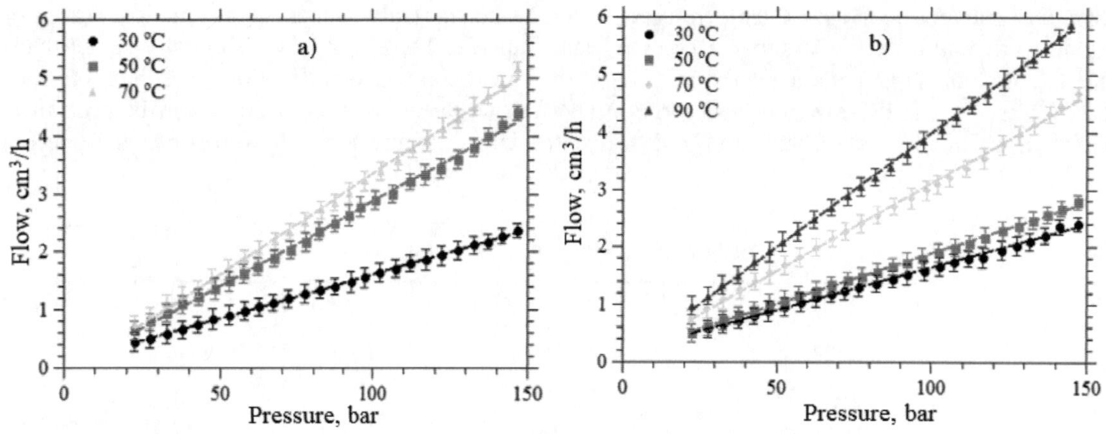

Figure 1. Flow of a) ethanol and b) propanol through PTMSP from pressure

The permeability coefficients of ethanol and propanol calculated by (1) are given in Fig. 2.

Apparently, at a pressure of 30 atm and temperature of 30°C, the permeability values measured by the DPD method and method of a filtration in a blind cell coincide [25]. As the temperature increases, permeability coefficients increase monotonously within the whole pressure range. So, the permeability of ethanol increases more than twice, as the temperatures increase from 30 to 70°C, the permeability of propanol increases by almost 3 times, as the temperatures increase from 30 to 90°C.

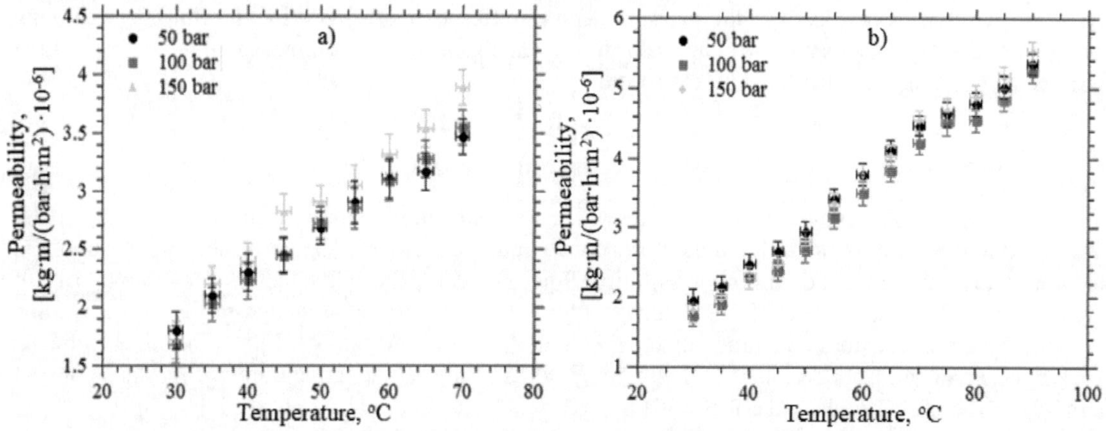

Figure 2. Permeability coefficient of a) ethanol and b) propanol through PTMSP from temperature

Figure 3a shows a change in pressure under water flowing through PTMSP. It can be seen that at a temperature of up to 40°C, the initial pressure of about 200atm did not change during the observation period, i.e. PTMSP is impermeable for water at such temperature and pressure. However at a temperature of 60°C and higher, there was a decrease in pressure related to the water flow through the membrane. A decrease in pressure during the first 30 minutes of measurements is related to the system relaxation and sorption processes in the membrane. The permeability values given in Fig. 3b were calculated for the pressure scaling-down area. It is seen that at a temperature of 60°C, water permeability achieves a value of $5 \cdot 10^{-9}$ kg·m/bar·h·m^2, i.e. comparable with diffusion flows in other polymeric membranes. Thus, one may suppose that at temperatures from 40 to 60°C, characteristic of liquid and its affinity to polymer become such that macroscopic diffusion of water is possible. As temperature further rises from 60 to 90°C, it results in the permeability increase by almost three times and achieves a value of ~ $15 \cdot 10^{-9}$ kg·m/bar·h·m^2. If we extrapolate linearly the permeability change, then the diffusion flow appears at a temperature of about 45°C.

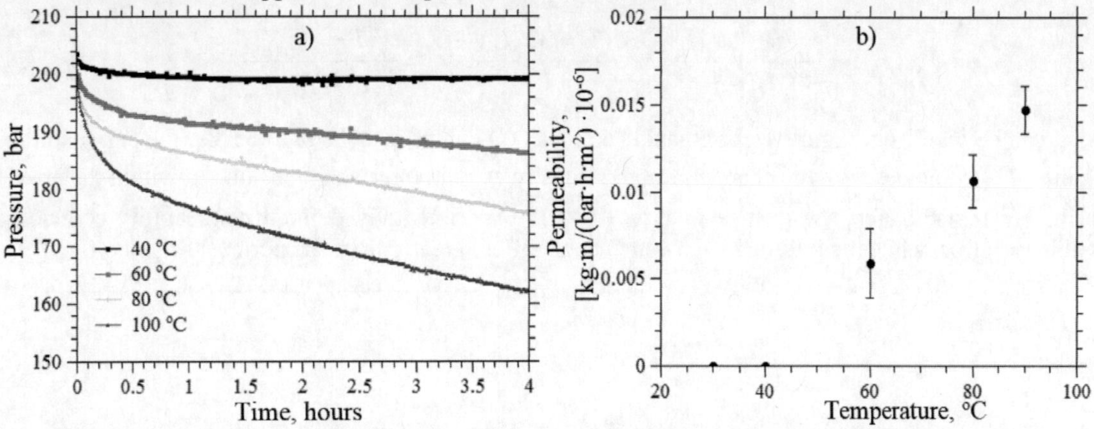

Figure 3. a) pressure change as water flows through PTMSP, b) water permeability coefficient through PTMSP

3. Results and discussion

In the nanofiltration mode through the nanostructured polymeric membranes, both convective and diffusion transfer of solvent [12,13,26-29] is possible. It is known that PTSMP has a non-uniform structure with areas of high and low density [32], porosity exceeds 25% and the structure of PTMSP non-relaxed free volume has bimodal distribution with the most probable diameter of "voids" of 0.66 and 1.36nm [33]. Respectively, related channels filled with liquid can occur in such free volume, through which there is a convective transfer goes and, in a general, a solvent flow through PTMSP can be described by the following equation:

$$J = A \cdot B \cdot \Delta p \tag{2}$$

where A – coefficient of liquid characteristics, B – coefficient defined by the polymer structure parameters, Δp – trans-membrane pressure [13]. If we consider a solvent flow through a membrane with a constant pressure, then the system temperature and interaction of liquid with polymer define coefficients A and B. On the one hand, in case of convective transport, the flow in channels in inverse proportion to liquid viscosity η(T). Also, in nanometer channels, the flow depends on the interfacial tension of liquid, which characterizes a sliing friction appearing because of interaction between liquid and material surface [6, 10]. As the temperature increases, the coefficient A and membrane permeability have to increase non-linearly as for the majority of liquids viscosity and interfacial tension decrease when heated and therefore, the coefficient A can be written down as follows:

$$A \sim \frac{1}{\eta(T) \cdot \gamma(T)} \tag{3}$$

On the other hand, the mobility of polymer chains and free volume can increase both due to the temperature change and in case of solvent sorption into polymer. The structural parameter B has to be as follows:

$$B = B(\Delta p, T, \phi) \tag{4}$$

where ϕ - sorption coefficient that depends on the affinity of liquid and polymer.

There is no and definite solution for the flow in the nanofiltration mode and the type of parameters A and B is not unambiguously defined. However, since the temperature changes of viscosity and interfacial tension of liquids are known, then using (3) it is possible to estimate the effect of these parameters on the permeability. Figure 5 shows the experimentally defined permeability coefficients for ethanol and propanol normalized to the corresponding values at 30 °C as well as the inverse values of viscosity and the interfacial tension:

$$P_{dn} \sim \frac{P_d}{P_d(30^0C)},\ \eta_n = \frac{\eta(T=30^0C)}{\eta(T)},\ \gamma_n = \frac{\gamma(T=30^0C)}{\gamma(T)} \tag{5}$$

As it can be seen from Figure 4, for ethanol (up to 45°C) and propanol (up to 65°C), the permeability change P_{dn} coincides with the change of both viscosity η_n and interfacial tension γ_n within the error. At higher temperatures, the change in viscosity and experimentally defined permeability coincides within the error, while the interfacial tension change differs greatly from the permeability one.

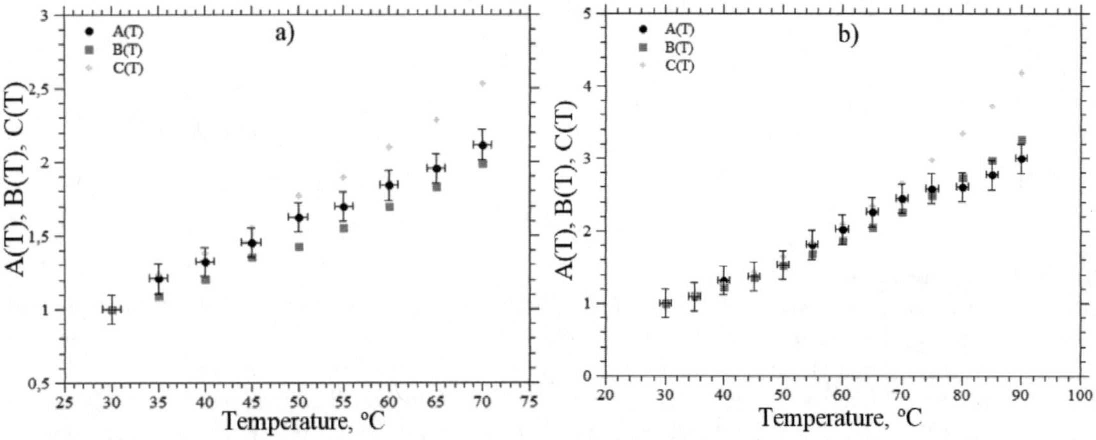

Figure 4. Normalized coefficients of permeability through PTMSP at a pressure of 100atm; viscosity and interfacial tension for a) ethanol and b) propanol.

This result corresponds to the model of liquid viscous flow in voids. At that, if the permeability temperature changes are related only to liquid viscosity, then the flow will be described by the following expression:

$$J = \frac{B(\phi)}{\eta(T)} \cdot \Delta p \tag{6}$$

Parameter $B(\phi)$ contains only structural characteristics of polymer, which depend only on the interaction of solvent with the membrane material and influence of pressure on the membrane structure can be neglected. Earlier it was shown that in case of sorption of alcohols, PTMSP swelling occurs and its volume increases by ~ 55% for ethanol and by ~ 110% for propanol [30]. Accounting swelling and free volume change, the following expression was proposed for structurál parameter $B(\phi)$:

$$B = \frac{a \cdot V_m}{l} \cdot \frac{\phi^3}{(1-\phi)^2} \cdot e^{1-\frac{\gamma_{SV}}{\gamma_L}} \tag{7}$$

where a –modeling parameter, ϕ – liquid volume fraction in mambrane, l – membrane thickness, V_m – molar volume of solvent, η – liquid viscosity, γ_{SV} and γ_L – surface energy of polymer and solvent [13,17]. Due to swelling resulted from solvent sorption and membrane compression under pressure, the geometry characteristics change:

$$\phi = \frac{\phi_0 + SD \cdot e^{-k\cdot\Delta p}}{1 + SD \cdot e^{-k\cdot\Delta p}} \text{ и } l = l_0 \cdot \frac{1 + SD \cdot e^{-k\cdot\Delta p}}{(1 + SD)^{\frac{2}{3}}} \tag{8}$$

where ϕ_0 and l_0 - membrane parameters without swelling, SD – swelling degree. Modeling parameter k – polymer compressibility coefficient – defines the membrane compression under pressure. As it can be seen from Figure 5, at temperatures up to 60°C for ethanol and 70°C for propanol, this model describes well the experimentally measured values of flows through PTMSP. At higher temperatures, the calculated values differ from the experimental ones.

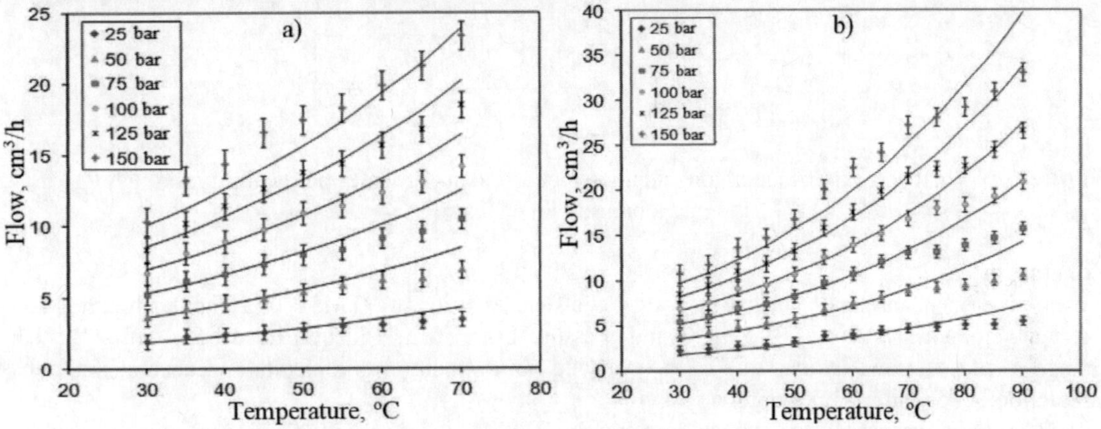

Figure 5. Solvent flow through PTMSP at different pressures and temperatures

a) ethanol, b) propanol

If solvent sorbs badly in polymer, then the transport channels filled with liquid cannot be formed and convective transfer of liquid is impossible. However, the diffusion transfer of solvent molecules in polymer may occur as a result of the transition between stable states, of which probability is defined by the activation energy and temperature of system. At that, the diffusion flow has a pronounced threshold character as the temperature rises, for example, for water in PTMSP. As it can be seen from Figure 3, at a temperature of up to 40°C and pressure higher than 150 atm, there is no water flow through PTMSP, i.e. its size is less than the measurement sensitivity $0.1 \cdot 10^{-9}$kg·m/bar·h·m^2. At a temperature of 60°C, the permeability coefficient becomes ~ $5 \cdot 10^{-9}$kg·m/bar·h·m^2, i.e. it increases almost by two orders of magnitude in the 20°C interval. Such essential increase in the water flow corresponds to the diffusion transport by the activation mechanism and the permeability coefficient can be described by the Arrhenius equation:

$$P = K_0 \cdot D_0 \cdot \exp\left(-\frac{E_a}{R \cdot T}\right), \tag{9}$$

where K_0 and D_0 – sorption and diffusion coefficients at the normalizing temperature, E_a – water activation energy, R – absolut gas constant, T – temperature. As it can be seen from Figure 6, the values of water permeability through PTMSP defined experimentally and calculated by (9) and normalized to the permeability at 60°C coincide within the error at an activation energy of E_a=31.1 kJ/mol. These activation energy values correspond to the diffusion characteristics of water in vitreous polymers. On the other hand, one may suppose that as the temperature rises and interfacial tension decreases, the sorption properties of water in PTMSP will change and the flow can be described within the convective diffusion model.

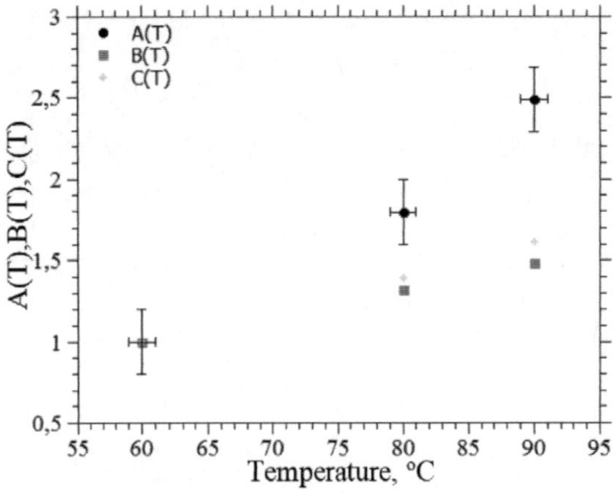

Figure 6. Normalized experimental and calculated coefficients of water permeability through PTMSP at a pressure of 150 atm

Conclusion

Flows have been measured and permeability coefficients through PTMSP of ethanol, propanol and water at temperatures from 20 to 90°C and pressure from 50 to 150 atm have been defined. Such change in pressure and temperature allows varying both structural parameters and characteristics of interaction of solvent with polymer.

Ethanol and propanol, for which sorption in PTSMP is observed, form a system of related channels, through which there is a convective diffusion of solvent. The permeability of these alcohols depends on the change in the PTMSP structural parameters under sorption and it is inverse proportion to viscosity. As for water, the thermo-activated transfer is observed at a temperature higher than 40°C. As the temperature rises, the mobility of water molecules and polymer chains increases as well as the permeability. At that, the activation energy and critical characteristics for water in PTMSP have been defined.

Acknowledgement

The authors gratefully acknowledge Prof. V S Khotimsky for providing of PTMPS samples.

References

[1] Volkov A, Korneeva G and Tereshchenko G 2008 *Russ. Chem. Rev.* **77** 1053
[2] Vandezande P,Geversb L and Vankelecom I 2008 *Chem. Soc. Rev.* **37** 365
[3] Volkov A, Parashchuk V, Stamatialis D, Khotimsky V, Volkov V and Wessling M 2009 *J. Membr. Sci.* **333** 88
[4] Fritsch D, Merten P, Heinrich K, Lazar M and Priske M 2012 *J. Membr. Sci.* **401-402** 222
[5] Robinson J, Tarleton Millington C and Niemeijer A 2004 *J. Membr. Sci.* **230** 29
[6] Geens J,Van der Bruggen B and Vandecasteele C 2006 *Sep. and Pur. Tech.* **48** 255
[7] Darvishmanesh S, Buekenhoudt A, Degrиve J and Van Bruggen B 2009 *J. Membr. Sci.* **334** 43
[8] Santos J, Hidalgo A, Oliveira R, Velizarov S, Crespo J. 2007 *J. Membr. Sci.* **300** 191
[9] Stefan 1872 *Sitzungsber. Kais. Akad. Wiss. Wien* **LXV** 323
[10] Grekhov A, Yushkin A, Tsar'kov S, Filippov A and Volkov A 2010 *Membrane* **47** 18
[11] Volkov A, Parashchuk V, Stamatialis D, Khotimsky V, Volkov V and Wessling M 2009 *J. Membr. Sci.* **333** 88
[12] Machado D, Hasson D and Semiat R 2000 *J. Membr. Sci.* **166** 63
[13] Volkov A, Yushkin A, Grekhov A, Shutova A, Bazhenov S, Tsarkov S, Khotimsky V, Vlugt T and Volkov V 2013 *J. Membr. Sci.* **440** 98
[14] Tsarkov S, Khotimsky V, Budd P, Volkov V, Kukushkina J and Volkov A 2012 *J. Membr. Sci.* **383** 241

[15] Khotimsky V, Tchirkova M, Litvinova E, Rebrov A and Bondarenko G 2003 *J. Polym. Sci. Part A: Polym. Chem.* **41** 2133
[16] Brunauer S, Emmett P and Teller E 1938 *J. Am. Chem. Soc.* **60** 309
[17] Yushkin A, Grekhov A, Matson S, Bermeshev M, Khotimsky V, Finkelstein E, Budd P, Volkov V, Vlugt T and Volkov A 2015 *Reac. and Func. Pol.* **86** 269
[18] William M, David R. and Thomas J 2012 *Handbook of chemistry and physics: A ready reference book of chemical and physical* data CRC
[19] Vazquez G, Alvarez E and Navaza J *J.* 1995 *Chem. Eng. Data*
[20] Shaoxian S and Changsheng P 2008 *J. of Disp. and Texn.* **29** 1373
[21] Pang F, Seng C, Teng T and Ibrahim M 2007 *J. of Mol. Liq.* **136** 71
[22] Bobylev V 2003 *Physical properties of most known chemical* (Moscow: RCTU)
[23] Volkov V, Tsar'kov S, Gokzhaev M, Bondarenko G, Legkov S, Kukushkin Yu and Volkov V 2012 *Membranes and membrane technology* **2** 1
[24] Grekhov A, Belogorlov A, Yushkin A, Volkov A 2012 *J. Membr. Sci.* **390-391** 160
[25] Volkov A, Tsarkov S, Gokzhaev M, Bondarenko G, Legkov S, Kukushkina Yu and Volkov V 2012 *Petr. Chem.* **52** 598
[26] Vandezande P, Geversb L, Vankelecom I. 2008 *Chem. Soc. Rev.* **37** 365
[27] Han S and Livingston A 2005 *J. Membr. Sci.* **262** 49
[28] Bhanushali D, Kloos S and Bhattacharyya D. 2002 *J. Membr. Sci.* **208** 343
[29] Muellera N, Van Der Bruggen B, Keuterc V, Luis P, Melind T, Pronke W, Reisewitzf R, Rickerbyg D, Riosh G, Wennekesi W and Nowack B 2012 *J. Haz. Mat.* **211–212** 275
[30] Tsar'kov S, Malakhov A, Litvinova E and Volkov A 2013 *Petr. Chem.* **53** 537
[31] Bhanushali D, Kloos S, Kurth C and Bhattacharyya D 2001 *J. Membr. Sci.* **189** 1
[32] Volkov V, Khotimsky V, Gokzhaev M, Litvinova E, Fadeev A and Kelly S 1997 *J. Phys. Chem.* **71** 1556
[33] Hofmann D, Heuchel M, Yampolskii Yu, Khotimskii V, Shantarovich V 2002 *Macrom,* **35** 2129